Ultra-Wideband, Short-Pulse Electromagnetics 8

Ultra-Wideband, Short-Pulse Electromagnetics 8

Edited by

Carl E. Baum

Department of Electrical and Computer Engineering
University of New Mexico
Albuquerque, NM, USA

Alexander P. Stone

Department of Mathematics
University of New Mexico
Albuquerque, NM, USA

J. Scott Tyo

Department of Electrical Engineering
University of New Arizona
Tucson, AZ, USA

Springer

Carl E. Baum
Department of Electrical
and Computer Engineering
University of New Mexico
EECE Bldg., MSC01 1100
Albuquerque, NM 87131-0001
USA

Alexander P. Stone
Department of Mathematics
University of New Mexico
Albuquerque, NM 87131
USA

J. Scott Tyo
Optical Sciences Center
Department of Electrical Engineering
University of Arizona
1630 E. University Blvd.
Tucson, AZ 85721
USA

ISBN 978-0-387-73045-5 e-ISBN 978-0-387-73046-2

Library of Congress Control Number: 2007932606

Printed on acid-free paper.

9 8 7 6 5 4 3 2 1

springer.com

PREFACE

The eighth conference on Ultra-Wideband, Short-Pulse Electromagnetics (UWBSP8) was held at the Convention Center of Albuquerque, New Mexico, USA, on 9-14 July 2006. This was part of AMEREM 2006. This in turn was part of a Joint Symposium including IEEE Antennas and Propagation Society International Symposium and USNC (US National Committee)/URSI (International Union of Radio Science) National Radio Science Meeting. This continues the tradition extending through Magdeburg, Germany (2004), on back to their beginning at Polytechnic University in Brooklyn, New York, USA (1992).

Like the previous conferences, the eighth in this series extends the earlier results. The subjects include pulse radiation and measurement, scattering theory, target detection and identification, antennas, signal processing, communications, and related subjects. It should be noted that, at this Joint Symposium, ultra-wideband was prominently recognized by the presentation of the John Kraus Antenna Award of the IEEE Antennas and Propagation Society to C.E. Baum, E.G. Farr, and D.V. Giri "For development of novel and innovative ultra-wideband antenna concepts that have enabled a new area of electromagnetics."

The photograph on the front cover is that of JOLT, an extremely powerful radiator of impulse-like electromagnetic waves. It was developed by the Air Force Research Laboratory, Directed Energy Directorate, on Kirtland AFB, adjacent to Albuquerque.

The editors wish to thank all of those involved in the Joint Symposium. The University of New Mexico, Department of Electrical and Computer Engineering, made an especially large contribution of personnel. The Three Dimensions company handled most of the administration. We would especially like to thank Chuck Reuben for handling many of the details in assembling this conference proceedings volume.

Carl E. Baum
Dept. of Electrical & Computer Engineering
The University of New Mexico
Albuquerque, NM 87131

Alexander P. Stone
Dept. of Mathematics & Statistics
The University of New Mexico
Albuquerque, NM 87131

J. Scott Tyo
College of Optical Sciences
The University of Arizona
Tucson, AZ 85721

TABLE OF CONTENTS

ANTENNAS

Development of the Impulse Slot Antenna 1
W.S. Bigelow, E.G. Farr, L.H. Bowen, W.D. Prather, and T.C. Tran

A High-Voltage Cable-Fed Impulse Radiating Antenna 9
L.H. Bowen, E.G. Farr, and W.D. Prather

Broadband Self-Complementary Antenna Arrays 17
M. Gustafsson

A Quasi-Planar Wide Band Conical Antenna 25
M. Lu, J.W. Bredow, S.Jung, and S. Tjuatja

Planar Tapered Loop Antennas for Ultra-Wideband Radio Systems 33
S.G. Mao and S.-L. Chen

An Omnidirectional and Low-VSWR Ultra Wideband Antenna for a Frequency Band of 6 to 40 GHz 41
A. Maeda and T. Kobayashi

Ultrawideband Band-Notched U-Shape Folded Monopole Antenna and its Radiation Characteristics 49
T.G. Ma and S.J.Wu

Comparison of UWB Antennas Considering Pattern Variation With Frequency 57
T. Dissanayake and K.P. Esselle

Dispersive Properties of Terminal Loaded Dipole Antennas in UWB Link 63
A.O. Boryssenko and D.H. Schaubert

An Approach to the Determination of the Phase Center of Vivaldi-Based UWB Antenna 69
B.-S. Jin, Q. Wu, Y.-M. Wu, L. Bian, and L.-W. Li

Application of UWB Antenna Descriptors to Antenna Performance Assessment 75
P. Miskovsky, J.M. Gonzalez-Arbesú, and J. Romeu

Broadband Tuning Limits on UWB Antennas Based on Fano's Formulation 83
M.C. Villalobos, H.D. Foltz, J.S. McLean, and I.S. Gupta

Electromagnetic Lens Design and Generalized E and H Modes 89
A.P. Stone and C.E. Baum

Producing Large Transient Electromagnetic Fields in a Small Region: An Electromagnetic Implosion 97
C.E. Baum

COMMUNICATION

Optimum Wireless Communication Through Unknown Obscuring Environments Using The Time-Reversal Principle: Theory and Experiments....................105
S. Jaruwatandilok, A. Ishimaru, and Y. Kuga

DATA PROCESSING

Correction of Time-Domain Data in Special Cases Where the Inverse Transfer Functions are Analytic Time Domain Operators....................113
C.E. Baum

Second Time Integral of the Impulse Response for Enhancing the Late-Time Target Response for Target Identification....................121
C.E. Baum

Blind Source Separation for Extraction of Target Scattering Centers....................127
I. Jouny

Extended Approaches for Integrated M-Sequence Based UWB Sensors 135
M. Kmec, R. Herrmann, J. Sachs, P. Peyrl, and P. Rauschenbach

Antenna Array Processing for Radar Applications with Support Vector Machines 143
M. Martínez-Ramón, N. Xu, C.G. Christodoulou

RADAR SYSTEMS

Portable Imaging UWB Radar System With Two-Element Receiving Array 153
A.O. Boryssenko, D.L. Sostanovsky, and Elen S. Boryssenko

On the Development of a Low-Cost Compact Planar Integrated-Circuit Sampling Receiver for UWB Sytems....................161
J. Han. R. Xu, and C. Nguyen

Time Reversal with Single Antenna Systems in Indoor Multipath Environments....................171
Z. Yun and M.F. Iskander

Further Developments in High Voltage UWB Directional Couplers....................179
L. Atchley, L. Bowen, E. Farr and W. Prather

SCATTERING

Resonance Behavior of a Dielectric Target in a Half-Space Using the CNR (Complex Natural Resonance) Method 187
S.K. Padhi, and N.V. Shuley

Analysis of the Late-Time Transient Field Scattered by a Line Source Above a Grounded Dielectric Slab 195
G.D. Dester and E. J. Rothwell

An Analysis of Time-Domain Dort Method for Ultrawideband Probing of Embedded Objects in Dispersive and Random Media 203
M.E. Yavuz and F.L. Teixeira

Application of the Method of Subregions to Measurement of Layered Materials 211
B. Perry, S. Cossmann, L. Kempel, and E. Rothwell

A Numerical Study on the Sensitivity of Time-Reversal Imaging Methods Against Clutter, Noise and Model Perturbations 219
M.E. Yavuz and F.L. Teixeira

Modelling of Reflection of UWB Pulses from Trapped Human Beings 227
A. Nezirovic, M Liu and A. Yarovoy

SOURCES

Global Limitation on Fast Switching by Semiconductor Devices 235
A.F. Kardo-Sysoev

A Low-Cost, Compact Planar Integrated-Circuit Tunable Multi-Pulse Transmitter for UWB Radar and Communication Systems 243
J. Han, R. Xu, and C. Nguyen

UWB Antenna Surrogate Design 251
P.R. Hayes, L. Hernandez and A. Austin

INDEX 259

ANTENNAS

DEVELOPMENT OF THE IMPULSE SLOT ANTENNA

W. Scott Bigelow, Everett G. Farr, Leland H. Bowen, William D. Prather, and Tyrone C. Tran*

1. INTRODUCTION

We describe the development of the Impulse Slot Antenna (ISA), which is a conformal ultra-wideband (UWB) tapered slot antenna, suitable for printing onto a non-conducting aircraft wing [1]. This antenna is designed to look out over the tip of a wing approximately 0.6 m wide by 1.5 m long. It is likely be useful in UWB radar applications having only limited space for an antenna. We describe the design, fabrication, and testing of the ISA; and compare its performance to that of a commercially available TEM horn.

The frequency range of interest for the ISA is from 250 MHz to 2 GHz, and we built $1/8^{th}$-scale antenna models operating in the 2–16 GHz range. We constrained these models to make maximum use of the assumed 2:5 wing aspect ratio, initially investigating tapered slot designs having 50 Ω input impedance. However, none of these antennas exhibited satisfactory performance. This experience led us to develop the ISA, a hybrid antenna consisting of 200-Ω flattened biconical coplanar plates near the feed and a spline-tapered slot of gradually increasing impedance toward the aperture. The antenna is fed through a 200-Ω twin-line by a 50-to-200 Ω splitter-balun.

The ISA out-performed its 50-Ω predecessor designs and performed nearly as well as the Farr Research Model TEM-1-50 sensor, which has a radiating element with nearly five times the area of the radiating elements of the scale model ISA. Moreover, the conformal ISA completely avoids the aerodynamic drag of a TEM horn, making it practical for use on an aircraft wing. We describe the $1/8^{th}$-scale ISA, and we compare its performance in terms of return loss, boresight gain, and antenna pattern to the Farr Research TEM sensor. Finally, we note some design improvements that should lead to improved performance in the next generation ISA.

* W. Scott Bigelow, Everett G. Farr, and Leland H. Bowen, Farr Research, Inc. Albuquerque, NM, 87123; William D. Prather and Tyrone C. Tran, AFRL/DE, Kirtland AFB, NM 87117.

2. BACKGROUND

In a survey of UWB antennas, Kraus and Marhefka [2] consider a simple V, a biconical V, and a curved biconical V, as shown in Figure 1. Of these three antennas, they claim that the curved biconical V has the most directionality and greatest bandwidth. With this insight, a design based on the curved biconical V appears most appropriate for our application.

To adapt either biconical design to a printable device, the cones can be flattened into equivalent coplanar plates, as derived by Farr and Baum in [3], and as shown in Figure 2. At a given location, the width of each plate would be approximately twice the diameter of the original cone. As described in [4], the aperture should be about one-half wavelength at the lowest frequency; the length of each element should be at least one wavelength; and, in order to produce similar E– and H–plane patterns, the opening angle should not exceed about 12 degrees. Thus, for operation as low as 250 MHz, one would want the antenna to have an aperture of about 0.6 m and a length of at least 1.2 m. To meet the 12-degree criterion, a much longer antenna (about 3 m) would be required.

The flattened curved biconical V antenna resembles a Vivaldi tapered slot antenna. Recent antipodal Vivaldi designs demonstrating as much as a 10:1 bandwidth and 10 dBi gain were described in [5]. However, neither the antipodal Vivaldi nor the flattened biconical designs can simultaneously fit our assumed 0.6 X 1.5-m wing and meet our 250 MHz low-end frequency requirement. The antipodal Vivaldi, with metallization on both sides of a supporting substrate, is additionally inconsistent with the desired capability to "paint" the antenna elements onto a single surface of a non-conducting wing.

The issues raised by the constraints led us to investigate design compromises in search of a workable antenna. To achieve the best possible low-end frequency response, the aperture was chosen to extend all the way across the short dimension of the aircraft wing; and we set the length of the antenna to fill the long dimension of the wing. Finally, portions of the flattened biconical design, exceeding the wing dimensions, were truncated. To facilitate construction and testing, we built our antennas as $1/8^{th}$-scale models. At this reduced size, the radiating elements occupied an area of 7.6 x 19 cm^2, and the low-end design frequency became 2 GHz.

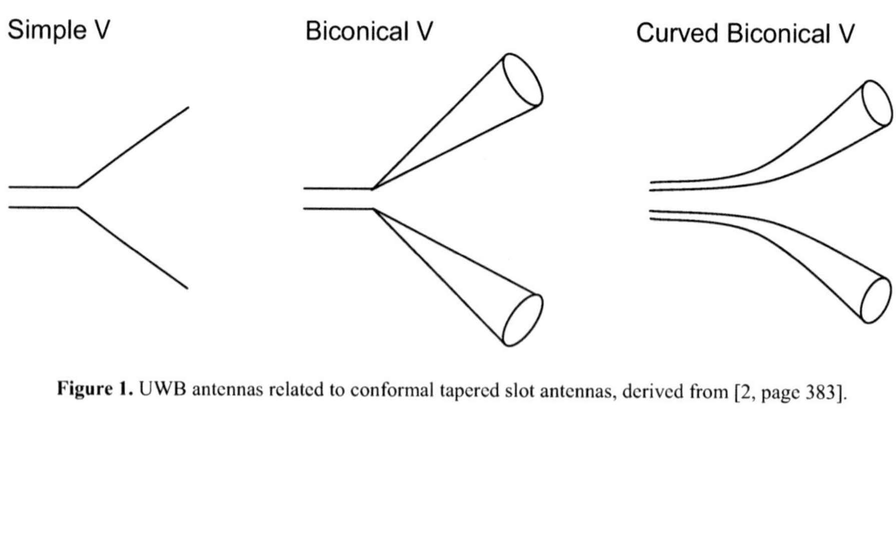

Figure 1. UWB antennas related to conformal tapered slot antennas, derived from [2, page 383].

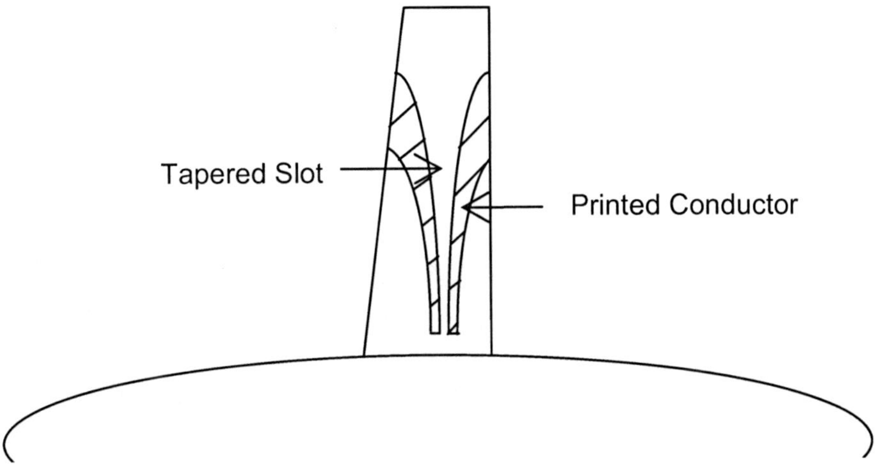

Figure 2. Wing-mounted tapered slot derived from the curved biconical V antenna.

We designed two 1/8th-scale-tapered-slot models, one approximating the flattened biconical V, and the other approximating the flattened curved biconical V of Figure 1. On the basis of quasi-static FEM analysis, the slot width at the feed was set to achieve an impedance of approximately 50 Ω. An SMA edge connector was used to mate a coaxial cable to the feed point. The radiating elements were fabricated from copper tape.

We built and tested these scale models with various substrates, but none were particularly successful. The antennas had more than −5 dB return loss, and realized gain was generally in the 0–5 dBi range. In response to these findings, we tried replacing the SMA edge connector feed with a cross-gap coaxial feed, as described in [6]. This new feed proved to be superior to the SMA edge connector, in that the maximum realized antenna gain reached 12 dB at 8 GHz. On the other hand, there was no significant reduction in return loss. Thus, while the cross-gap feed was an improvement over previous designs, the return loss remained high enough to keep us searching for a better solution.

3. IMPULSE SLOT ANTENNA

We developed the ISA to address the large return losses of earlier tapered-slot designs. The antenna, shown in Figure 4, is a hybrid antenna, somewhat similar to a combination of flattened versions of the biconical V antennas shown in Figure 1. It consists of a 200-Ω flattened biconical region near the feed and a spline-tapered slot region farther out. To adapt the impedance to a 50-Ω cable, we used a 50-to-200-Ω splitter-balun, similar to those commonly used in reflector impulse radiating antennas (IRAs). We connected the balun to the antenna with a short length of homemade 200-Ω twinline, following an approach developed for an earlier effort [7].

The design of the ISA's radiating arms close to the feed point follows the same approach used to design the feed arms of an IRA in a coplanar plate configuration, as described in [3]. From the feed point, as shown in Figure 4, the edges of the radiating arms emerge at angles corresponding to a 200-Ω pair of flattened cones. This is a region of constant impedance. As the radiating arms approach the outer edge of the wing, the outer edges follow circular arcs until they become parallel with the antenna centerline. At the same time, the inner edge of each arm gradually tapers away from the centerline at an increasing rate, until it intersects with the outer edge at the aperture. This is a spline-taper region of increasing impedance.

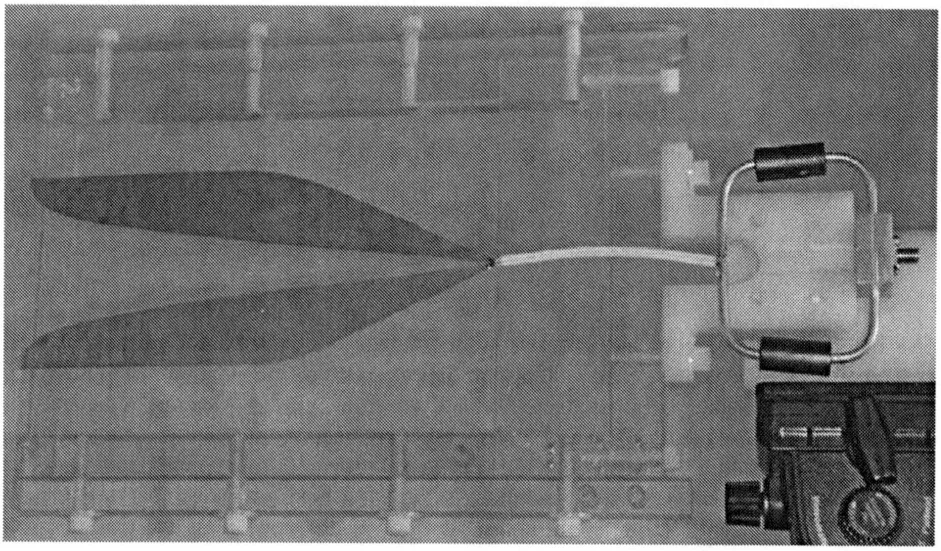

Figure 3. Impulse slot antenna (ISA) with Mylar® substrate, twinline feed, and splitter-balun.

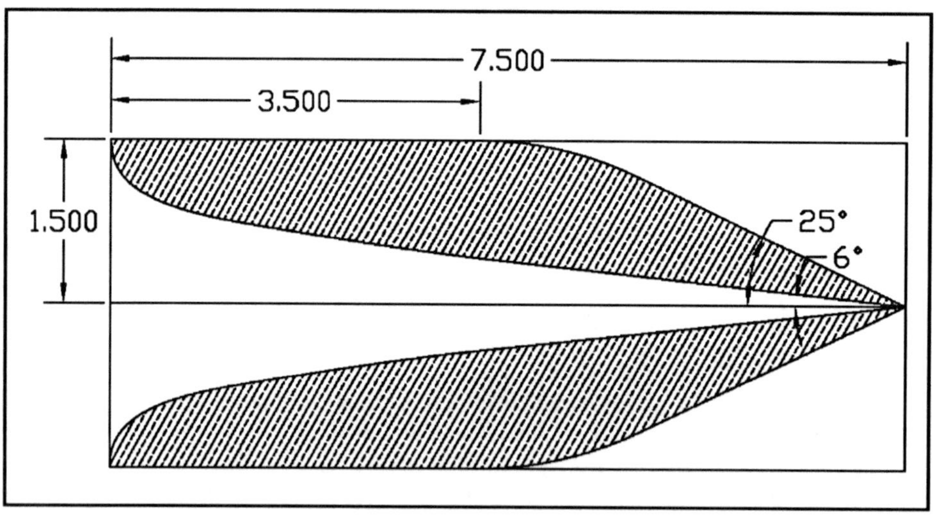

Figure 4. Radiating elements of the 1/8th-scale ISA. Dimensions are in inches.

By eliminating large feed point impedance discontinuities, we hoped the ISA design would reduce the return loss of previous designs. In fact, the impedance discontinuities associated with the feed point, twinline, and splitter-balun are all 10 Ω or less. This is apparent from the TDR shown on the left in Figure 5. On the right in the same figure, we also observe that the return loss (or S_{11}) of the ISA has been reduced to about −10 dB.

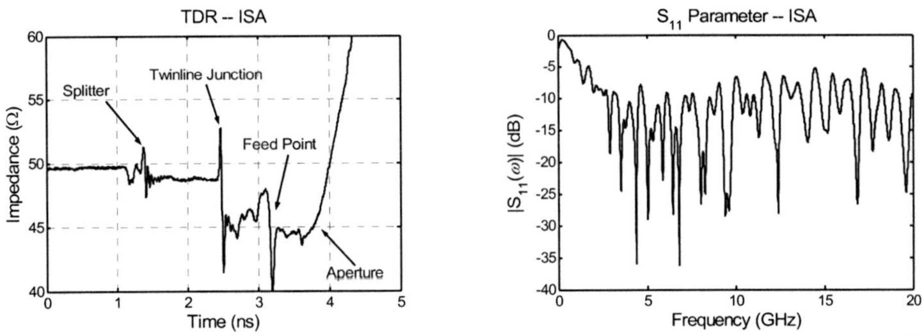

Figure 5. TDR (left) and associated return loss (right) of the ISA.

The impulse response of the ISA was measured on boresight and in H– and E–planes, as shown in Figure 6 and Figure 7. The peak response on boresight is 1.46 m/ns, with a full-width-at-half-maximum (FWHM) of only 35 ps. When we integrate the response, we find that the effective aperture height of the ISA, h_{eff}, is 1.9 cm, as defined by $V_{rec} \approx h_{eff} \cdot E_{inc}$, where V_{rec} is received voltage, and E_{inc} is the incident electric field. The maximum realized gain is 15 dB at 13 GHz. These results represent improved high-frequency response and a 3-dB increase in gain over the tapered slot antenna with cross-gap feed. The maximum response of the antenna is on physical boresight in both H–plane and E–plane, and the patterns are somewhat broader in the H–plane than in the E–plane.

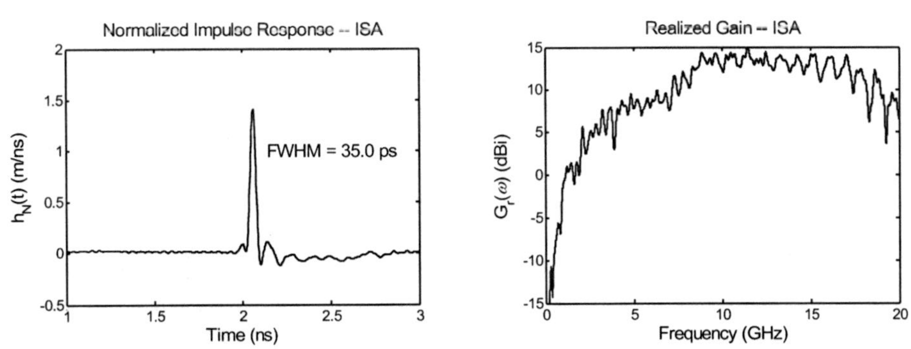

Figure 6. Boresight impulse response (left) and realized gain (right) of the ISA.

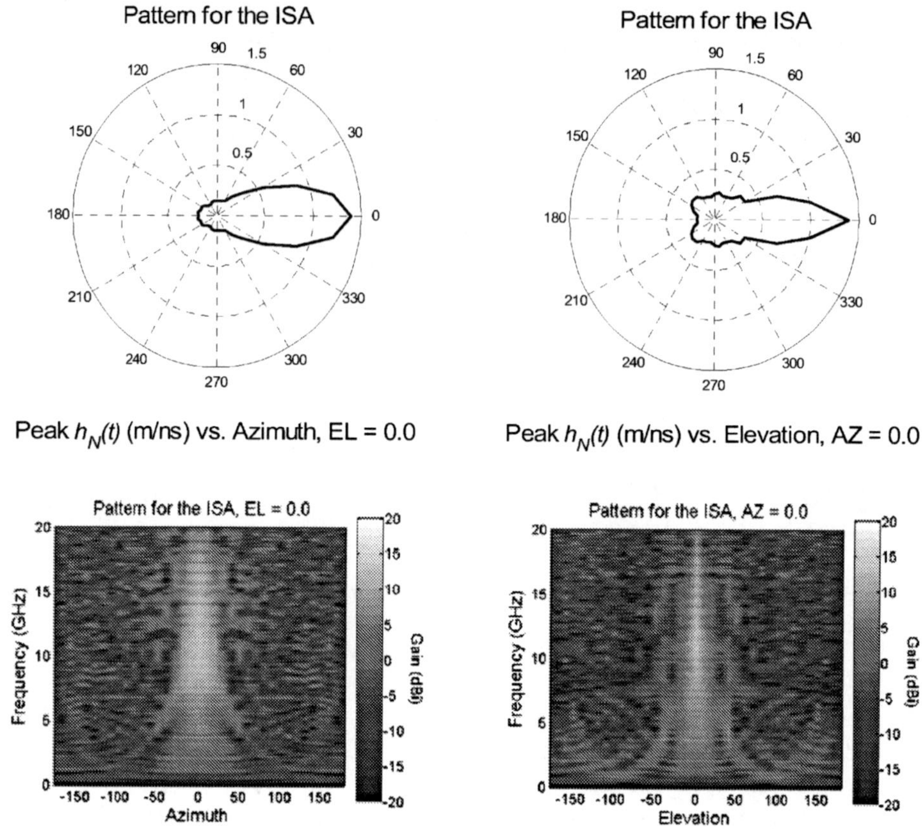

Figure 7. Patterns of peak normalized impulse response (top) and realized gain (bottom) for the H–plane (left) and E–plane (right) of the ISA. The plane of the antenna was vertically aligned for these pattern measurements.

4. COMPARISON OF IMPULSE SLOT ANTENNA AND FARR RESEARCH TEM-1-50 SENSOR

Upon examining the performance of the ISA, we found it to be remarkably similar to that of the Farr Research 50-Ω TEM sensor, model TEM-1-50, shown in Figure 8, and described in [8]. When we compare the TDR and S_{11} of the ISA (Figure 5) with those of the TEM-1-50 (Figure 9), we observe that the TDR of the ISA is less smooth than that of the TEM-1-50, and the return loss for the ISA is about 5 dB higher than that of the TEM-1-50.

Next, we compare the impulse response and realized gain of the ISA (Figure 6) with those of the TEM-1-50 (Figure 10). We observe that the impulse responses of the two antennas are remarkably similar. The ISA has a peak response of 1.46 m/ns and a FWHM of 35 ps, while the TEM-1-50 has a peak response of 1.55 m/ns and a FWHM of 30 ps. We have already seen that the integral of the impulse response, h_{eff}, is 1.9 cm for the ISA. For the TEM-1-50, it is 1.8 cm.

Figure 8. The Farr Research model TEM-1-50 sensor.

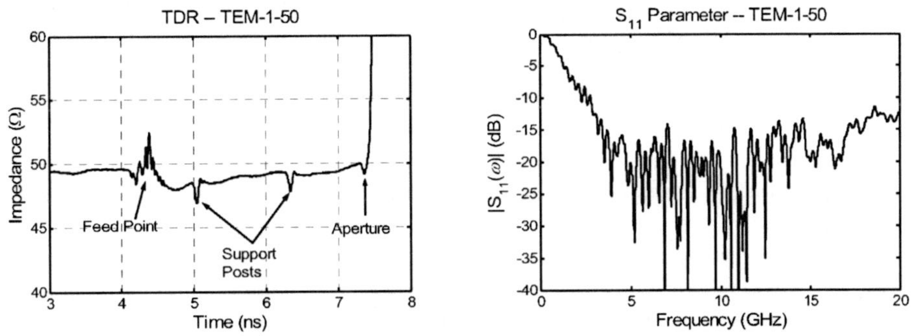

Figure 9. TDR (left) and associated return loss (right) of the TEM-1-50 sensor.

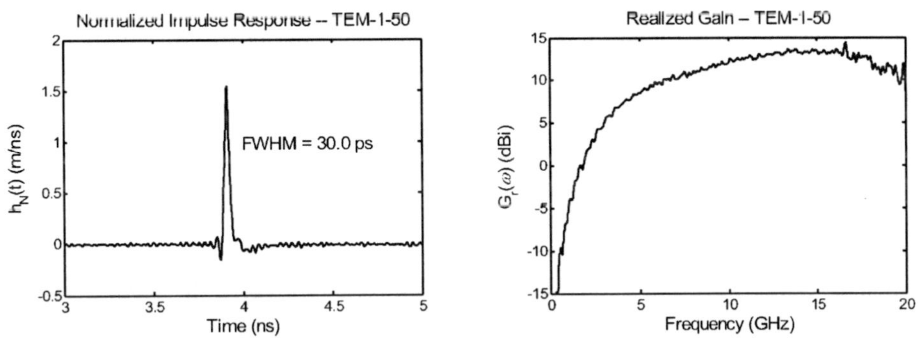

Figure 10. Impulse response (left) and realized gain (right) for the TEM-1-50 sensor.

As for realized gain, both antennas have essentially the same boresight gain across most of the bandwidth. Recall that the maximum for the ISA was found to be 15 dB at 13 GHz. For the TEM sensor, the maximum is 14 dB at 17 GHz, a result that reflects the sensor's somewhat better high frequency response. To put the relative performances of ISA and TEM sensor in perspective, we note that the ISA, with radiating elements that occupy an area of 7.6 x 19 cm^2, is only about 20% the size of the sensor, whose radiating element occupies about 15 x 46 cm^2. Thus, it may be argued that the ISA is about five times more efficient than the TEM sensor.

6. CONCLUSIONS

We designed, built, and tested a $1/8^{th}$-scale model of a new conformal UWB antenna, the Impulse Slot Antenna (ISA). We observed that the performance of the ISA is remarkably similar to that of the Farr Research TEM-1-50 sensor. This is an important result, because the TEM sensor has a radiating element with nearly five times the area (length × width) of the radiating elements of the ISA. Furthermore, unlike the larger TEM sensor, the ISA is a conformal design with little or no aerodynamic drag. Thus, with refinement, the ISA may become an alternative to the TEM sensor for applications requiring a conformal design.

Several modifications to the ISA may warrant future investigation in order to increase its gain and further reduce its return loss. The first is resistive loading of the outer radiating arm surfaces. The second is termination of the aperture at less than a 90-degree angle with the centerline, to smooth the transition to air. Other possibilities open up if we relax the requirement that the antenna conform to a 2:5 aspect ratio.

Finally, we note that we may expect even better performance in a full-scale ISA than in the $1/8^{th}$-scale model built here. This arises because the balun and twinline used here will perform much better at lower frequencies.

ACKNOWLEDGEMENTS

We are pleased to acknowledge many helpful discussions on this topic with Dr. Carl E. Baum, formerly with the Air Force Research Laboratory, Directed Energy Directorate, and currently with the University of New Mexico. We also wish to thank AFRL/DE for funding this work.

REFERENCES

1. W. S. Bigelow, E. G. Farr, *et al.*, "Development of the Impulse Slot Antenna (ISA) and Related Designs," Sensor and Simulation Note 505, December 2005.
2. J. D. Kraus and R. F. Marhefka, *Antennas for all Applications*, Third Edition (McGraw-Hill, 2002), p. 383.
3. E. G. Farr and C. E. Baum, "Prepulse Associated with the TEM feed of an Impulse Radiating Antenna," Sensor and Simulation Note 337, March 1992.
4. R. Garg, *et al.*, "Tapered Slot Antennas," in *Microstrip Antenna Design Handbook*, (Artech House, Boston, 2001).
5. J. Noronha, *et al.*, "Designing Antennas for UWB Systems," *Microwaves and RF*, June 2003.
6. K. C. Gupta, *et al.*, *Microstrip Lines and Slotlines*, Second Edition, (Artech House, Boston, 1996).
7. L. H. Bowen, *et al.*, "Fabrication and Testing of a Membrane IRA," Sensor and Simulation Note 464, January, 2002.
8. Farr Research, Inc., *Catalog of UWB Antennas and HV Components* (Albuquerque, July 2006); http://www.Farr-Research.com.

A HIGH-VOLTAGE CABLE-FED IMPULSE RADIATING ANTENNA

Leland H. Bowen, Everett G. Farr, and William D. Prather*

1. INTRODUCTION

We describe here a high-voltage cable-fed Impulse Radiating Antenna (IRA) that was developed as part of an Ultra-Wideband (UWB) radar system using a single antenna and a directional coupler [1]. In this configuration, it is not possible to use a sharpening switch at the apex to handle high voltages, so we investigated using higher voltages on a standard IRA with splitter balun. The resulting antenna, referred to as the IRA-6, was built and tested at three different voltage levels. Note that a more complete version of this paper appeared in [2].

This antenna is intended to be operated with a pulser with 30 kV peak voltage, 150 ps risetime, 3 ns pulse duration (at 1/*e* of peak), and 1 kHz maximum pulse repetition frequency (PRF) (manufacturer's specs.). To prevent flashover at this voltage level, we incorporated a number of compromises into the antenna design at the splitter and feed point (focus), which led to somewhat larger reflections in the TDR than we are accustomed to seeing. These compromises also led to reduced realized gain, especially at higher frequencies. Nevertheless, the IRA-6 performed quite well as high as 4 GHz, which met the requirements for the radar system under study.

2. DESCRIPTION OF THE IRA-6

The IRA-6 is 1.52 m (5 ft.) in diameter, with a focal length of 0.56 m (F/D = 0.37). It has feed arms positioned at ±45° to the vertical, and it is manufactured from a spun aluminum reflector. It includes a ground plane in the horizontal plane of symmetry, which adds structural rigidity and reduces crosspol, as described in [3]. A photo of the IRA-6 is shown in Figure 1.

---

* Leland H. Bowen and Everett G. Farr, Farr Research, Inc. Albuquerque, NM, 87123; William D. Prather, AFRL/DE, Kirtland AFB, NM 87117.

Figure 1. Front view of the IRA-6.

The feed arms and ground plane were fabricated from aluminum honeycomb sheets produced by Teklam to reduce weight. This reduced the stress on the cable connections at the feed point (focus), which are sensitive to shock and vibration. The honeycomb sheets are 6.3 mm (0.25 in) thick with 0.5 mm (0.020 in) thick surfaces on each side of the honeycomb material. This provides approximately 80% of the strength of solid aluminum at approximately 20% of the weight. The edges of the honeycomb material were covered with copper tape to maintain electrical continuity between the two sides.

A new feature of the antenna is the high-voltage splitter balun, show in Figure 2. This uses a custom cable that is roughly equivalent to RG-213 in diameter, with a narrower center conductor to realize a 100 Ω impedance. An HN-Type panel-mount connector is used at the input port. The insulation on connector and the ends of the 100-Ω cables are both tapered, which is apparent on the right in Figure 2. The walls of the aluminum housing are also tapered, but in the reverse direction, to increase the high-voltage standoff between the center conductor and the housing wall. The minimum distance between the center conductors and the housing walls is 3.6 mm and the path length along the surface of the tapered dielectric is 19 – 21 mm.

The void inside the splitter is filled with Jet-Lube silicon dielectric compound DM-3, which has a dielectric strength of 20 kV/mm (500 V/mil). This compound was selected because it is available in a cartridge for a standard automotive type grease gun. This allows one to inject the compound under considerable pressure to completely fill the void. The tapers on the insulators and the inside walls of the splitter were designed to maintain a constant impedance, based on dielectric constants of 2.9 for the DM-3, 2.25 for the polyethylene insulation in the 100 Ω cable, and 2.1 for the TFE insulation in the HN-Type connector. We were unable to locate an accurate dielectric constant for the DM-3, so we used the value of 2.9, because this is typical for silicone compounds at low frequency. Note that dielectric constants are seldom specified at frequencies above 1 MHz, so accurate information is generally difficult to obtain. We show the splitter and feed cables on the back side of the IRA-6 in Figure 3.

Figure 2. High voltage splitter, front (left), back removed (right).

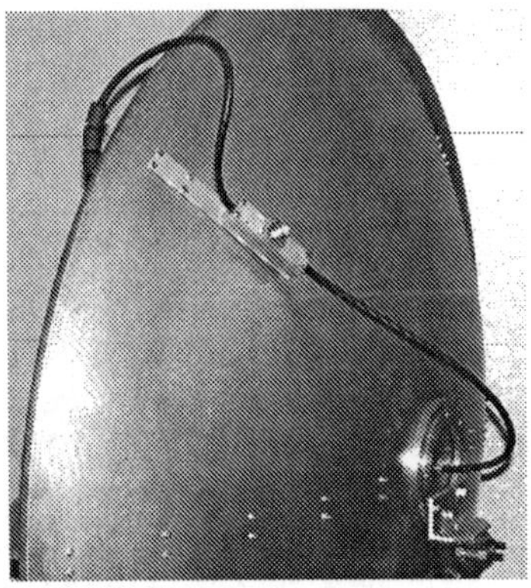

Figure 3. Splitter with feed cables mounted on back side of IRA-6.

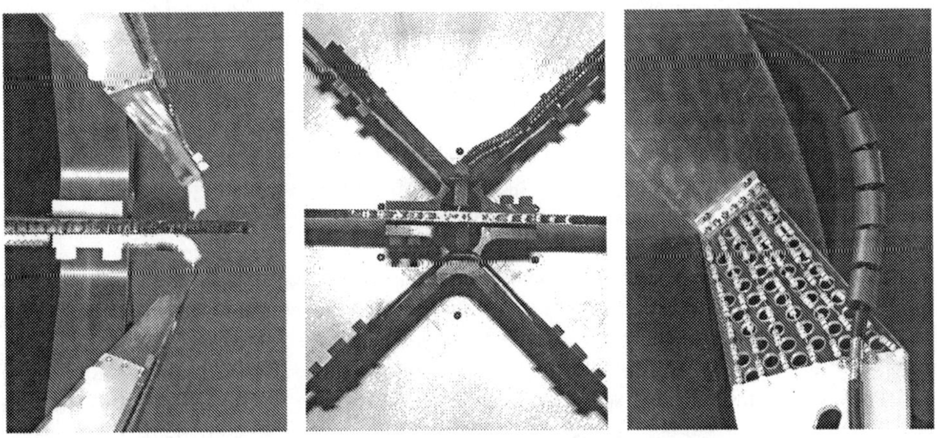

Figure 4. IRA-6 feed point from the left (left) and front (center), and detail of the load resistors and ferrite beads on feed cable (right).

At the feed point are two cable stubs that are 19 mm (0.75 in.) in length, to prevent flashover. Two close-up views of the feed point are shown in Figure 4. The tight bend at the tip of the bottom cable was formed by warming the dielectric with a heat gun, bending the insulation to the required shape, and allowing the plastic to cool.

The high-voltage splitter consists of two cables of equal length, connected in parallel at the connector and in series at the focus. We found it necessary to tune the cable lengths by measuring the time of arrival of the pulse on either side of the ground plane with a home-made sensor. The sensor was a semi-rigid cable with center conductor protruding and bent at an angle of 90 degrees. It was carefully placed at the same position on either side of the ground plane, to obtain an accurate time of arrival for the main pulse.

The resistive load consists of seven strings of 2-watt metal film resistors, as shown in Figure 4. The material supporting the resistors at the end of the feed arms is polycarbonate (Lexan®) with a thickness of 5 mm (3/16 in.) that has been drilled with many holes to reduce the capacitance at the end of the feed arms. The feed cable is routed away from the resistors and has five ferrite beads to prevent shorting out the resistors.

3. LOW-VOLTAGE MEASUREMENTS

We first characterized the IRA-6 using a low voltage pulser on the Portable Automated Time-domain Antenna Range (*PATAR*®) developed by Farr Research. The pulser, a Picosecond Pulse Labs model 4015C, has a negative 4 V peak voltage with a rise time of approximately 20 ps. The pulser drove a Farr Research model TEM-1-50 sensor to form the transmit portion of the antenna range. The receive section of the range consisted of the IRA-6 connected to a Tektronix model TDS8000 digital sampling oscilloscope with a model 80E04 sampling head. The TEM-1-50 and the IRA-6 were supported approximately 3 m above the ground with the apertures spaced 20 m apart.

In Figure 5 we show the TDR of the IRA-6 on the left. The reflections at the splitter and feed point are due to the compromises we have made to accommodate high voltage standoff. The TDR of the splitter is not quite as smooth as expected, possibly due to voids in the dielectric gel. The TDR at the feed point was initially much worse, but we were able to improve the configuration considerably by tuning the cable lengths and bending the cable stubs. Reflections at the feed point may be a concern; because they may cause ghost images in a radar measurement.

The impedance match in the vicinity of the feed arm loads is quite good. We experimented with various cable placements near the resistors, testing whether the TDR could be improved by placing the cable farther away from the resistors. However, we found this made little difference.

In order to provide a point of comparison, we also provide data on the Farr Research model CIRA-2, which is a collapsible man-portable IRA, shown in Figure 6. The CIRA-2 is only a little smaller than the IRA-6 (1.22 m diameter instead of 1.5 m), so the comparisons should be useful. Additional information on the CIRA-2 may be found in [4].

The TDR of the CIRA-2 is shown on the right in Figure 5. The CIRA-2 has no high voltage requirement, so the splitter and feed point are much smaller than those on the IRA-6, resulting in a much flatter TDR. At the load, however, there is a larger dip than in the IRA-6. This is due to the fact that IRAs with feed arms positioned at $\pm 30^{\circ}$ should have loads concentrated at the edge of the feed arms, instead of being evenly distributed, as shown in [5]. However, distributed resistors are probably necessary in a collapsible IRA for practical reasons.

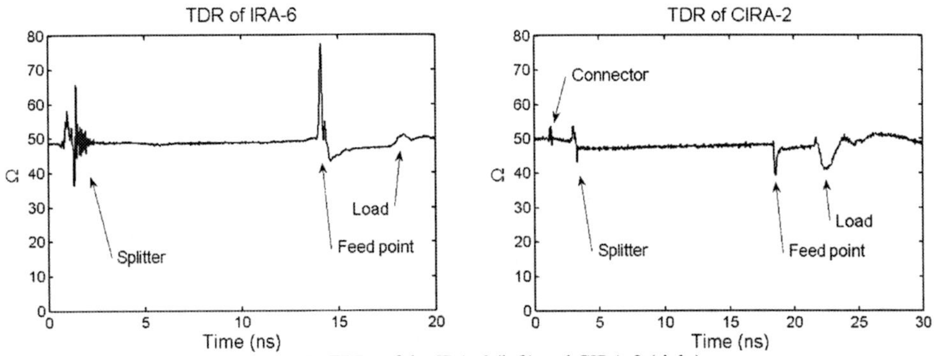

Figure 5. TDRs of the IRA-6 (left) and CIRA-2 (right).

Figure 6. Farr Research model CIRA-2.

Next, we show the normalized impulse responses for the IRA-6 and CIRA-2 in Figure 7. The FWHM for the IRA-6 is much greater than that of the CIRA-2, due to the insulated stubs required for high voltage operation. It may be possible to improve the high frequency response of the IRA-6 by using shorter cable stubs at the feed point. One might also try machining ribs into the stubs to decrease the stub length while maintaining voltage standoff. However, this would eliminate the possibility of tuning the cable lengths to synchronize the response of the two antenna halves.

We show the realized gains for the IRA-6 and CIRA-2 in Figure 8. The gain of the IRA-6 drops off above 4 GHz due to the high-voltage design. However, below 2 GHz the gain of the IRA-6 is higher than that of the CIRA-2. This is expected, since the IRA-6 is larger in diameter by a factor of 1.25. The peak realized gain of the IRA-6 is approximately 20 dBi at 3.12 GHz.

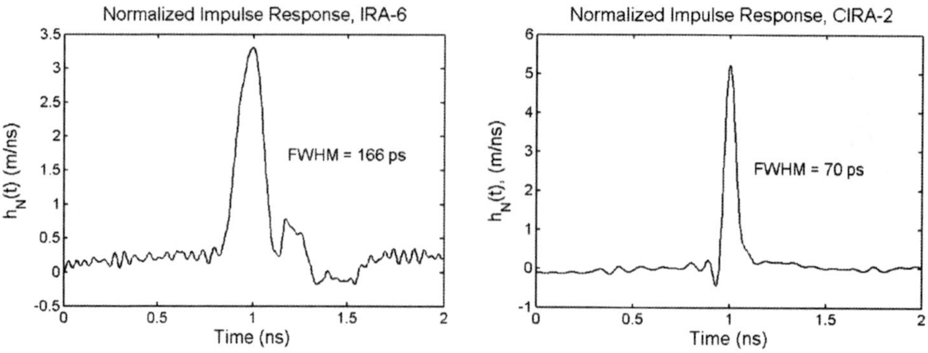

Figure 7. Normalized impulse response for the IRA-6 and the CIRA-2.

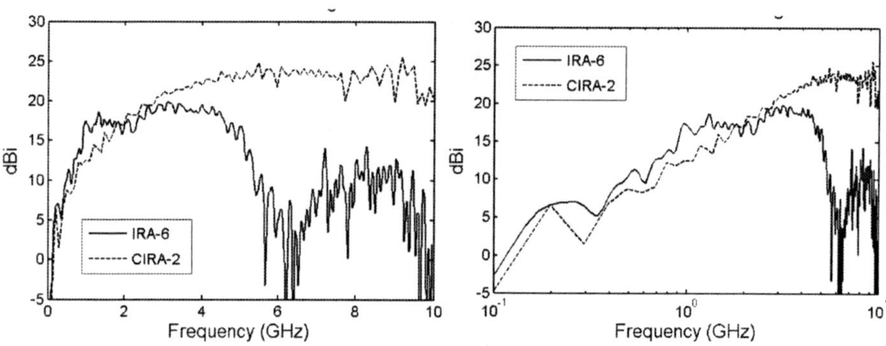

Figure 8. Realized gain on boresight for the IRA-6 and CIRA-2.

4. MID-VOLTAGE MEASUREMENTS

Next, we tested the IRA-6 with a Grant Applied Physics model HYPS pulser. The output of this pulser is a negative voltage pulse with a peak of –4.5kV. We show the output of the HYPS in Figure 9, on the left. We characterize the pulser in terms of the FWHM of the derivative of the output voltage, which we find to be 139 ps.

We drove the IRA-6 with the HYPS pulser and measured the field at a distance of 20.4 m. The electric field was measured using a Farr Research model TEM-1-50 sensor connected to a Tektronix model TDS8000 digital sampling oscilloscope. The received voltage was converted to incident electric field using the effective height of the sensor h_{eff} and the equation [6, eqn. A.3]

$$V_{rec}(t) \approx h_{eff}\, E_{inc}(t)$$

For the TEM-1-50, h_{eff} is 0.018 m. The E-field is shown in Figure 9, on the right, where we see that the peak field is approximately 600 V/m.

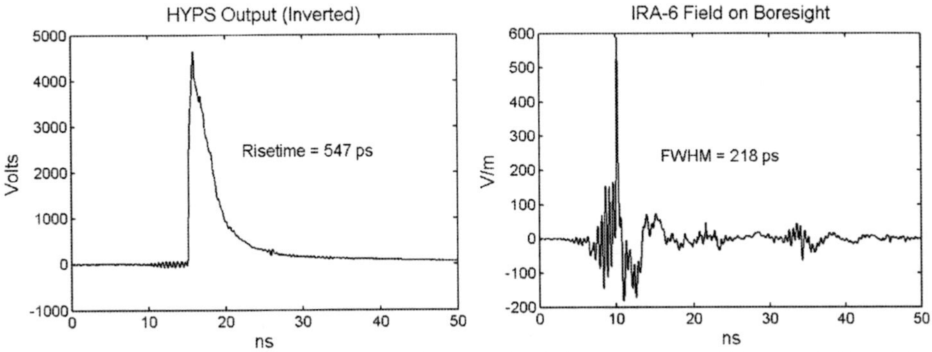

Figure 9. Derivative of the voltage output from the Grant Applied Physics model HYPS pulser (left), and the radiated field on boresight from the IRA-6 when driven by the HYPS pulser.

5. HIGHER VOLTAGE MEASUREMENTS

Finally, we tested the IRA-6 when driven by a higher voltage pulser, the FID model FPG 30-1KM.

To characterize the FID pulser, we fired it into a dummy load, and measured the voltage using a V-dot cable sensor, the Farr Research model VDC-1. The resulting waveform had to be integrated, and the result is shown in Figure 10, on the left. We found the FWHM of the derivative waveform to be 170 ps, which is close to the manufacturer's specification of a 150 ps 10-90% risetime. The peak voltage is 25 kV.

Having characterized the source, we then measured the radiated field when the FID pulser drives the IRA-6. We measured the field using a Farr Research model TEM-1-50 sensor (h_{eff} = 0.018 m) at a distance of 20.4 m, using a Tektronix model TDS6804 oscilloscope. There was a 2-meter cable of type RG 58 A/U between the sensor output and source, and the effect of this cable was not deconvolved from the data. The sensor was accounted for by a scalar correction factor.

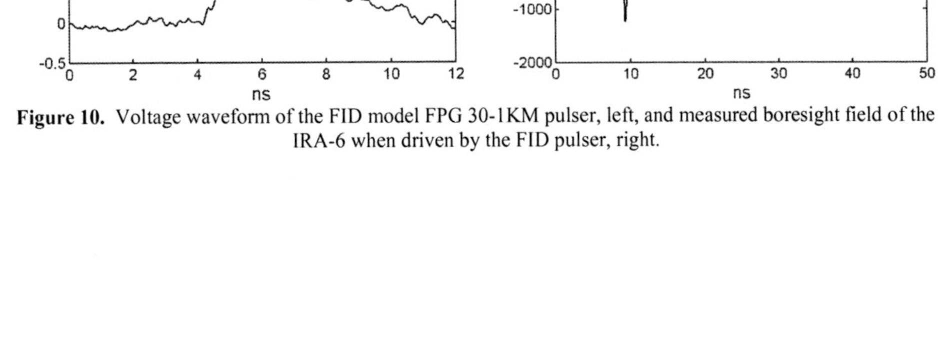

Figure 10. Voltage waveform of the FID model FPG 30-1KM pulser, left, and measured boresight field of the IRA-6 when driven by the FID pulser, right.

The radiated field on boresight is shown in Figure 10, on the right, where we observe a peak radiated field of 2.8 kV/m, and a FWHM of 254 ps. Since we had already characterized the IRA-6 at low voltage, there was little new to be learned from this data. However, it was important to demonstrate the ability to drive the IRA-6 at 25 kV without any apparent flashover. We can observe from the data this was indeed the case.

6. CONCLUSIONS

We were successful in developing the IRA-6, which is a higher-voltage version of a cable-fed IRA with a Type-HN input connector, and with a diameter of 1.5 meters. This antenna was able to tolerate a peak voltage of 25 kV without flashover, and it performed well up to 4 GHz. It is possible that the IRA-6 might be able to tolerate even higher voltages, but this was not tested. These characteristics were sufficient for the intended application of a UWB radar with directional coupler.

Further refinement of the splitter and feed point may lead to a flatter TDR, and improved risetime and realized gain. In radar applications, it is important that the TDR of an antenna be very flat, however, it becomes more challenging to achieve this at higher voltages. The TDR was very flat at the load resistors.

The peak realized gain of the IRA-6 was 20 dBi at 3.1 GHz. The peak measured E field at 20 m on boresight of the IRA-3 was 600 V/m when driven by a 4.5-kV pulser, and 2.8 kV/m with the 25-kV pulser.

Finally we found that the aluminum honeycomb material used in the feed arms and ground plane was very effective in reducing the weight of the antenna while maintaining strength and rigidity. The original design of the IRA-6 included some additional structural members that were not used because the honeycomb performed so well.

7. ACKNOWLEDGEMENTS

We are pleased to acknowledge many helpful discussions on this topic with Dr. Carl E. Baum, formerly with Air Force Research Laboratory, Directed Energy Directorate, and currently with University of New Mexico. We also wish to thank AFRL/DE for funding this work.

8. REFERENCES

1. L. M. Atchley, E. G. Farr, D. E. Ellibee, and D. I. Lawry, "A High-Voltage UWB Coupled-Line Directional Coupler," Sensor and Simulation Note 489, April 2004.
2. L. H. Bowen, E. G. Farr, and W. D. Prather, A High-Voltage Cable-Fed Impulse Radiating Antenna, Sensor and Simulation Note 507, December 2005.
3. L. H. Bowen, E. G. Farr, *et al*, Results of Optimization Experiments on a Solid Reflector IRA, Sensor and Simulation Note 463, January 2002.
4. L. H. Bowen, E. G. Farr, C. E. Baum, and W. D. Prather, "Experimental Results of Optimizing the Location of Feed Arms in a Collapsible IRA and a Solid IRA," Sensor and Simulation Note 450, November 2000.
5. E. G. Farr and L. H. Bowen, "The Relationship Between Feed Arm Position and Input Impedance in Reflector Impulse Radiating Antennas," Sensor and Simulation Note 499, April 2005.
6. L. M. Atchley, E. G. Farr, *et al*, "Characterization of a Time Domain Antenna Range," Sensor and Simulation Note 475, June 2003.

BROADBAND SELF-COMPLEMENTARY ANTENNA ARRAYS

Mats Gustafsson[1]

1. INTRODUCTION

Self-complementary antennas are basic prototypes of frequency independent antennas[1,2]. They exist both as single port antennas and antenna arrays. It is well known that planar self-complementary antennas have a constant impedance of $Z_0/2 \approx 188\,\Omega$, *i.e.*, half the intrinsic impedance of vacuum[1,2]. The situation changes for antenna arrays where it is common and often necessary to place a ground plane under the array. The destructive interference from the reflection of the ground plane limits the bandwidth[3]. These effects can be reduced by radar absorbing material between the antenna elements and the ground plane. This gives a broadband array at the expense of half the power is absorbed in the radar absorbing material.

In this paper, it is shown that stacking of dielectric slabs above planar self-complementary antenna elements can reduce the degrading effect of the ground plane and hence be used to design broadband antenna arrays. The dielectric slabs act as filters and transform the impedance of the antenna elements. The slabs are chosen to be of equal optical thickness, and, hence, resembling the use of quarter-wave length transformers in broadband matching[4–6]. The Bode-Fano theory is used to derive limitations on the bandwidth as a function of the Q-factor. Although the limitations are based on an approximate circuit model the results are useful and enhance our understanding of the planar array. Moreover it is shown that the self-complementary patch array offers the possibility to design an in principle grating lobe free broadband array. Of course, this requires a very short inter element spacing and hence a very complex feeding network. Numerical results are presented for the infinite antenna array with broadside bandwidths of 4.7:1 at -13dB and of 5.5:1 at -17dB for the cases of two and three dielectric slabs, respectively.

[1]Department of Electroscience, Lund University, Box 118, Lund, Sweden

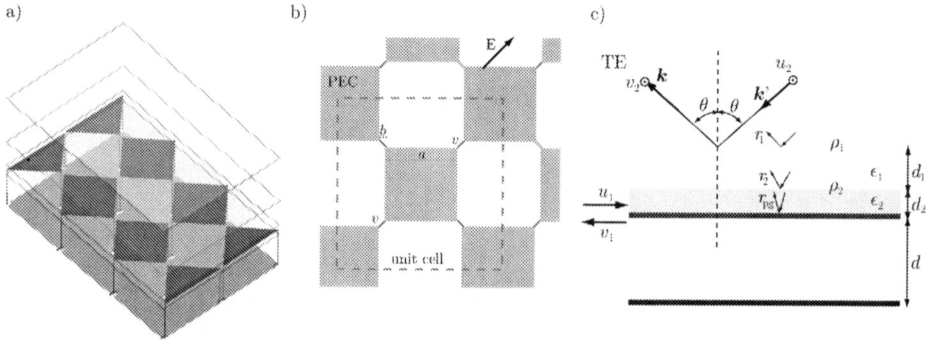

Figure 1: a) The array geometry. b) top view. The infinite array consists of a periodic repetition of square perfectly electric conductor (PEC) patches fed at the corners. Use of the feed points labeled with a v give a linear polarized field in the 45° direction as illustrated by the arrow. c) side view. Dielectric slabs are stacked above the patches. Scattering parameters of the array antenna, where the input signal, u_1, and output signal, v_1, relates to the antenna elements and the incident wave, u_2, and scattered wave, v_2 are the electromagnetic waves above the array.

The use of dielectric slabs to improve antenna performance is not new. A dielectric slab can be used for wide-angle impedance matching of planar arrays[7]. It has also been shown that dielectric slabs can be used to improve the bandwidth of an array composed of closely spaced dipoles[4]. The bandwidth can also be increased by the use of high impedance material under the antenna elements[4].

2. SELF-COMPLEMENTARY PATCH ELEMENTS

Here, we consider an infinite antenna array consisting of PEC patches as depicted in Figure 1. The patches are fed at the corners of each patch[8] giving a linearly polarized field in the ±45° directions depending on the used feed points. The patch array is almost self complementary, *i.e.*, the PEC structure is almost identical to its complement. The infinite antenna array can be simulated with either the FDTD, MoM, or FEM methods as long as the code can handle periodic boundary conditions[4,9,10]. Here, the code periodic boundary FDTD (PB-FDTD) developed by H. Holter[10] is used. Numerical simulations, using PB-FDTD, verify that the impedance is frequency independent and equal to $Z_0/2$ in the considered frequency range at broad-side radiation. The input impedance normalized to 189 Ω for the frequency range 1 GHz to 30 GHz is seen as the dot in the center of the Smith chart in Figure 2a. For larger scan angles, θ, and low frequencies, the impedance is approximately half the transverse impedance of free space η_{T}, *i.e.*,

$$Z_{\mathrm{TE}} \approx \frac{\eta_{\mathrm{TE}}}{2} = \frac{\eta_0}{2\cos\theta} \quad \text{and } Z_{\mathrm{TM}} \approx \frac{\eta_{\mathrm{TM}}}{2} = \frac{\eta_0\cos\theta}{2}. \tag{1}$$

Observe that the H- and E-planes correspond to the TE and TM cases, respectively. The impedance does not depend strongly on the frequency as long as we are far below the onset of grating lobes. For the case of a load $Z = R$, we get the active

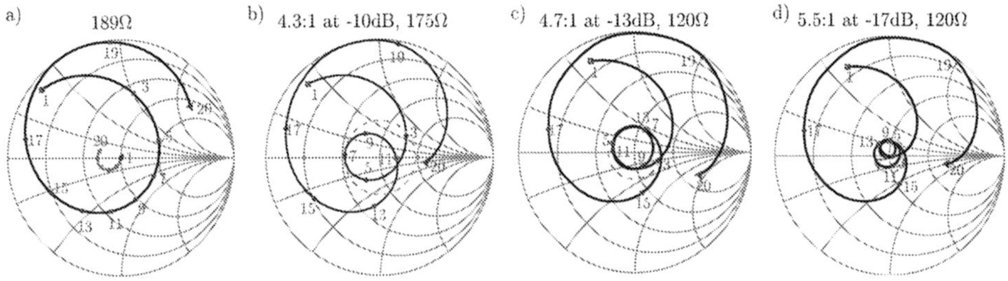

Figure 2: FDTD simulated impedance at broadside scan. The frequencies are given in GHz. a) the patch array as the dot in the center, patch array together with the environmental protection as the short arc leaving the center, and the patch array with environmental protection and the ground plane at the distance $d = 8\,\mathrm{mm}$. The ground plane transforms the impedance to rotate around $Z_0/2$. b) impedance normalized to $175\,\Omega$ for the single dielectric slab with $\epsilon_1 = 4$ giving a $-10\,\mathrm{dB}$ bandwidth of 4:1. c) impedance normalized to $120\,\Omega$ for the two dielectric slabs case with $d = 8\,\mathrm{mm}$, $\epsilon_1 = 7$, and $\epsilon_2 = 3$ giving a $-13\,\mathrm{dB}$ bandwidth of 4.7:1. d) impedance normalized to $120\,\Omega$ for the three dielectric slabs case with $d = 8\,\mathrm{mm}$, $\epsilon_1 = 7.2$, $\epsilon_2 = 3.4$, and $\epsilon_3 = 1.8$ giving a $-17\,\mathrm{dB}$ bandwidth of 5.5:1.

antenna reflection coefficient $\rho = \frac{\eta_{\mathrm{T}}-2R}{\eta_{\mathrm{T}}+2R}$. These are effective parameters of the self-complementary patch array. A homogeneous resistive sheet with conductivity $\sigma \to \infty$ and thickness $d \to 0$ such that $\sigma d = R^{-1}$ has similar scattering properties as the patch array in the low-frequency limit[11].

A thin dielectric slab ('dielectric underwear'[4]) is used as an environmental protection of the patch array. From the results in Figure 2a, it is observed that the thin dielectric slab ($1\,\mathrm{mm}$ with $\epsilon = 2.33$) hardly changes the impedance at all. The effect is seen as the small arc going from the center in the Smith chart. Due to the constant impedance character of the self-complementary array, the effect of a ground plane, here at the distance $d = 8\,\mathrm{mm}$, is profound as seen in Figure 2a. The impedance grazes the rim of the Smith chart at approximately 18 GHz, corresponding to the destructive interference of a ground plane distance of half a wavelength[3]. The effect of changing the ground-plane distance is mainly a rotation and stretching of the impedance in the Smith chart, *i.e.*, a frequency scaling.

We now consider the patch array together with its environmental protection as fixed and improve the bandwidth by placing dielectric slabs above the elements. The dielectric slabs act as a filter matching the antenna for a range of frequencies $f_{\mathrm{l}} \leq f \leq f_{\mathrm{u}}$. The upper frequency f_{u} is limited by the onset of grating lobes and the destructive interference from a ground plane at half a wavelength distance. In analogy with quarter-wave transformers in broadband matching, the ground plane distance and the slabs are chosen to be of equal optical thickness, *i.e.*, a slab thickness of $d/\sqrt{\epsilon_i}$ is used[5]. The case with a single dielectric slab is easily analyzed with a parametric study. The result with a single dielectric slab is seen in Figure 2b. In this case the dielectric slab can be designed to give one single loop in the center of the Smith chart. The $-10\,\mathrm{dB}$ bandwidth of approximately 4:1 is comparable to the case of wire dipoles above a ground plane without dielectric slabs[4].

It is reasonable that the bandwidth can be improved by stacking more dielectric

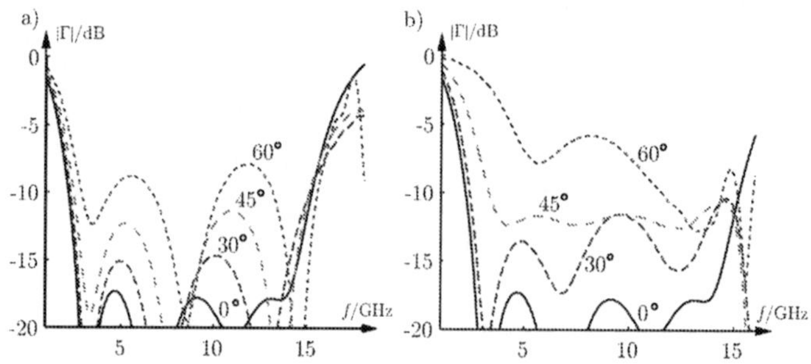

Figure 3: Simulated reflection coefficients normalized to 120 Ω for the three slab case for the scan angles of 30°, 45°, and 60°. a) H plane. b) E plane.

slabs above the patch array. As the number of slabs increases, the parametric study gets more involved. The effect of stacking several dielectric slabs above the patch array can be analyzed with a global optimization algorithm, *e.g.*, the Genetic Algorithm [12]. However, empirical studies have showed that the permittivities can be chosen from a parametric study of a set of slabs generated by a constant reflection coefficient between two slabs, *i.e.*, $\epsilon_i = \epsilon_{i+1}(1+\rho)^2(1-\rho)^{-2}$ for $i = 1, .., N$ where N is the number of slabs (here $N = 2$ or $N = 3$) and $\epsilon_{N+1} = 1$. The parametric study (or line search) in ρ gives good initial values of the permittivities. These values are easily improved by the use of a parametric study.

The loops are centered in the Smith chart with a normalization of 120 Ω as seen in Figure 2cd. This gives a −13 dB bandwidth of 4.7:1 and a −17 dB bandwidth of 5.5:1 for the cases of two and three dielectric slabs, respectively. As seen in Figure 2c, the impedance makes two overlaying loops in the Smith chart with two slabs. The third slab adds a loop and hence increases the bandwidth and tightens the impedance to the center of the Smith chart. The property of adding loops in the center of the Smith chart is very favorable as it gives an almost constant magnitude of the reflection coefficient over the matched frequency range. It is also interesting to observe that the property of adding loops in the Smith chart is similar to the result of Chebyshev transformers where each quarter-wave transformer adds an approximate loop in the Smith chart [5].

The magnitude of the reflection coefficient $|\Gamma|$ is used to illustrate the behavior versus the scan angle. The effect of increasing scan angles are shown in Figure 3 for the three slab case, see [13,14] for the corresponding two slab case. The scan angles 30°, 45°, and 60° are considered in both the H-plane and E-plane, where the H-plane and E-plane are the ±45° diagonal planes, see Figure 1b. As seen in Figure 3, the reflection coefficient increases with increasing scan angle as expected. This corresponds to input impedance loops with an increased radius in the Smith chart. Hence, the bandwidth reduces as the scan angle increases.

As the impedance of a self-complementary structure is independent of the geometry of the structure it is reasonable that the input impedance of the patch array does not depend strongly on the dimensions of the patch elements. In Figure 4a, the

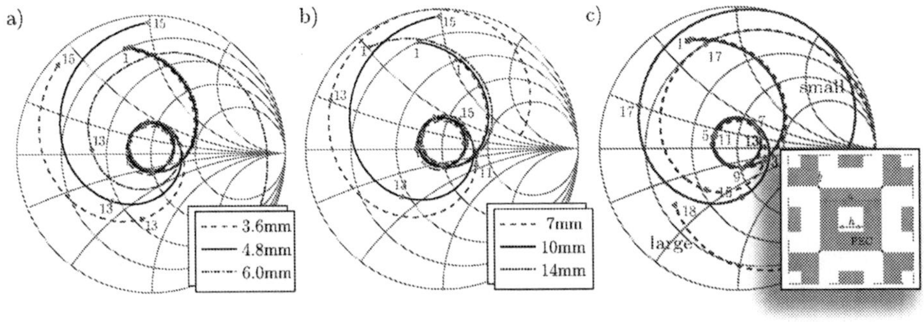

Figure 4: Patch array with two dielectric slabs. The impedance is simulated and normalized to 120 Ω. a) variation of the patch width a for a fixed ground plane distance $d = 10\,\text{mm}$ and fixed dielectric slabs. b) variation of the ground plane distance d and slab thicknesses for a fixed patch width of $a = 4.8\,\text{mm}$. c) Self-complementary arrays with small and large holes.

input impedance, normalized to 120 Ω, of the two slab case is shown for the patch widths 3.6 mm, 4.8 mm, and 6.0 mm. The ground plane distance is 10 mm and the slab parameters are as in Figure 2c. As seen in Figure 4a, the input impedance is almost independent of the patch width up to 12 GHz. For higher frequencies the input impedance start to differ as the distance between two feed points approach half a wavelength and hence the onset of grating lobes. The onset of grating lobes at 15 GHz corresponds to a patch width of just above 6 mm. The frequency independent property of the patch array can also be seen in Figure 4b, where the the ground plane distance is changed to 7 mm and 14 mm.

To illustrate how the self-complementary property of the patch array is utilized in the antenna, the patch array is reshaped into a new self-complementary structure with square holes, Figure 4c. The simulated input impedance for the cases with small ($h = a/3$) and large holes ($h = 2a/3$) are seen in Figure 4c. As seen in the figure, the input impedance of the two cases is similar for low frequencies. As the frequency increases the input impedances start to differ and for high frequencies (above 15 GHz) they are quite different. However, the considered patch array is only approximately self-complementary, *i.e.*, the PEC patches are a little smaller then their complementary structure, due to the feed-point distance $b = a/12$. As the properties of the patch array do not depend strongly on the patch width it is possible to design an antenna with grating lobes far above the operational band. Of course it is necessary to consider the effect of the feeding structure for this case.

3. SCATTERING MODEL

To analyze the effect of the antennas elements on the array, we simplify the array in several steps. To start we assume that the array is large enough such that the array can be approximated with an infinite array. This gives a unit cell analysis. A voltage gap model is used to feed the antenna elements. Moreover, a transmission

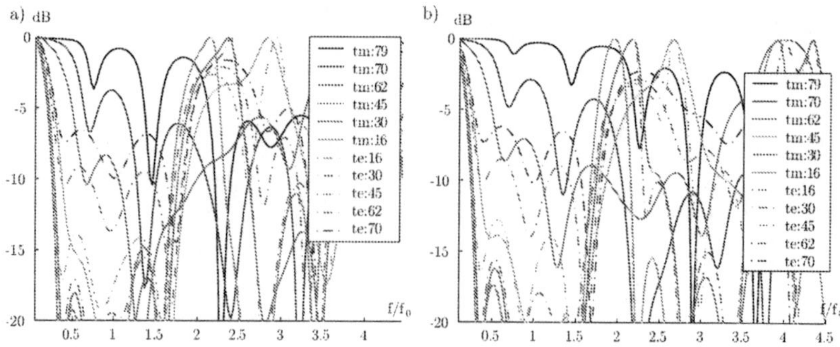

Figure 5: The reflection coefficient of the effective patch array with $R = 120\,\Omega$ for several scan angles, where f_0 is the resonance frequency. a) PB-FDTD. b) scattering matrix approximation.

line model is used to get a circuit model that can be analyzed to characterize the properties of the antenna elements. We use a scattering matrix to model the scattering properties of the element array

$$\begin{pmatrix} \rho_{\mathrm{p}} & \tau_{\mathrm{p}} & \tau_{\mathrm{p}} \\ \tau_{\mathrm{p}} & r_{\mathrm{p}} & 1 + r_{\mathrm{p}} \\ \tau_{\mathrm{p}} & 1 + r_{\mathrm{p}} & r_{\mathrm{p}} \end{pmatrix} \begin{pmatrix} u_1 \\ u_2 \\ u_3 \end{pmatrix} = \begin{pmatrix} v_1 \\ v_2 \\ v_3 \end{pmatrix} \tag{2}$$

where u_i and v_i are the incident waves and reflected waves, respectively, see Figure 1c. The scattering parameters depend on the frequency, scan angle, and polarization. We assume that the cross polarization is negligible. The scattering matrix is reciprocal and lossless. Conservation of power simplifies the relations between the scattering coefficients. We assume that the scattering parameters are real-valued in the low frequency limit. This gives $r_{\mathrm{p}} = -\frac{1}{2} - \frac{\rho_{\mathrm{p}}}{2}$ where ρ_{p} is given by the miss match of the feed point impedance.

The patch array can be tuned to resonance at an arbitrary frequency by placing a ground plane under the patch array at approximately quarter of a wavelength distance. The reflection coefficient of the ground plane is $r_{\mathrm{g}} = -\exp(-\mathrm{i}2kd\cos\theta)$. It is easiest to determine the scattering properties of the element array together with ground plane and dielectric sheets from the reflection of a set of dielectric sheets. The successive reflection coefficients of the dielectric slabs are determined with recursions[5]. The approach with the scattering matrix approximation gives a good understanding of the basic physics of the antenna array. A comparison between FDTD simulations and the scattering matrix approximation for the two dielectric slab case is shown in Figure 5. We observe that the basic structure of them is similar up to twice the resonance frequency, $f = 2f_0$.

The basic limitations of the antenna design is analyzed with the 1D or transmission line model[4] corresponding to the scattering problem in Figure 1c. Here, the dielectric slabs are modeled as a lossless matching network. The case of a resonance

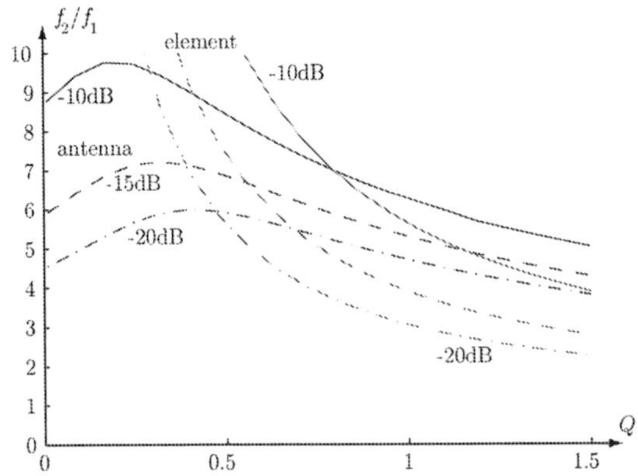

Figure 6: Bode-Fano limit on the bandwidth as a function of the Q-factor for the reflection coefficients $-10\,\mathrm{dB}$, $-15\,\mathrm{dB}$, and $-20\,\mathrm{dB}$.

circuit is considered as a model of the antenna elements, *i.e.*,

$$Z_{\mathrm{A}}(s) = Z_{\mathrm{A}}(\mathrm{i}\omega) = R + \mathrm{j}X_{\mathrm{A}}(s) = R(1 + Z_{\mathrm{Q}}(s)) = R + RQ\left(\frac{\omega_0}{s} + \frac{s}{\omega_0}\right). \tag{3}$$

The Bode-Fano matching theory is used to get bounds on the reflection coefficient[15]. The Fano equations are solved as a constrained optimization problem. The bandwidths are shown in Figure 6 as a function of the Q-factor for the reflection coefficients $-10\,\mathrm{dB}$, $-15\,\mathrm{dB}$, and $-20\,\mathrm{dB}$. We observe that bandwidth is maximal for a small but non-zero Q-factor, *i.e.*, the dipole array is better than the self-complementary array in this respect. However, observe that the above analysis is performed for fixed impedance. It is obviously advantageous to increase the ground plane impedance, *i.e.*, by the use of a high permeability between the antenna elements and the ground plane[4].

4. CONCLUSIONS

In this paper it has been shown that infinite self-complementary antenna arrays above a ground plane together with dielectric slabs above the antenna elements can be used to design broadband antennas. The dielectric slabs match the impedance of the antenna elements to free space. It is shown that, at least for the first three slabs, each slab adds one loop to the input impedance in the Smith chart. Moreover, the radius of the loops reduces with increasing number of slabs, and hence reduces the reflection coefficient over a large bandwidth. It is interesting to observe that the circular loop pattern gives an almost constant reflection factor over the matched frequencies. The presented results based on an infinite antenna and a simple feed model indicates that dielectric slabs are useful in the design of broadband arrays based on self-complementary structures. When realizing an antenna design it is of course necessary to improve the model of the feeding network, analyze finite arrays, and obtain experimental verification.

It is interesting to compare the performance of the self-complementary array presented here with results for arrays composed of closely spaced wire dipoles presented in[4]. In free space, the dipole array is broadband but not frequency independent as the self-complementary array. This is utilized in the dipole case by carefully balancing the reactive effects between the dipoles and the ground plane, and hence increasing the bandwidth[4]. This is also shown by a simple Q-factor model, where the Q-factor is the Q-factor of the infinite planar array in free space. The ground plane and the dielectric slabs are included with a transmission line model or equivalently cascade coupling between appropriate scattering matrices. The Bode-Fano matching theory shows that it is advantageous to have small but non zero Q-factor, and hence the analysis confirms the results in[4].

5. ACKNOWLEDGMENTS

I thank Anders Höök at Ericsson Microwave systems (EMW), Sweden, for valuable discussions. The support by EMW and FMV is also gratefully acknowledged.

References

1. J. D. Kraus and R. J. Marhefka, Antennas, 3rd ed. New York: McGraw-Hill, 2002.
2. Y. Mushiake, "Self-complementary antennas," IEEE Antennas and Propagation Magazine, vol. 34, no. 6, pp. 23–29, 1992.
3. X. Dardenne and C. Craeye, "Simulation of the effects of a ground plane on the radiation characteristics of self-complementary arrays," IEEE Antennas and Propagation Society International Symposium, vol. 1, pp. 383–386, 2003.
4. B. Munk, Finite Antenna Arrays and FSS. New York: John Wiley & Sons, 2003.
5. S. J. Orfanidis, "Electromagnetic waves and antennas," 2002, www.ece.rutgers.edu/ orfanidi/ewa, revision date June 21, 2004.
6. D. M. Pozar, Microwave Engineering. New York: John Wiley & Sons, 1998.
7. E. G. Magill and H. A. Wheeler, "Wide-angle impedance matching of a planar array antenna by a dielectric sheet," IEEE Trans. Antennas Propagat., vol. 14, no. 1, pp. 49–53, 1966.
8. D. McGrath and C. Baum, "Scanning and impedance properties of TEM horn arrays for transient radiation," IEEE Trans. Antennas Propagat., vol. 47, no. 3, pp. 469–473, 1999.
9. A. F. Peterson, S. L. Ray, and R. Mittra, Computational Methods for Electromagnetics. New York: IEEE Press, 1998.
10. H. Holter and H. Steyskal, "Infinite phased-array analysis using FDTD periodic boundary conditions—pulse scanning in oblique directions," IEEE Trans. Antennas Propagat., vol. 47, no. 10, pp. 1508–1514, 1999.
11. M. Gustafsson, "RCS reduction of integrated antenna arrays with resistive sheets," J. Electro. Waves Applic., vol. 20, no. 1, pp. 27–40, 2006.
12. J. Johnson and Y. Rahmat-Samii, "Genetic algorithms in engineering electromagnetics," IEEE Antennas and Propagation Magazine, vol. 39, no. 4, pp. 7–21, 1997.
13. M. Gustafsson, "Broadband array antennas using a self-complementary antenna array and dielectric slabs," Lund Institute of Technology, Department of Electroscience, P.O. Box 118, S-221 00 Lund, Sweden, Tech. Rep. LUTEDX/(TEAT-7129)/1–8/(2004), 2004, http://www.es.lth.se.
14. ——, "Broadband array antennas using a self-complementary antenna array and dielectric slabs," in Antennas and Propagation Society International Symposium. IEEE., 2006.
15. R. M. Fano, "Theoretical limitations on the broadband matching of arbitrary impedances," Journal of the Franklin Institute, vol. 249, no. 1,2, pp. 57–83 and 139–154, 1950.

A QUASI-PLANAR WIDE BAND CONICAL ANTENNA

Mingyu Lu, Jonathan W. Bredow, Sungyong Jung, and Saibun Tjuatja*

1. INTRODUCTION

The conical antenna has wide input impedance bandwidth and omni-directional radiation pattern (Kawakami and Sato, 1987; Liang and Wah, 2000). This special property makes it irreplaceable in some modern applications, for instance, in portable or mobile communications requiring very wide-bandwidth channels or where continuous frequency coverage is needed. The prototype of the conical antenna has a three-dimensional structure. It is mechanically difficult to fabricate and integrate. In (Gentili et al., 2004), a dielectric loaded conical antenna is reported, where the dielectric loading mechanically stabilizes the conical antenna while maintaining wide band input impedance characteristics. However in (Gentili et al., 2004), the loading material is magnetic, hence not practical; and the antenna still has a three-dimensional configuration. In this study, a conical antenna with metallic cones coated on non-magnetic dielectric slab is proposed. The proposed antenna has a quasi-planar structure. It is mechanically stable, and easy to build and integrate with planar circuits. A full wave analysis code is developed to simulate this novel conical antenna. It is shown that the input impedance remains close to a constant value when the loading material's dielectric constant is chosen within a wide range. A quasi-planar conical antenna is fabricated with high density polyurethane foam as the loading material. The measurement data verifies the simulation results. Techniques to reduce the antenna size and adjust the radiation pattern are also discussed.

2. ANTENNA DESIGN AND ANALYSIS

The proposed antenna is illustrated in Figure 1(a), and its cross section is shown in Figure 1(b). A cylindrical dielectric slab is made of homogeneous material with permittivity $\varepsilon_r \varepsilon_0$ and permeability μ_0, where ε_0 and μ_0 are the permittivity and

* M. Lu, J. W. Bredow, S. Jung, and S. Tjuatja, Wave Scattering Research Center, Department of Electrical Engineering, University of Texas at Arlington, Arlington, TX 76019.

permeability of the free space, respectively. One side of the slab is coated by metal and behaves as the ground plane. A conical cavity is etched in the slab, and metal is coated on the cone wall. The antenna is fed at the tip of the cone. The dielectric material loading makes the antenna mechanically stable, quasi-planar, and easy to fabricate. Furthermore, if any planar circuits are on the other side of the ground plane, they can be simply connected to the antenna through a via hole.

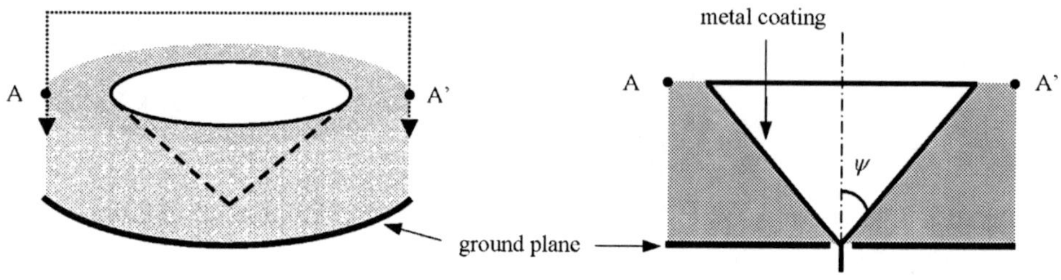

(a) Antenna configuration (b) Cross section (A-A' view)

Figure 1. The quasi-planar conical antenna

The radiation mechanism of this quasi-planar conical antenna (Figure 1) is similar to the classical conical antenna (Kawakami and Sato, 1987; Liang and Wah, 2000). Since the feed is located at the center of a revolutionarily symmetric structure, a spherical transverse electromagnetic (TEM) wave is launched in the dielectric material. The cone and the ground plane can be considered constituting a piece of transmission line with characteristic impedance (Kawakami and Sato, 1987),

$$Z_0 = \frac{1}{\pi}\sqrt{\frac{\mu_0}{\varepsilon_r \varepsilon_0}} \ln\left[\cot\left(\frac{\psi}{2}\right)\right]. \tag{1}$$

In Eq. (1), ψ denotes the angle of the cone (Figure 1(b)). When the TEM wave hits the end of the cone, it is reflected and scattered. With the increase of frequency, the reflection and scattering attenuate, and the antenna behaves more and more like an infinitely long transmission line. Compared with the conventional conical antennas, the additional dielectric loading makes the wave's behavior more complicated. Firstly, the wavelength within the dielectric material is shorter than that in the air. As a result, the electrical length of the transmission line is enlarged due to the dielectric loading, which is helpful to the high frequency input impedance performance. However secondly, because of the dielectric-air interface, the reflection and scattering at the end of the cone is larger in the dielectric loaded antenna, making the antenna less matched to free space. Thirdly, the dielectric material forms a cavity, like that underneath microstrip patch antennas (Richards et al., 1981). This cavity tends to store energy, hence would reduce the antenna's bandwidth. Fourthly, a dielectric material's electrical properties usually vary with respect to frequency. Specifically, the dielectric constant ε_r depends on frequency (this dependence is implicit in Eq. 1). Such variations may deteriorate the antenna's wide band feature if not carefully taken into account. Therefore, it is not easy to predict the effect of dielectric loading in a straightforward manner.

To analyze the dielectric loaded conical antenna, a full wave simulation is carried out. The numerical code is based on the method of moments (MOM) following (Medgyesi-Mitschang et al., 1994). The antenna geometry is illustrated in Figure 2(a). The whole space is divided into two regions: Region 1 is filled with air; and Region 2 is the dielectric region. Parts of the surface between the two regions are made of metal (depicted by solid lines in Figure 2(a)); the other parts are dielectric-air interfaces (dashed lines). Two equivalent problems are established in Figure 2(b) and Figure 2(c), respectively. In Figure 2(b), there is air in the whole space; the exterior fields are the same as those in Figure 2(a); and the interior fields are zero. In Figure 2(c), the whole space is filled with the dielectric material ($\varepsilon_r \varepsilon_0$, μ_0); the interior fields are the same as those in Figure 2(a), and the exterior fields are zero. In both equivalent problems, electric currents ($\mathbf{J}_1$ and $\mathbf{J}_2$) and magnetic currents ($\mathbf{M}_1$ and $\mathbf{M}_2$) exist on the surface between the two regions. Since the excitation is at the origin, all the fields and currents have revolutionary symmetry. Then, integral equations can be constructed and numerically solved as detailed in (Medgyesi-Mitschang et al., 1994).

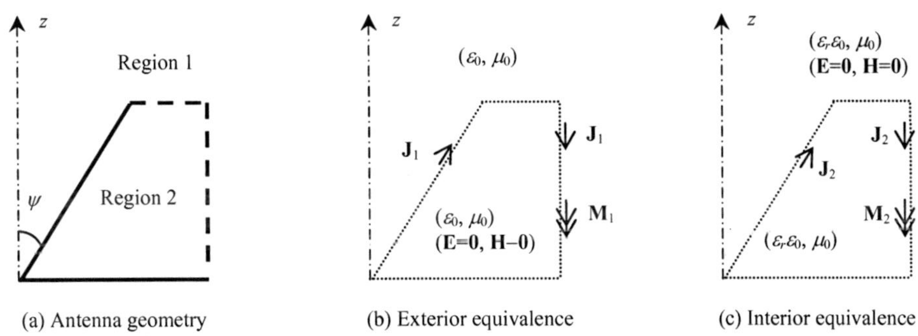

Figure 2. Illustration of the numerical modeling of the antenna

Numerical simulation indicates that the choice of loading material is critical to the antenna's electrical performance. If the dielectric constant is too large, much power is stored within the dielectric material instead of being radiated. The antenna then becomes narrow band due to the high quality factor. On the other hand, when ε_r is reasonably small (less than 4 as shown in the next section), the antenna's input impedance remains close to a constant within a wide frequency band. Of greater importance, the wide-band characteristics are not sensitive to the variation of ε_r. In other words, the value of ε_r does not have to be exactly known to design a dielectric loaded conical antenna; and as long as ε_r's dependence on frequency is not too strong, it does not degrade the antenna's wide band feature. As a result, a wide range of materials can be chosen for the proposed quasi-planar conical antenna. Obviously, low conductivity should be another criterion for the loading material to minimize the dielectric loss.

3. NUMERICAL AND EXPERIMENTAL RESULTS

In this section, some numerical and experimental results for the proposed quasi-planar conical antenna are presented. It is assumed that the antenna is to be matched to characteristic impedance of $50\,\Omega$.

The antenna configuration illustrated in Figure 3(a) is simulated, fabricated, and measured. The fabricated antenna is shown in Figure 3(b). The dielectric slab is made of high density polyurethane foam. Copper sheet is attached to the lower side of the slab as ground plane. The conical wall is coated by silver paint. The antenna is fed by an SMA coaxial connector. Figures 4 and 5 show the simulated voltage standing wave ratio (VSWR) at the antenna feed obtained from the MOM solver. In Figure 4, the dielectric constant of the loading material is assumed to be 1, 2, 3, and 4, respectively; and the cone angle is dictated by ε_r through Eq. (1) with $Z_0 = 50\,\Omega$. When $\varepsilon_r = 1$, the antenna reverts to the conventional conical antenna. Its input impedance approaches Z_0 when l, the side length of the cone, is larger than quarter wavelength. In practice, the high end of the frequency band is restricted by the feed structure. It is observed in Figure 4 that, with the increase of ε_r, there are more and more ripples in the VSWR curves. When ε_r is too large, the antenna loses its wide band feature. Indeed, too large value of ε_r is not practical for fabrication either. From Eq. (1), large ε_r results in small value of ψ. Since the function $\cot(\psi/2)$ changes very fast when its argument is small, a tiny error in the antenna fabrication could produce a large error in the input impedance. The data plotted in Figure 5 are similar to those in Figure 4. However in Figure 5, the cone angle ψ is fixed while ε_r changes. Specifically, ψ is determined by assuming $\varepsilon_r = 2$ in Eq. (1). The three curves in Figure 5 correspond to three loading materials with ε_r = 1, 2, and 3, respectively. Figure 5 indicates that, the quasi-planar conical antenna has wide band input impedance behavior even when the cone angle is not matched to the dielectric material. Based upon this observation, when the antenna is designed and fabricated, inaccuracy of the knowledge about ε_r or ε_r's variation with respect to frequency can be tolerated. A simple wideband measurement with a network analyzer shows that the dielectric constant of the high density polyurethane foam is between 1.4 and 1.9 in the frequency band of our concern. The antenna in Figure 3(b) is then fabricated with angle $\psi = 36.2°$. In the numerical simulation, ε_r is assumed to be 1.8. The simulated and measured VSWR results agree well with one another as shown in Figure 6. The VSWR stays below 2 over more than ten octaves. Because of the mechanical imperfection at the feed, at high frequencies, the measured input impedance has more impedance mismatching than the numerical one. The simulated and measured antenna gains are compared with each other in Figure 7, at frequencies 3 GHz and 6 GHz, respectively. The discrepancies between the two sets of results are largely caused by the metal loss of the silver paint used in the antenna in Figure 3(b). It is expected that the conductivity of the metal coating would be greatly improved when a plating approach is used.

The proposed quasi-planar conical antenna allows more flexible designs than the regular conical structure in Figure 3: the cone radiator's shape can be easily changed. For instance, a deformed conical antenna is illustrated in Figure 8, where the cone's shape is tilted upward. The deformation increases the antenna aperture, hence could result in size reduction (Liang and Wah, 2000). When the ground plane is also deformed, a dipole configuration is arrived at, as in Figure 9. By adjusting the angles of the dipoles, radiation pattern can be tuned while the input impedance is kept unchanged. The simulated VSWR results of the deformed monopole conical antenna in Figure 8 with $\varepsilon_r = 2$ are plotted in Figure 10, and compared with those for the antenna geometry in Figure 3 with $\varepsilon_r = 1$ and $\psi = 47°$. When VSWR<2 criterion is assumed, the tilted configuration has a smaller low end of the frequency band, as expected.

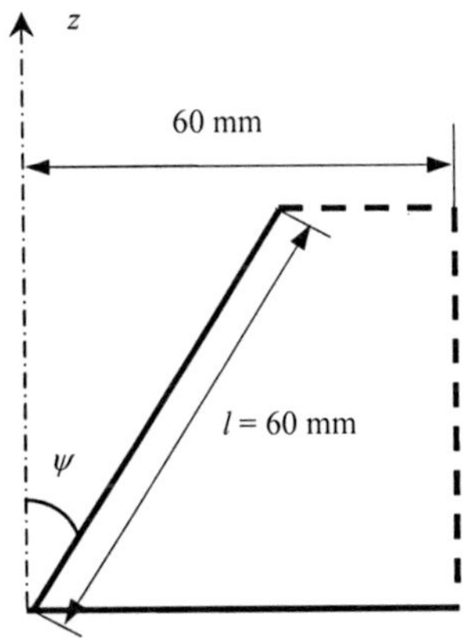

(a) Antenna geometry

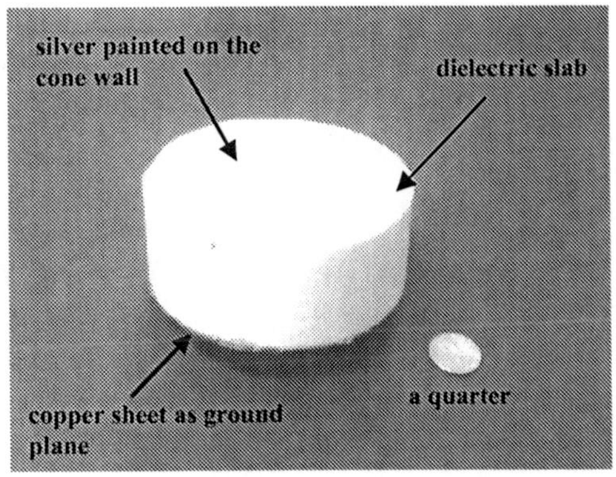

(b) Antenna picture

Figure 3. A quasi-planar conical antenna fabricated and tested

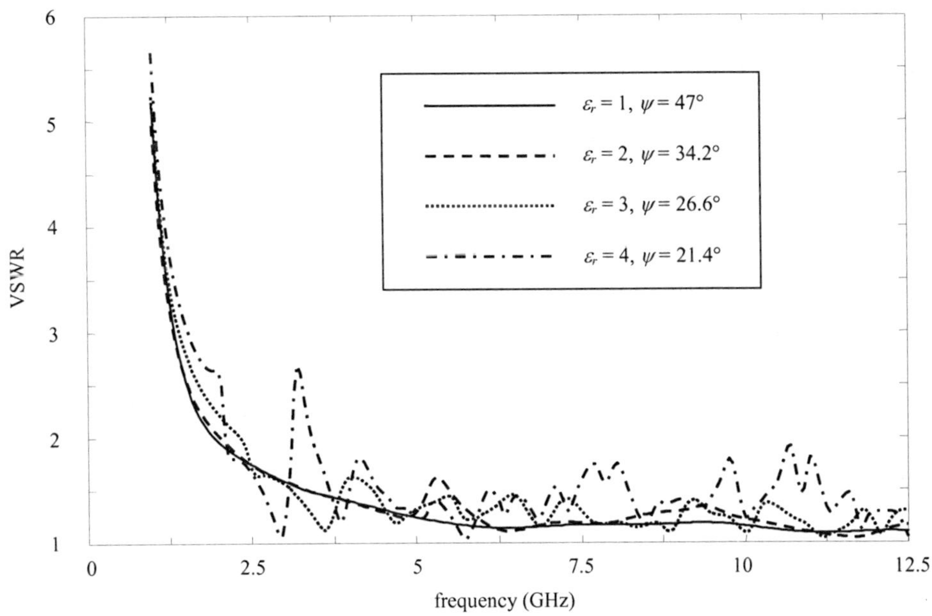

Figure 4. Simulated VSWR with various dielectric constants

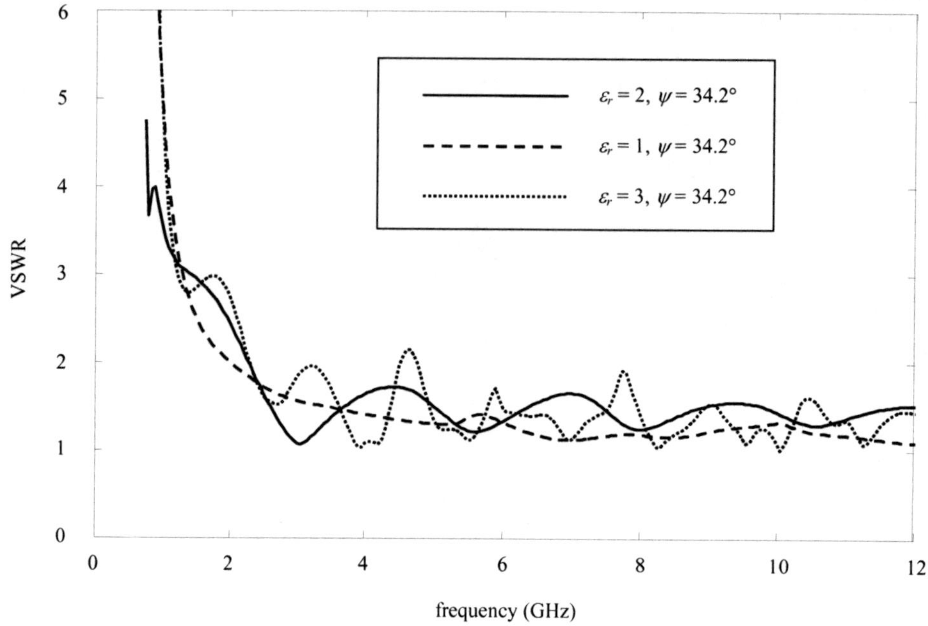

Figure 5. Simulated VSWR with various dielectric constants (cone angle fixed)

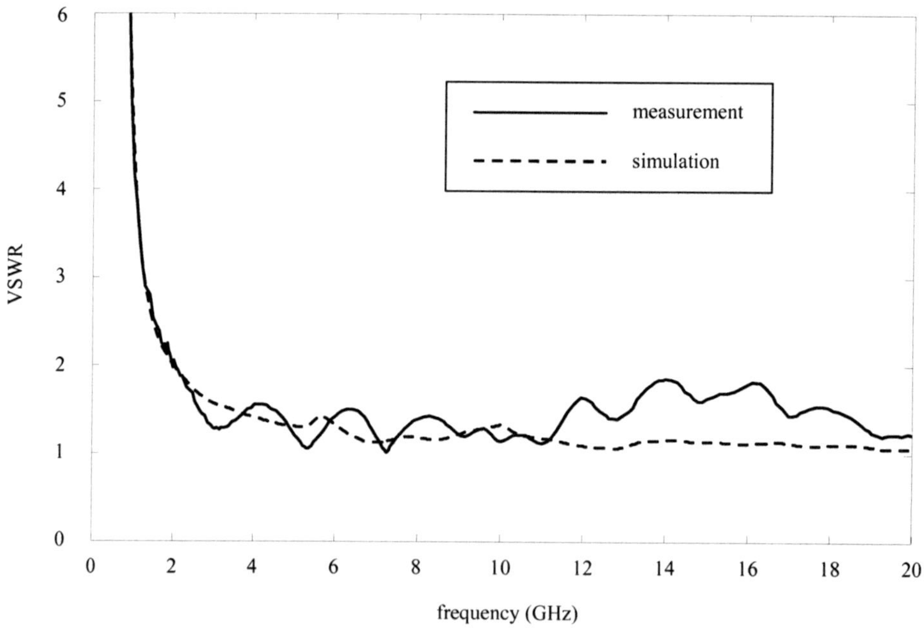

Figure 6. Simulated and measured VSWR for the antenna in Figure 3

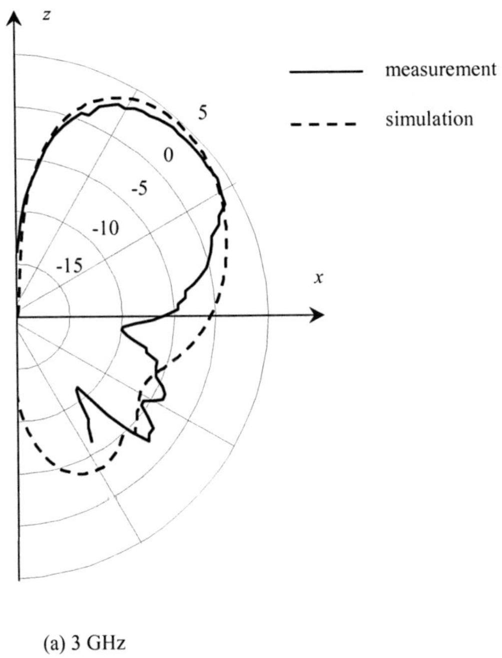

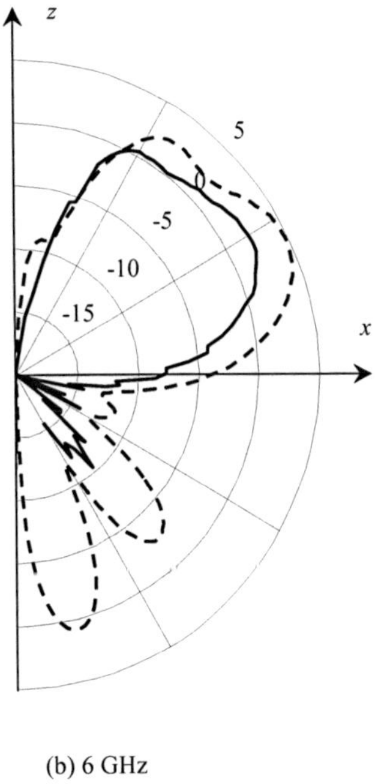

Figure 7. Simulated and measured gains (with unit dBi) for the conical antenna in Figure 3

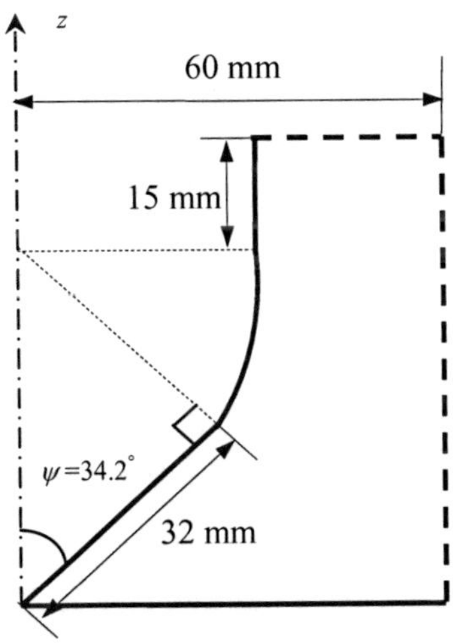

Figure 8. A tilted monopole conical antenna

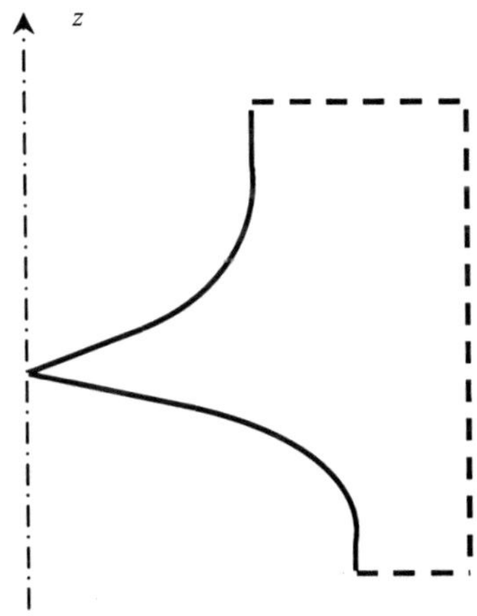

Figure 9. A tilted dipole conical antenna

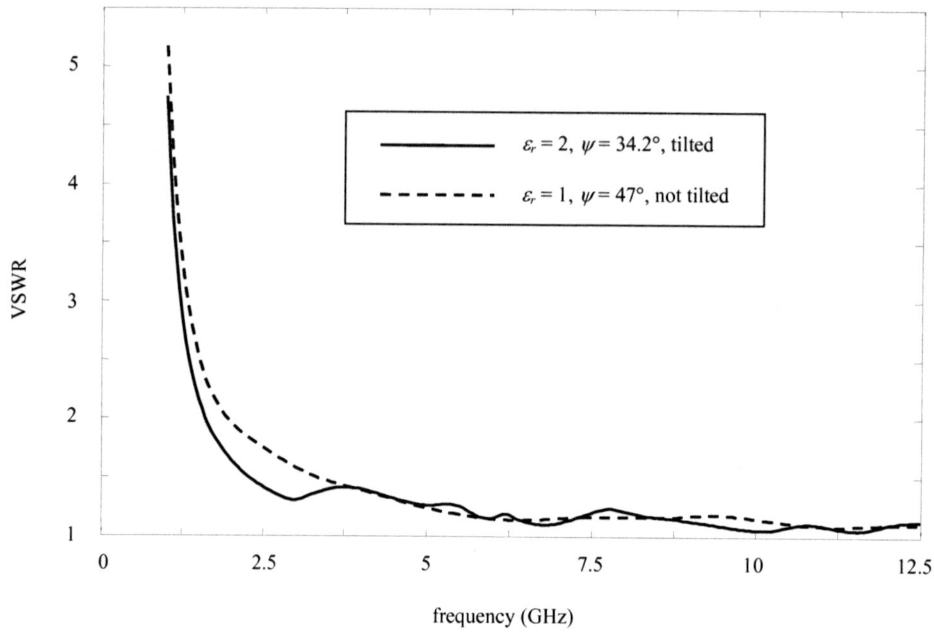

Figure 10. Simulated VSWR for the tilted monopole structure in Figure 8

4. CONCLUSIONS

In this study, a novel quasi-planar conical antenna is proposed. Compared to the traditional conical antennas, it is mechanically stable and easy to fabricate while maintaining nice wide-band characteristics. The proposed antenna is validated by both numerical simulation and experimental measurement. Research is currently going on to investigate various geometrical configurations of this antenna.

5. REFERENCES

Gentili G. B., Cerretelli M., and Cecchi L., 2004, Coated conical antennas for automotive application, *Journal of Electromagnetic Waves and Applications* **18**:85.

Kawakami H. and Sato G., 1987, Broad-band characteristics of rotationally symmetric antennas and thin wire constructs, *IEEE Transactions on Antennas and Propagation* **35**:26.

Liang X. and Wah M. C. Y., 2000, Low-profile broadband omnidirectional monopole antenna, *Microwave and Optical Technology Letters* **25**:135.

Medgyesi-Mitschang L. N., Putnam J. M., and Gedera M. B., 1994, Generalized method of moments for three-dimensional penetrable scatterers, *Journal of the Optical Society of America A* **11**:1383.

Richards W. F., Lo Y. T., and Harrison D. D., 1981, An improved theory for microstrip antennas and application, *IEEE Transactions on Antennas and Propagation* **29**:38.

PLANAR TAPERED LOOP ANTENNAS FOR ULTRA-WIDEBAND RADIO SYSTEMS

Shau-Gang Mao and Shiou-Li Chen

1. INTRODUCTION

Recently, ultra-wadeband (UWB) radio technology has found a variety of applications in science and engineering communities, such as imaging, location tracking, ground penetration radar, and low cost and short-range communications. One of the challenges of the implementation of UWB radio systems is the development of a suitable antenna that can be operated over an ultra wide bandwidth allocated by the Federal Communication Commission[1]. When considering the design of UWB antennas, not only the usual specifications over the entire frequency range, such as input impedance, gain, polarization, and radiation pattern, etc., but also a good time-domain performance, i.e. the impulse response with minimal distortion, are required[2].

In this paper, a novel UWB tapered loop antenna fed by the composite right/left-handed coplanar waveguide-to-coplanar stripline transition[3] is proposed and investigated based on the time- and frequency-domain analysis. Compared with the conventional dipole-like UWB antennas[4], this balanced-fed planar antenna configuration shown in Fig. 1 is easily integrated with other circuit components and is less dispersive in pulse response. Moreover, when considering the proximity effect, the antenna performance would not be degraded because it reduces the induced currents on the conducting environment[5]. The features of UWB tapered loop antenna in time- and frequency-domain are studied including input impedance, gain, radiation pattern, impulse response, and cross-correlation and group delay of impulse radiation system. To further explore pulse propagation effects in UWB radio such as scattering at boundaries and wave dispersion caused by a frequency-dependent interaction around the environment, the time-frequency analysis (TFA) to retrieve simultaneously both the time and frequency information of the pulse at low frequency and a good time resolution at high frequency, signal is proposed. The most popular method of the TFA is the short-time Fourier transform. However, this method is restrictive due to its fixed resolution in both time and

The authors are with the Graduate Institute of Computer and Communication Engineering, National Taipei University of Technology, Taipei 106, Taiwan, R.O.C. (e-mail: sgmao@ieee.org).

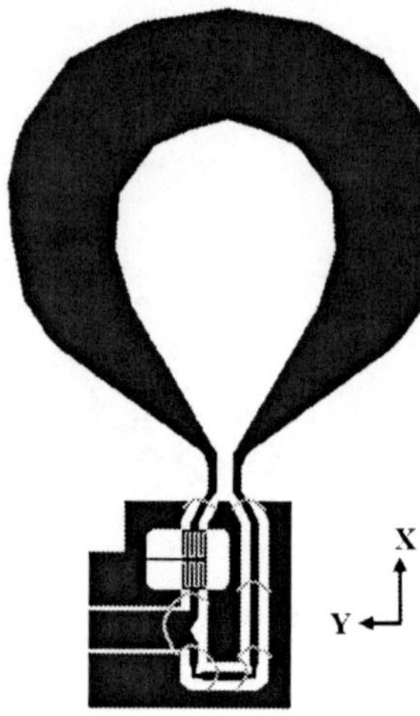

Figure 1. Geometry of the tapered loop antenna fed by the CPW-to-CPS transition. The dimensions are listed in [3].

frequency domains. Hence, the wavelet transform, which has a good frequency resolution is applied to analyze the UWB impulse signals. First, the impedance bandwidth, gain, and radiation pattern of the UWB antenna are measured in frequency domain. Second, the characteristic of UWB antenna is described by a transfer function and the time-domain impulse response of the antenna can then be obtained. Third, the pulse distortion in the transmitting and receiving antenna systems is evaluated in terms of the group delay and fidelity of radiated signals. Finally, the mutual coupling and multipath reflection of UWB antennas are examined based on the wavelet transform.

2. CHARACTERIZATION OF UWB ANTENNAS IN FREQUENCY AND TIME DOMAINS

A. Antenna Transfer Function

The transient properties of the UWB antennas can be characterized by using antenna transfer function, in which both the magnitude and phase of the radiated signal are included. The normalized complex antenna transfer function ($H_{TX/RX}$), derived from the two-port S-parameter measurement made with a pair of antennas and a vector network analyzer[6], can be used to obtain the time-domain impulse response ($v_{in/RX}(t,\theta,\phi)$) via the inverse Fourier transform techniques.

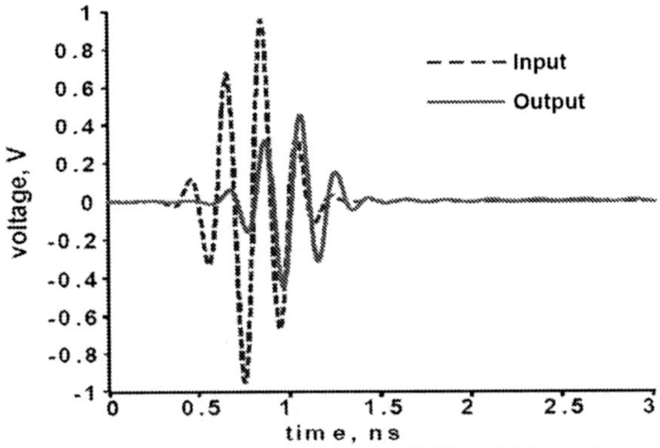

Fig. 2 Simulated input and output of pulse responses of the CPW-to-CPS transition.

$$V_{RX}(f,\theta,\phi) = V_{in}(f,\theta,\phi) H_{TX}(f,\theta,\phi) \left(\frac{j\lambda}{4\pi R_1} e^{-j2\pi\frac{R_1}{C}} + \Gamma \frac{j\lambda}{4\pi R_2} e^{-j2\pi\frac{R_2}{C}} \right) H_{RX}(f,\theta,\phi), \quad (1)$$

where Γ is the reflection coefficient of the dielectric interface, and R_1 and R_2 are the path lengths from the transmitting antenna to the receiving antenna for direct and reflected rays, respectively. The fourth-derivative Gaussian pulse $V_{in}(t)$ is given by

$$V_{in} = A\left(1 - 6\pi\left(\frac{t-0.9}{T_{au}}\right)^2 + \frac{6\pi^2}{2}\left(\frac{t-0.9}{T_{au}}\right)^4 - \frac{2\pi^2}{64}\left(\frac{t-0.9}{T_{au}}\right)^{16}\right) \cdot e^{-\pi\left(\frac{t-0.9}{T_{au}}\right)^2} (V), \quad (2)$$

where A=1 V and T_{au}=180 ps.

B. Group Delay and Fidelity of Impulse Signals

The variation in the group delay is the factor to distort the signals in UWB systems. In contrast to the narrowband antennas usually having linear phase response across the operating band, the pulse input to the UWB antenna system has an extremely large bandwidth and hence any variation of the group delay within the frequency band of interest is likely to distort the pulse. The standard deviation of the group delay is

$$\sigma_{gd} = \sqrt{\frac{1}{f_2 - f_1} \int_{f_1}^{f_2} (\tau_g - \overline{\tau}_g)^2 df}, \quad (3)$$

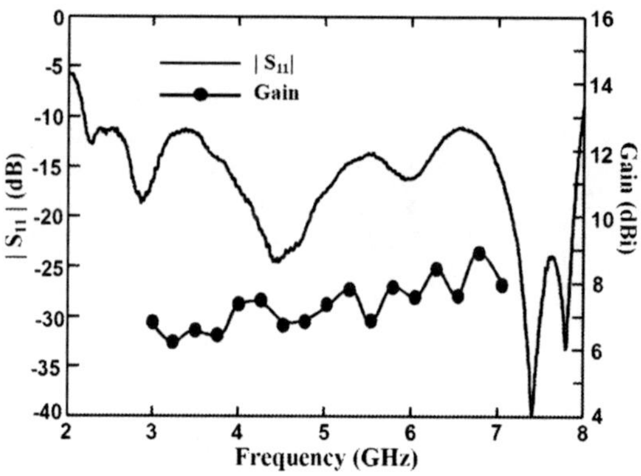

Figure 3. Measured $|S_{11}|$ and gain of the tapered loop antenna fed by the CPW-to-CPS transition.

where f_2 and f_1 are the upper and lower bounds of impedance bandwidth and $\overline{\tau_g}$ is the average group delay.

Another important characteristic of the radiated pulses is the similarity in shape. To quantitatively determinate how similar the received pulse and the selected detection template, the fidelity of the signal in the transmitting and receiving antenna systems is calculated. The pulse fidelity is the maximum cross-correlation of the normalized received pulse v_{RX} and the reference input pulse v_{in}.

$$\text{Fidelity} = \frac{\max_{\tau}\left[\int_{-\infty}^{\infty} v_{in}(t)\cdot v_{RX}(t-\tau)dt\right]}{\sqrt{\int_{-\infty}^{\infty}|v_{in}(t)|^2 dt\cdot\int_{-\infty}^{\infty}|v_{RX}(t-\tau)|^2 dt}} \tag{4}$$

C. Time-Frequency Analysis

The continuous wavelet transform of time-domain signal $v(t)$, which uses short windows at high frequencies and long windows at low frequencies to achieve a better resolution in time-frequency domain, is defined as follows:

$$W_t(a,b) = \frac{1}{\sqrt{|a|}}\int_{-\infty}^{\infty} v(t)\Psi^*\left(\frac{t-b}{a}\right)dt \qquad a,b\in R, \qquad a\neq 0, \tag{5}$$

where $\Psi(t)$ is the mother wavelet, *a* reflects the flexibility scale of Ψ, and *b* represents the shift position of wavelet on time axis.

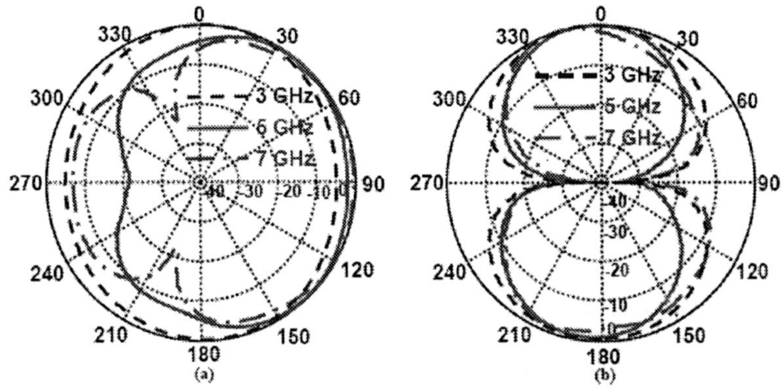

Figure 4. Measured E_ϕ-component radiation pattern of the tapered loop antenna at 3, 5, and 7 GHz. (a) XZ-plane and (b) YZ-plane.

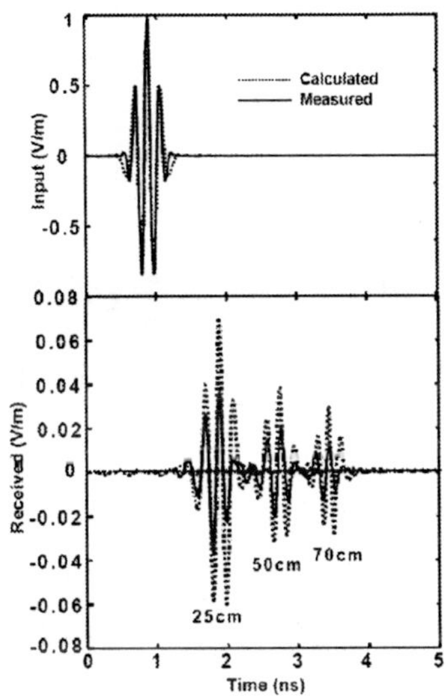

Figure 5. Measured and calculated input and received pulses with various antenna separations.

3. RESULTS AND DISCUSSION

To obtain a better performance of the loop antenna, the unbalanced-to-balanced line transition, i.e. balun, is necessary. The proposed balun is comprised of the conventional CPW and the CRLH CPW to achieve the exact 180° phase difference that CPW-to-CPS transition required, as shown in Fig. 1. To examine the time-domain characteristic of the

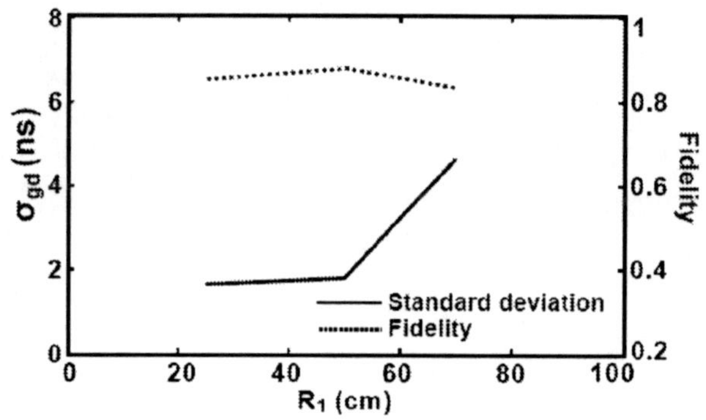

Figure 6. Measured standard deviation of group delay and fidelity of impulse responses of transmitting and receiving UWB systems along Z-axis with various separations R_1.

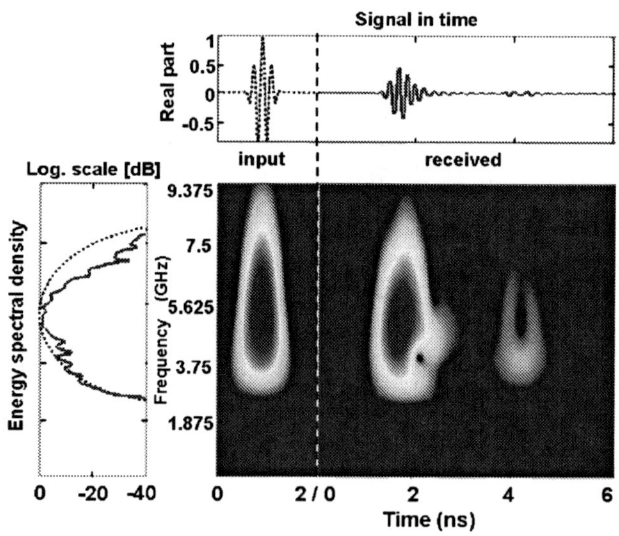

Figure 7. Time-frequency analysis of input and received pulses of UWB antennas with 27 cm-height above a dielectric interface.

balun, the fourth-derivative Gaussian pulse $V_{in}(t)$ with spectrum from 2.5-7.5 GHz is fed into the CPW port. The S-parameters of the CPW-to-CPS transition is obtained by using full-wave simulator IE3D and converting this frequency-domain data to the time domain, and then the output pulse $V_{out}(t)$ in CPS port can be obtained, as shown in Fig. 2. By using (4), the calculated fidelity of the pulse response of balun is 0.9976, which represent that the $V_{in}(t)$ and $V_{out}(t)$ of the CPW-to-CPS transition are very similar in shape and thus the proposed balun structure is less dispersive corresponding to the 2.5-7.5 GHz frequency spectrum.

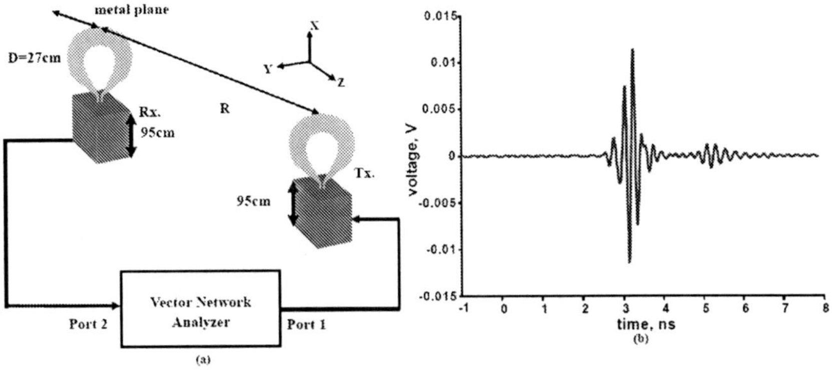

Figure 8. (a) Measurement setup for the time-domain response with a metal plane. (b) Received pulse of the two-antenna system separated by R=70 cm and a metal plane located at D=27 cm.

The measured $|S_{11}|$ and gain of the proposed antenna shown in Fig. 1 are depicted in Fig. 3, which indicates that the 10-dB return loss bandwidth of 114% and 6-9 dBi antenna gain are obtained in the operating band. In Fig. 4, the similar E_ϕ -component radiation patterns of the antenna on XZ- and YZ-plane at 3, 5, and 7 GHz are presented. In addition to these frequency-domain parameters, the response of the antenna to the UWB signals should be further described in terms of its temporal characteristics. Two identical tapered loop antennas with 105 cm-height above ground were used for measurement, which point each other along Z-axis with distances 25, 50, and 70 cm. The measured pulse responses are shown in Fig. 5 and compared with the calculated results obtained from (1) by the inverse Fourier transform. Fig. 6 illustrates the measured standard deviation of group delay and fidelity as a function of transmitting and receiving antennas separation. The antenna exhibits a low standard deviation of group delay ($\leq$ 4.67ns) and a high fidelity ($\geq$ 0.86) over the antenna separations from 25 to 70 cm. The environment influence in UWB antenna systems is analyzed by using wavelet transform. Fig. 7 depicts the time-frequency representation of mutual coupling within two side-by-side 6-cm neighboring tapered loop antennas above a dielectric plane. Compared with the time-frequency characteristic of the input pulse, the resonance effect caused by the interaction of two adjacent antennas represents a vertical expansion in the time-frequency diagram, and the attenuated and delayed pulse due to the reflection on a planar dielectric interface is easily interpreted. To further investigate the influence of environment in UWB antenna performance, the metal plane behind the received antenna with D=27 cm is considered into the two identical UWB antennas separated by R=70 cm, as shown in Fig. 8(a). The received pulse response of this configuration is measured and depicted in Fig. 8(b). Results show that the reflected pulse coming from the metal plane is observed with a delayed time t_{delay}=1.8 ns, which agrees well with the time duration of the reflected pulse by calculating $t_{delay} = 2D/c$, where c is the velocity of light. Therefore, the immunity to multipath interference can be achieved by maintaining precise timing at the receiver.

4. CONCLUSION

The time-domain as well as frequency-domain characteristics of the UWB tapered loop antennas have been investigated theoretically and experimentally. A good agreement between the measured and simulated results is observed and thus validates the proposed impulse response technique for the joint antenna-channel system. This antenna offers high-fidelity and low-σ_{gd} transmission and reception of UWB impulse signals with minimal distortion. The wavelet transform is used to determinate the dispersion and transient behavior of the radiated signal, which is useful for the UWB antenna array and receiver design.

5. ACKNOWLEDGMENTS

This work was supported in part by the National Science Council of Taiwan, R.O.C., under Grant NSC 95-2221-E-027-018, NSC 95-2221-E-027-031, and NSC 95-2752-E-002-002-PAE.

References

1. J. H. Reed *ed.*, *An introduction to ultra wideband communication systems*, NJ : Prentice Hall PTR, 2005.
2. J. S. Tyo, E. G. Farr and D. I. Lawry, "Effect of defocus on the prompt response of a reflector IRA," *IEEE Trans. Antennas Propag.*, vol. 53, pp. 3247-3254, Oct. 2005.
3. S.-G. Mao and Y.-Z. Chueh, "Broadband Composite Right/Left-Handed Coplanar Waveguide Power Splitters With Arbitrary Phase Responses and Balun and Antenna Applications," *IEEE Trans. Antennas Propag.*, vol. 54, pp. 243-250, Jan. 2006.
4. J.-P Zhang, Y.-S. Xu, W.-D. Wang,"Ultra-wideband microstrip-fed planar elliptical dipole antenna," *Electrom. Lett.,* vol. 42, pp. 144-145, Feb. 2006.
5. H. Morishita, H. Furuuchi, and K. Fujimoto, "Characteristics of a balance-fed loop antenna system for handsets in the vicinity of human head or hand," in *Proc. IEEE AP-S. Int. Symp.*, vol. 4, pp. 2254-2257, July 2000.
6. Reed J. H. ed.: 'An introduction to ultra wideband communication system', NJ : Prentice Hall PTR, 2005.

AN OMNIDIRECTIONAL AND LOW-VSWR ULTRA WIDEBAND ANTENNA FOR A FREQUENCY BAND OF 6 TO 40 GHz

Akihide Maeda* and Takehiko Kobayashi*

Abstract—We have proposed an omnidirectional-in-azimuth, low voltage-standing-wave-ratio (VSWR), and easy-to-construct antenna for ultra-wideband (UWB) systems. Omnidirectional and low-VSWR requirements are essential for applications such as UWB channel sounding. The proposed antenna consists of a circular ground plane and a teardrop that is defined as a combination of a finite cone and a sphere inscribed inside the cone at the cone's base. This antenna can be considered either as a rounded finite monocone antenna or as a simplified volcano smoke antenna. Assuming 50-Ω excitation, the optimum half-cone angle is 48° which gives the lowest maximum VSWR. Following a prototype in the frequency band of 3 to 20 GHz, we constructed the 6 - 40 GHz version. The teardrop was 12.5 mm high (the quarter wavelength at 6 GHz) and ground plane was 50 mm in diameter (significantly larger than any quarter wavelength in the designated bandwidth for 6 - 40 GHz). The teardrop is connected to the inner conductor of the coaxial cable that penetrates the ground plane. The tip of the inner conductor of a K connector was inserted into the apex of the teardrop.

1. INTRODUCTION

Recently, encouraged by deregulation, ultra-wideband (UWB) technology has attracted attention for use in communication and sensing applications in the commercial domain. In February 2002 the Federal Communications Commission (FCC) of the United States conditionally liberalized unlicensed operation of personal UWB devices in the private sector. In the meantime, various antennas have been developed to be used for UWB systems, such as double-ridged waveguide horn, log-periodic, biconical, and monocone antennas. None of these antennas, however, simultaneously meet omnidirectional and low voltage-standing-wave-ratio (VSWR) requirements, essential for some measurement applications such as UWB channel sounding [1]. Double-ridged

* Wireless System Laboratory, Tokyo Denki University, 2-2 Kanda-nishiki-cho, Chiyoda-ku, Tokyo 101-8457, Japan.

waveguide horn and log-periodic antennas are directional. Infinite biconical antennas (Fig. 1(a)) have a self-similar structure and are, therefore, frequency-independent, and omnidirectional in azimuth. By replacing one of the cones with the infinite ground plane, the infinite biconicals are extended to infinite monocones (Fig. 1(b)), which are also frequency-independent and omnidirectional. By truncating the ideal infinite monocone, a real finite monocone (Fig. 1(c)), known as a discone, can be obtained. This structure is omnidirectional in azimuth, but not frequency-independent anymore since the truncation causes reflected current from the base of the cone. A volcano smoke antenna [2] (Fig.1 (d)) fulfills these two requirements simultaneously, but its shape is not well-defined. Whereas rotationally symmetric monopole antennas were analyzed by means of an extended method of moment, parameter optimization was not performed. Taxonomy of other wide variation of UWB antennas and intensive references are given in [3].

We proposed an antenna that is both omnidirectional in azimuth and nearly frequency-independent and defined as a combination of simple structures. So far, we have constructed and measured this shape antenna in the frequency band of 3 - 20 GHz [4], [5], [6], [7]. The shape and size were optimized to yield the lowest maximum VSWR between 3.1 and 10.6 GHz. This antenna was utilized in an array antenna formation in spatio-temporal UWB channel sounding to achieve 10° and 0.67-ns resolution in spatial and temporal domains [1]. This system was also successfully deployed in various propagation environments [8]. Following the prototype in the frequency band of 3 to 20 GHz, we constructed a 6 - 40 GHz version. This paper will describe the design and the measured performance of the 6 - 40 GHz antenna, in terms of the VSWR, gain, and group delay.

2. THE PROPOSED ANTENNA AND ITS OPTIMIZATION

The proposed antenna consists of a circular ground plane and a teardrop that is defined as a combination of a finite cone and a sphere inscribed inside the cone at the cone's base, as shown in Fig. 1(e). The teardrop is excited with a coaxial cable penetrating the ground plane. This antenna can be considered either as a rounded finite monocone antenna (Fig. 1(c)) or as a simplified form of the Volcano smoke (Fig. 1(d)). Input impedances of the infinite biconical and infinite monocone antennas are governed by the half-cone angle ψ [2].

We calculated the input impedance of the proposed antenna using finite Integration method. Assuming 50-Ω excitation, the maximum VSWR between 3.1 and 10.6 GHz is shown in Fig. 2, from which the optimum half-cone angle ψ was found to be 48°. This formation was found to outperform the finite monocone in the VSWR through calculation and the first prototype experiment [4].

The lowest operable frequency (the lowest frequency that meets VSWR < 2) of the proposed antenna is determined by the height of teardrop, because a finite monocone antenna is also determined by the height of the cone. We computed the input impedance of the proposed antenna that changed the height h (Fig. 1(e)) of teardrop. The lowest operable frequency is nearly inversely proportional to the height h of the teardrop, and

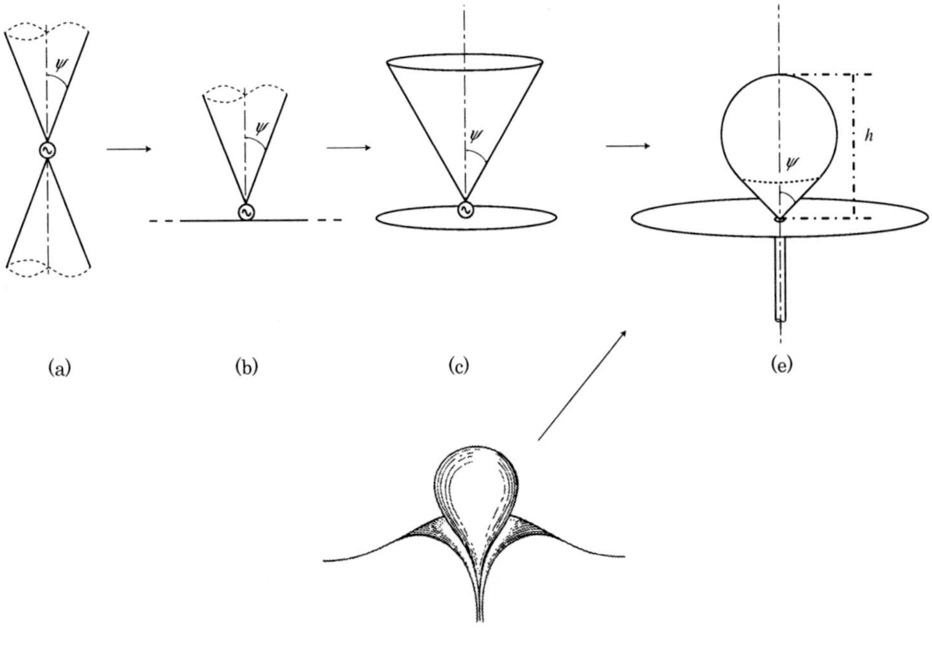

Fig. 1. Broadband omnidirectional antennas: (a) infinite biconical, (b)infinite monocone (c)finit monocone (discone), (d) volcano smoke (in cutaway view), and (e) proposed antenna.

the h is found at about a quarter wavelength of the lowest operable frequency, as shown in Fig. 3. Prototype antennas (3 - 20 GHz and 6 - 40 GHz versions) are shown in Figs. 4 and 5. Height of teardrop h and the diameter of the ground plane are optimized to 1/4 and 1 wavelength of the lower operable frequency: the teardrop was 12.5 mm high, the ground plane was 50 mm in diameter. The coaxial cable connection was K-type and its inner conductor was 0.3 mm in diameter. The eardrop element was mated with and supported by the inner conductor of the connector.

3. MEASURED AND CALCULATED RESULTS OF THE PROTOTYPE ANTENNA

The antennas used in these systems should meet several requirements: namely, constant directivity over the UWB bandwidth, low VSWR, and low frequency dependencies of the gain and the group delay. All of these affect the resulting delay spread, which should be minimized to suppress the signal distortion. The VSWR, gain, and group delay of the 6 – 40 GHz prototype antenna were measured and calculated.

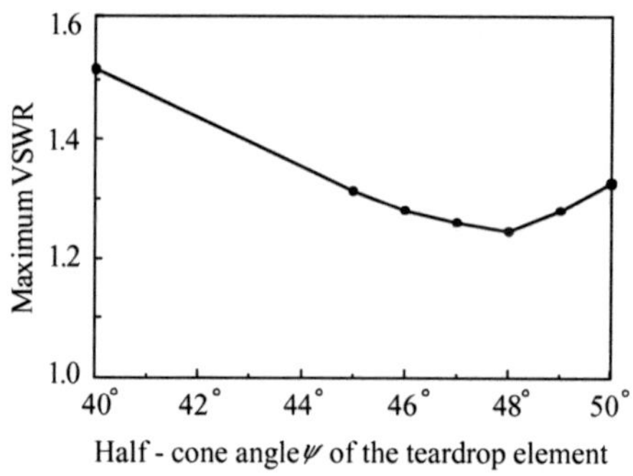

Fig. 2. Computed maximum VSWR versus half cone angle of the first and second prototype between 2 to 10 GHz.

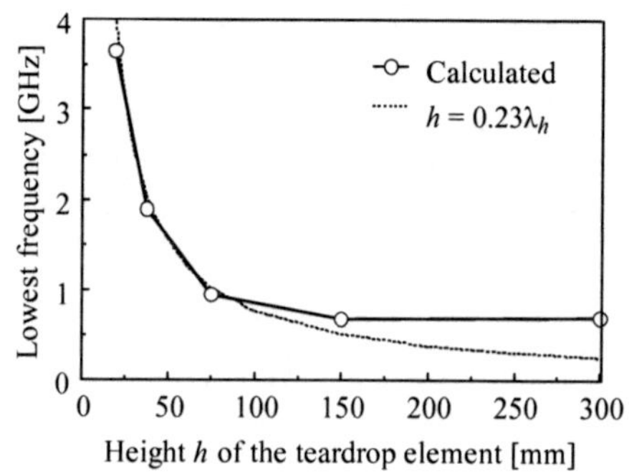

Fig. 3. Relation of height h of the teardrop and lowest available frequency.

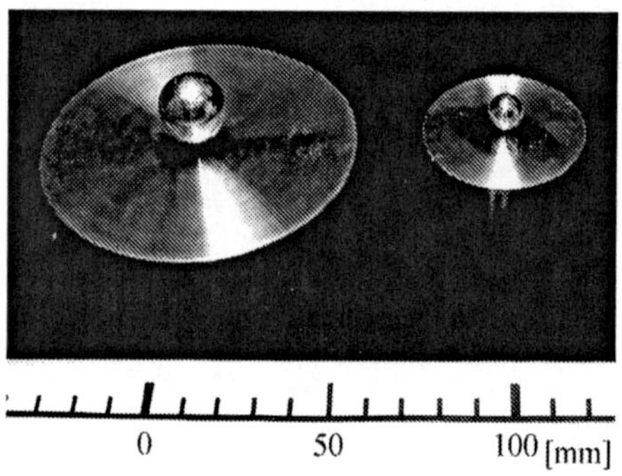

Fig. 4. Prototype UWB antennas (left: 3 - 20 GHz, right: 6 - 40 GHz).

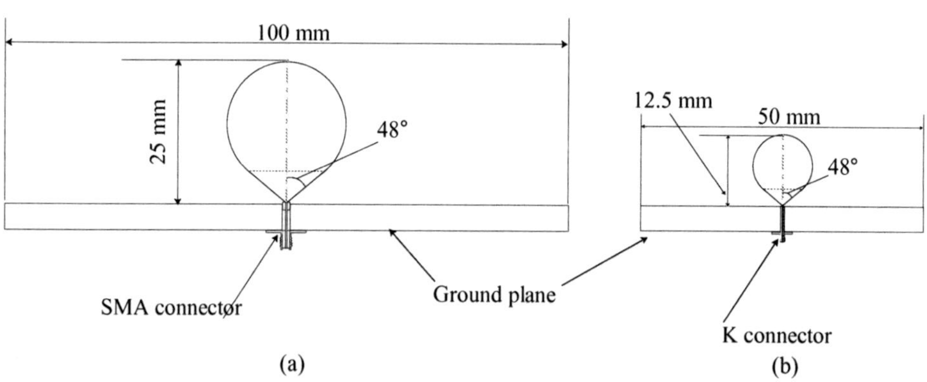

Fig. 5. Cross section of the prototype antenna: (a) 3 - 20 GHz and (b) 6 - 40 GHz versions.

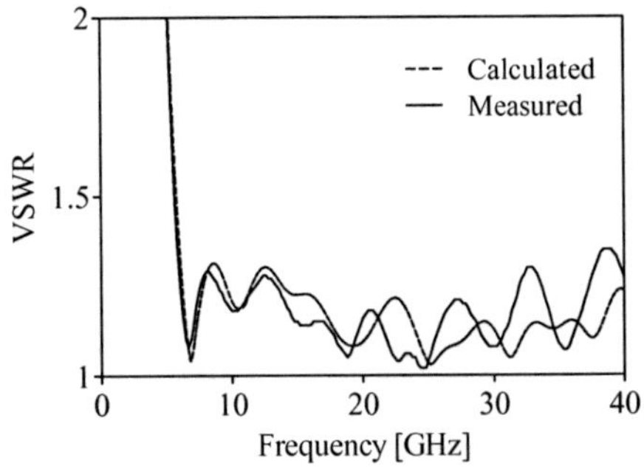

Fig. 6. Measured and Calculated VSWR of the prototype antenna.

All the measurements were carried out in a radio anechoic chamber. For VSWR measurements, the reflecting coefficients S_{11} at antenna inputs were measured with a vector network analyzer (VNA). For the other measurements, the transmitting and receiving antennas were placed 1 meter apart, which fulfills the far-field condition for all the antennas under test; and the transmission coefficients S_{21} from the transmitting antenna inputs to the receiving antenna outputs were measured with the VNA. Calibration was performed with a coaxial calibration kit between the transmitting antenna input and the receiving antenna output (namely, the ends of the antenna feeding cables). The VSWR and radiation patterns were calculated with CST MICROWAVE STUDIO^{TM}.

3.1. VSWR

The measured and calculated VSWRs are as shown in Fig. 6. The calculated and Measured VSWRs are less than 1.35 between 6 and 40 GHz.

3.2. Absolute Gain and Radiation Pattern

The absolute gain of the prototype antenna measured with the 3-antenna method is shown in Fig. 7 along with the calculation. The absolute gain was from -6 to 1 dBi in the bandwidth. The root-mean-square difference between the calculated and measured gain was below 1.2 dB between 6 and 40 GHz, which was a reasonable value, taking into account ±1.5 dB amplitude of uncertainty of the VNA. The absolute gain was measured at the direction included in the ground plane (horizontal), which was not necessarily the antenna maximum gain direction.

The calculated and measured E-plane radiation patterns of the proposed antenna at 6, 23 and 40 GHz are shown in Figs. 8 and 9. In these graphs, 0° corresponds to the zenith, and 90° and 270° correspond to the horizontal direction. There was a clear correlation between measured and the calculated results. The antenna radiated the maximum power

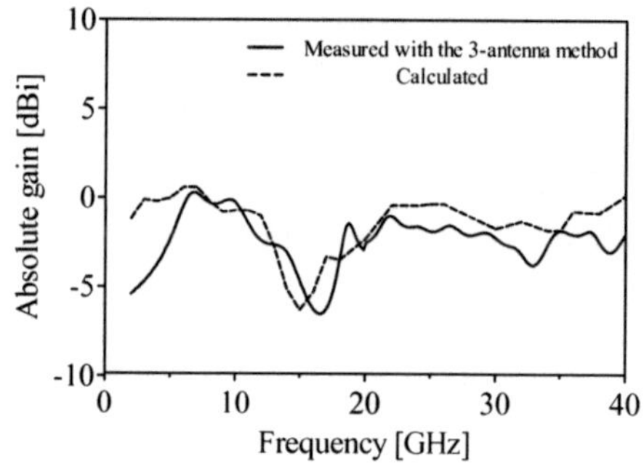

Fig. 7. Measured and Calculated Absolute gain of the prototype antenna.

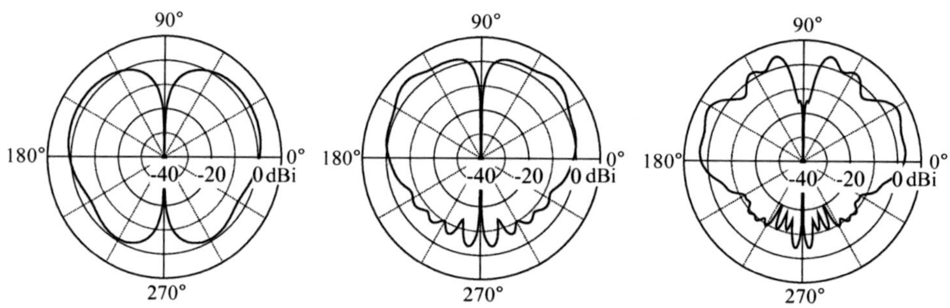

Fig. 8. Calculated E-plane radiation patterns: (a) 6 GHz, (b) 23 GHz, and (c) 40 GHz.

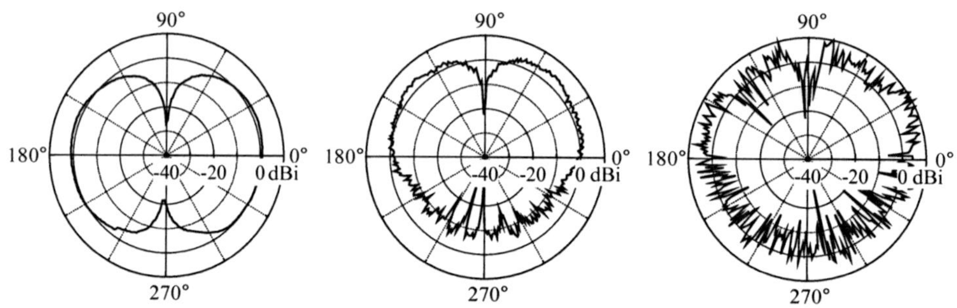

Fig. 9. Measured E-plane radiation patterns: (a) 6 GHz, (b) 23 GHz, and (c) 40 GHz.

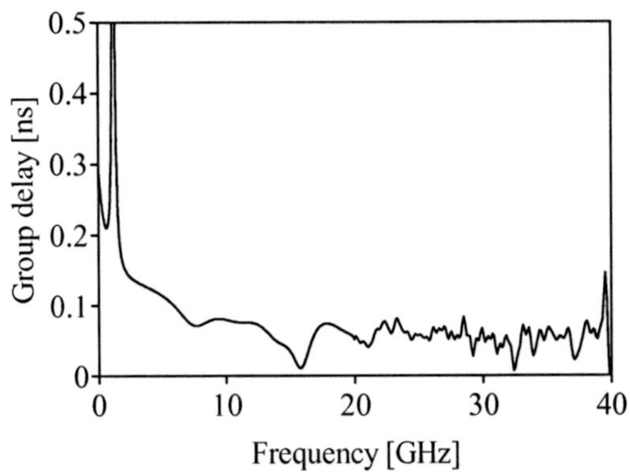

Fig. 10. Measured Group delay of the prototype antenna.

in the region from 30° to 70°. The maximum gain was between 5 ± 2 dBi from our calculated model. The gain in the horizontal direction was between 0 ±1.5 dBi from our calculated model. These values are in a reasonable agreement with the measurements.

3.3. Group Delay

Figure 10 shows the group delay measurement results. The group delay of the antenna under test was measured with the VNA by placing a pair of identical antennas face-to-face and 1-m apart, and by halving the group delay to obtain the value for one antenna. The maximum and average group delay was 0.1 and 0.09 ns though 6 to 40 GHz. Low dependence of the group delay on the frequency was observed, as shown in Fig. 10.

4. CONCLUSION

We constructed a new design of omnidirectional and low-VSWR antenna in the frequency band of 6 – 40 GHz version, which consists of a circular ground plane and a teardrop defined as a combination of a cone and a sphere. We optimized its parameters to realize very low VSWR.

The new prototype antenna demonstrated VSWR < 1.35 and group delay < 0.1 ns between 6 and 40 GHz, while the former one's VSWR < 1.3 and group delay < 0.2 ns between 3 and 20 GHz. This constructed antenna has already been used for propagation measurements in Ka frequency band.

Acknowledgement

This work has been supported in part by a grant for Top Priority Research and Development to be Focused (Frequency Resources Development) from the Ministry of Public Management, Home Affairs, Posts and Telecommunications of Japan within its R & D Support Scheme for Funding Selected IT Proposals, and in part by the Research Institute for Science and Technology,

Tokyo Denki University, under the grant Q02J–08. The antennas should be greatly indebted to K. Tahara at KEAD Co., Tokyo, Japan for fabricating the prototypes. The antennas described in this present paper are available from KEAD (http://www.kead.co.jp).

References

1. K. Haneda, J. Takada, and T. Kobayashi, A parametric UWB propagation channel estimation and its performance validation in an anechoic chamber, *IEEE Trans. Microwave Theory and Tech.*, 54(4) pp.1802-1811, (2006).
2. J. D. Kraus, Antennas, 2nd ed. (McGraw-Hill, New York, 1988) pp.340-358 and 692-694.
3. H. Schantz, *The Art and an Sience of Ultrawideband Antennas*, (Artech House, Boston, 2005).
4. T. Taniguchi, and T. Kobayashi, An Omnidirectional and low-VSWR antenna for ultra-wideband wireless systems, *Proc. of 2002 IEEE Radio and Wireless Conf. (RAWCON 2002)*, pp.145-148, (2002).
5. T. Taniguchi, A. Maeda, and T. Kobayashi, Development of An omnidirectional and low-VSWR Ultra wideband antenna, *Int. Journal on Wireless and Optical Comms.*, to be published.
6. T. Taniguchi, and T. Kobayashi, An omnidirectional and low-VSWR antenna for the FCC-approved UWB frequency band, *Proc. of IEEE Int. Antennas and Propag. Symp. Digest*, 3, pp.460-463, (2003).
7. A. C. Newell, C. F. Stubenrauch, and R. C. Baird, Calibration of microwave antenna gain standards, *Proc. IEEE*, 74, pp.129-132, (1986).
8. K. Haneda, J. Takada, and T. Kobayashi, Double directional ultra wideband channel characterization in a line-of-sight home environment, *IEICE Trans. Fundamentals*, E88-A(9) pp.2264-2271, (2005).
9. A. Maeda, and T. Kobayashi, A performance comparison of various ultra wideband antennas using the 3-antenna method, *Proc. of 2004 Antenna Applications Symp.*, (2004).
10. T. Barts, et al., Maxwell's grid equations, *Frequenz*, 44(1), pp.9-16, (1990).

ULTRAWIDEBAND BAND-NOTCHED U-SHAPE FOLDED MONOPOLE ANTENNA AND ITS RADIATION CHARACTERISTICS

Tzyh-Ghuang Ma*, and Sung-Jung Wu

1. INTRODUCTION

The demands for high-speed, high-capacity, low-cost indoor wireless communications have been raised rapidly in recent years. While the IEEE 802.11a/b/g/n standards are proposed for the wireless local-area networks (WLANs), the IEEE 802.15.3a standard is now developed to fulfill the needs of wireless personal-area networks (WPANs). The WPANs, mainly serving as cable eliminators in personal multimedia equipments, feature much higher data rates but shorter transmission range than the WLANs, which aim at managing the outgoing data communications over the Internet. Instead of competing with, the WPANs actually complement the WLANs to satisfy the diverse requirements in indoor wireless applications.

Ultrawideband (UWB) technology has become the most promising candidate for future WPAN communications. In 2002, the Federal Communication Commission (FCC) in United States officially allocated the spectra from 3.1 to 10.6 GHz for unlicensed UWB applications. The antenna implemented in an UWB radio plays a unique role than it does in conventional narrowband systems. In such a system, the antenna behaves like a filter in both spatial and frequency domains, and tends to introduce unpleasant signal distortion and degradation. In addition, it is noticed that the WLAN band from 5.15 to 5.825GHz is unfortunately overlapped within this allocated spectra. To avoid the potential interference, the antenna community has proposed several new UWB antenna designs with fairly good band-notched properties at the WLAN band[1-3]. These new designs have alleviated the need for an additional bandstop filter in UWB radios and thus significantly reduce the circuit complexity.

Planar monopole antenna has the merits of planar structure, compact size, and consistent radiation characteristics. It has become one of the most attractive designs for

---

*T.-G. Ma and S.-J. Wu are with the Department of Electrical Engineering, National Taiwan University of Science and Technology, Taipei 10607, Taiwan.

UWB antennas[4-7]. In this paper, we discuss a microstrip-fed folded U-shape planar monopole antenna with band-rejected characteristics. This antenna features compact size, wide impedance bandwidth, almost omni-directional H-plane patterns, and the favorite band-notched properties. The study reveals a gain suppression of as high as 10 dB at the center frequency of the rejection band. The antenna parameters in a traditional sense like the return losses, radiation patterns, and averaged gain are first introduced in Section 3 while the wideband antenna transfer functions and their corresponding group delays are addressed in Section 4. The detailed derivations of the dimensionless normalized antenna transfer functions[8] are discussed as well.

2. ANTENNA CONFIGURATION

The geometry of the proposed antenna is shown in Fig. 1. This antenna was fabricated on a 1.6mm-FR4 substrate with an overall size of 3 by 3 cm^2. The dielectric constant and loss tangent are 4.4 and 0.022, respectively. As shown the figure, the antenna consists of a feeding microstrip line, a U-shape radiator, and a pair of capacitive loadings. Key parameters to improve the antenna impedance matching are summarized as below:

1) The tapered profile between the microstrip feeding line and the radiator, i.e. the ratio of L_2 to W_2, plays an important role on the antenna impedance matching.
2) The feeding gap L_2 itself has significant effects on the impedance matching..
3) The ground plane size can serve as a tuning parameter for further improving the in-band antenna matching.

To achieve the desired band-rejected properties, a pair of capacitive loadings are added to the end of the radiating strips. These capacitive loadings are achieved by folding the radiating strips to the bottom layer of the substrate. Together with the radiator, which behaves like inductors due to the high current density, the capacitive loadings act as narrowband series resonators at the targeted rejection band, and thus provide the favorite band-notched properties. The length of radiating strips at this frequency is about $\lambda_g/4$,

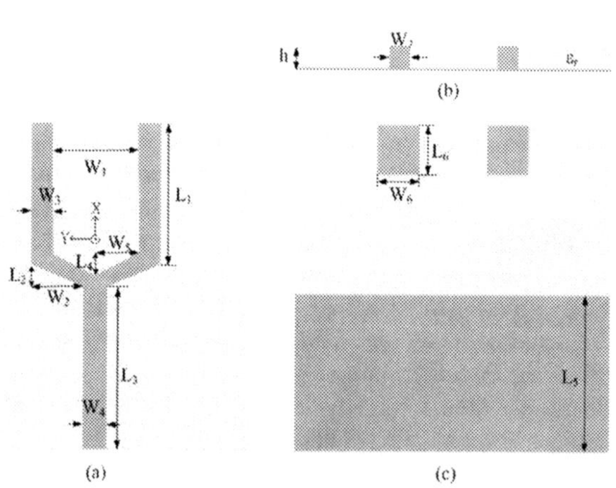

Figure 1 Geometry of the antenna. (a) Top view. (b) Cross-sectional view. (c) Bottom view.

and the loaded capacitance values are determined by the areas of the pads, i.e. W_6 by L_6.

3. ANTENNA PERFROAMNCE

Following the design rules given in Section 2, the proposed antenna is designed for UWB radios from 3.1 to 10.6 GHz together with a notched band from 5 to 6 GHz. The design parameters are W_1=12.3mm, W_2=5mm W_3=2mm, W_4=2.3mm, W_5=1.8mm, W_6=4.5mm, W_7=2mm, L_1=12.5mm, L_2=1.5mm, L_3=15mm, L_4=4.15mm, L_5=14.4mm, L_6=4.5mm, h=1.6mm. The propose antenna was simulated using Ansoft HFSS 9.2 and measured by an Agilent E8362B vector network analyzer. Figure 2 illustrates the simulated and measured return losses. The impedance bandwidth with $|S_{11}|< -9.5$ dB covers almost the whole UWB spectra and a rejection band with $|S_{11}| > -5$ dB is observed from 4.925 to 5.9 GHz. The somehow discrepancy between the simulated and measured results at lower frequency range can be mostly attributed to the coupling effects of the connecting cable in the measurement. The antenna radiation patterns were measured in a 6×4×3m^3 anechoic chamber in NTUST. The measurement was again performed by an Agilent E8362B network analyzer along with the NSI 2000 far-field measurement software. In the measurement the connecting cables were carefully shielded by absorbers to reduce the potential multiple-reflection interferences and the antenna absolute gains were calibrated by a standard EMCO 3115 double-ridged horn antenna. Figure 3 illustrates the measured radiation patterns in both H-(yz-) and E-(xz-) planes, and reasonable omni-directional patterns can be observed in the H-plane. The average gain in dBi, which is the average of the radiation patterns taken in the H-plane, is also shown in Fig. 2. From the figure we notice a gain suppression of as high as 10 dB at the center frequency of the rejection band.

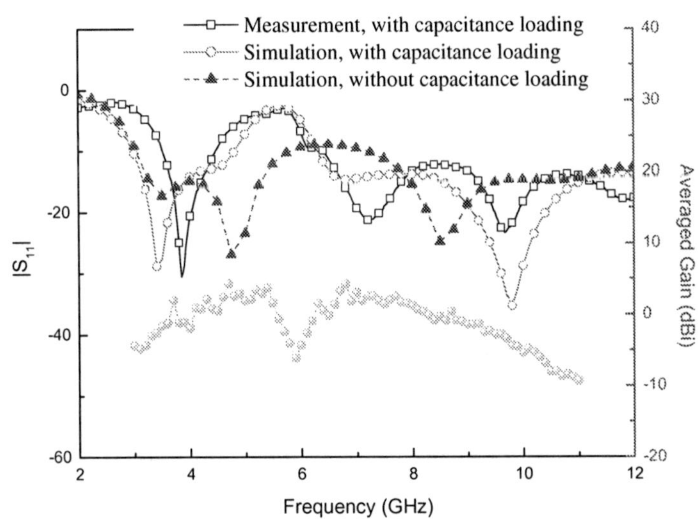

Figure 2 The reflection coefficients and average gain of the proposed antenna.

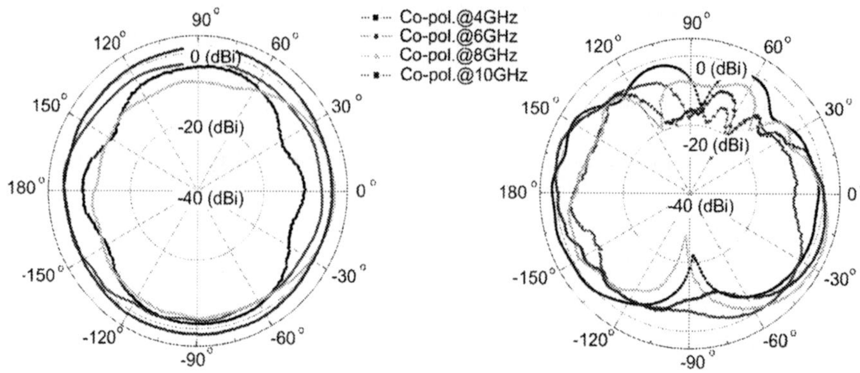

(a) (b)

Figure 3 Measured radiation patterns in (a) H-plane (yz-plane) and (b) E-plane (xz-plane).

4. THE ANTENNA TRANSFER FUNCTIONS

Due to the large operating bandwidth of an UWB antenna, the antenna parameters evaluated only in the traditional sense like the return loss, radiation patterns, etc. are generally insufficient. Instead, many researchers devote themselves to investigating the antenna transfer functions or the corresponding impulse responses to give a complete understanding of the antenna performance[8-11]. C. E. Baum et al. extend the definitions of antenna gain and radiation pattern into time domain, and discuss the relationship between the antenna transient gain and the conventional CW gain definition[9,10]. B. Scheers et al. propose a normalize impulse response (IR) for analyzing the transient behavior of a TEM horn in various problems[11].

Recently, we re-defined the normalized antenna transfer function by Scheers[11] to take full advantage of the conventional two antenna gain measurement arrangement to evaluate the wideband response of an UWB antenna[8]. The derivation of this transfer function is detailed as follows. First of all, the normalized impulse responses (IRs)[11] of the transmitting and receiving antennas in a transceiving antenna system are given by

$$h_{N,Tx}(t) = (\sqrt{Z_c/Z_a})\cdot(\tau_{Tx}/\sqrt{f_g})h_{Tx}(t) \tag{1}$$

$$h_{N,Rx}(t) = (\sqrt{Z_c/Z_a})\cdot(\tau_{Rx}/\sqrt{f_g})h_{Rx}(t) \tag{2}$$

$$\tau_{Tx} = 2Z_a/Z_c + Z_a\,,\quad \tau_{Rx} = 2Z_c/Z_c + Z_a \quad \text{and} \quad f_g = Z_a/Z_o\,. \tag{3}$$

In Eqs. (1)-(3), $h_{TX}(t)$ and $h_{RX}(t)$ are the impulse responses (IRs) of the transmitting and receiving antennas, Z_c is the characteristic impedance of the measurement system, Z_a is the antenna input impedance, and Z_o is the intrinsic impedance of free space. With the normalized IRs, the output voltage at the receiving antenna terminal is expressed by

$$V_{rec}(t) = \frac{1}{2\pi \text{R}c} h_{N,Tx}(\tau,\theta,\phi) * \frac{d}{d\tau} V_S(\tau) * h_{N,Rx}(\tau,\theta,\phi) * \delta(\tau - \frac{\text{R}}{c}), \tag{4}$$

where $V_s(t)$ is the applied voltage at the transmitting antenna terminal, and R is the distance between the virtual sources of the antennas.

Now converting Eq. (4) to frequency domain and using $S_{21}(f) = V_{rec}(f)/V_S(f)$, we have

$$S_{21}(f) = j\omega\Im\{h_{N,Tx}(\tau,\theta,\phi)\}\Im\{h_{N,Rx}(\tau,\theta,\phi)\}e^{-j2\pi fR/c}/2\pi \text{ R}c$$

$$= j\lambda H_{DN,Tx}(f,\theta,\phi) H_{DN,Rx}(f,\theta,\phi) e^{-j2\pi fR/c}/4\pi\text{R}. \tag{5}$$

In (5), $H_{DN,Tx}(f,\theta,\phi) = \sqrt{4\pi}/\lambda \cdot \Im\{ h_{N,Tx}(t,\theta,\phi) \}$ and $H_{DN,Rx}(f,\theta,\phi) = \sqrt{4\pi}/\lambda \cdot \Im\{ h_{N,Rx}(t,\theta,\phi) \}$ are defined as the *dimensionless normalized antenna transfer functions* of the transmitting and receiving antennas, respectively, which are modifications of the definitions by Scheers[11] with a normalization factor $\sqrt{4\pi}/\lambda$.

To experimentally evaluate the dimensionless normalized antenna transfer function, the conventional two-antenna gain measurement arrangement is modified to add the required phase information. The measuring arrangement is shown in Fig. 5. The position and polarization of the transmitting antenna were fixed during the measurement. The antenna under test (AUT) and the standard antenna were in turn mounted on a rotator as the receiving antenna. The virtual source of the antenna, which is regarded as the origin of radiated impulse, was evaluated in advance and aligned with the rotation center of the rotator. The distance between the transmitting and receiving antennas was 3.6 m in the measurement. The transmission scattering parameter $S_{21}(f)$ of this transceiving antenna system was then measured using a vector network analyzer (VNA), whose reference planes have been calibrated to the antenna terminals in advance. The dimensionless normalized antenna transfer function of the AUT (antenna under test) can be evaluated by

$$H_{DN,AUT}(f,\theta,\phi) = \frac{S_{21,AUT}(f,\theta,\phi)}{S_{21,STD}(f)} H_{DN,STD}(f), \tag{6}$$

where $S_{21,AUT}(f,\theta,\phi)$ is the measured transmission scattering parameter of the AUT at a specific angle (θ,ϕ), $S_{21,STD}(f)$ is that of the standard antenna in its maximum gain direction, and $H_{DN,STD}(f)$ is dimensionless the normalized antenna transfer function of the standard antenna. To improve the measuring accuracy of the antenna group delays, the intermediate-frequency (IF) bandwidth of a VNA is reduced from 3kHz to 700Hz and the necessary time-gating technique is adapted as well.

In Eq. (6), the dimensionless normalized antenna transfer function of the standard antenna is yet determined. To simplify the evaluation of $H_{DN,STD}(f)$, we first assume that the standard antenna is well matched to the measuring system, i.e. $Z_a = Z_c = f_g Z_o$. With this assumption, Eqs. (1), (2) can be reduced to

$$h_{N,STD}(t) = h_{STD}(t) \Big/ \sqrt{f_g} \ . \tag{7}$$

Furthermore, we assume that the standard antenna is with constant group delay over the frequency band of interest. With these assumptions, the dimensionless normalized antenna transfer function of the standard antenna can be expressed as

$$H_{DN,STD}(f) = \left|H_{DN,STD}(f)\right| \cdot e^{j2\pi f t_d} = \sqrt{4\pi}/\lambda \cdot \left|\Im\{h_{N,STD}(t)\}\right| \cdot e^{j2\pi f t_d}$$

$$= \sqrt{4\pi} \Big/ \sqrt{f_g}\,\lambda \cdot \left|\Im\{h_{STD}(t)\}\right| \cdot e^{j2\pi f t_d} = \sqrt{4\pi} \Big/ \sqrt{f_g}\,\lambda \cdot |\, H_{STD}(f)\,| \, e^{j2\pi f t_d} \tag{8}$$

where t_d is a constant group delay and $H_{STD}(f) = \Im\{h_{STD}(t)\}$ is the effective height of the standard antenna. According to Farr[9], as the input impedance of the standard antenna is purely resistive and well matched to the measurement system, we have

$$G_{STD}(f) = (\sqrt{4\pi}/\lambda)^2 \left|H_{STD}(f)\right|^2 \Big/ f_g \,, \tag{9}$$

where $G_{STD}(f)$ is the absolute power gain of the standard antenna. By substituting Eq. (9) into Eq. (8), the dimensionless normalized antenna transfer function of the standard antenna can now be simplified as

$$H_{DN,STD}(f) = \sqrt{G_{STD}(f)} \cdot e^{j2\pi f t_d} \,. \tag{10}$$

In this paper, an EMCO 3115 double-ridged horn antenna is chosen to be the standard antenna. This antenna has proved to be well matched to the measurement system

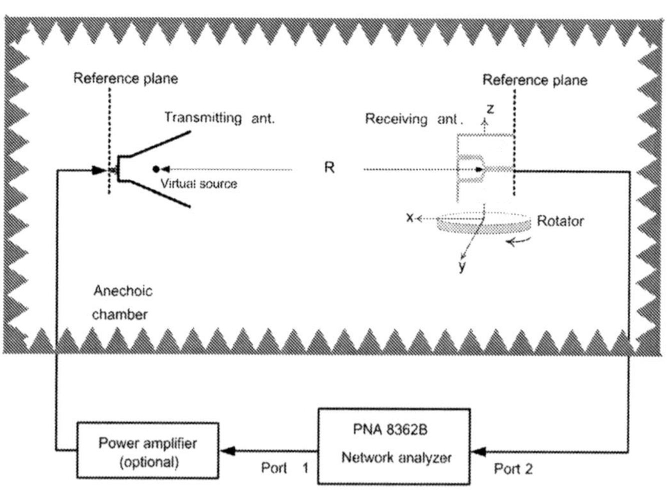

Figure 5 Arrangement for measuring the dimensionless normalized antenna transfer functions

and possess constant group delay over the frequency band of interest. The group delay of this standard antenna can be readily determined since the position of the virtual source has been specified. Refer to the data sheet, the absolute gain of the standard antenna is obtained as well. After the antenna transfer function of the standard antenna has been properly evaluated, the desired dimensionless normalized antenna transfer function of the AUT can be readily achieved by substituting Eq. (10) back into Eq. (6). It should be emphasized that the magnitude in dB of this antenna transfer function is exactly the absolute gain of an antenna, and therefore will be referred to as the gain response in the following discussion for simplicity.

The measured gain responses of the proposed antenna at four reception angles in the H-plane are shown in Fig. 6 whereas the antenna group delays at the same reception angles are illustrated in Fig. 7. Referring to the Fig. 6, it demonstrates that the proposed antenna experiences a gain variation of less than 5 dB in both UWB low (mandatory) and high (optional) bands. In addition, it is shown that the WLAN band has been well excluded from the operating spectra expect that at boresight direction (θ =0°). The degradation of the band-notched properties at the boresight direction is believed to be caused by some constructive interference of the current flowing on the radiating strips. To tackle the problem, antenna with stripline feeding scheme is now under investigation to guarantee consistent band-rejected characteristics over all spatial directions. The antenna group delays, on the other hand, reveals fairly stable responses over the frequency band of interest.

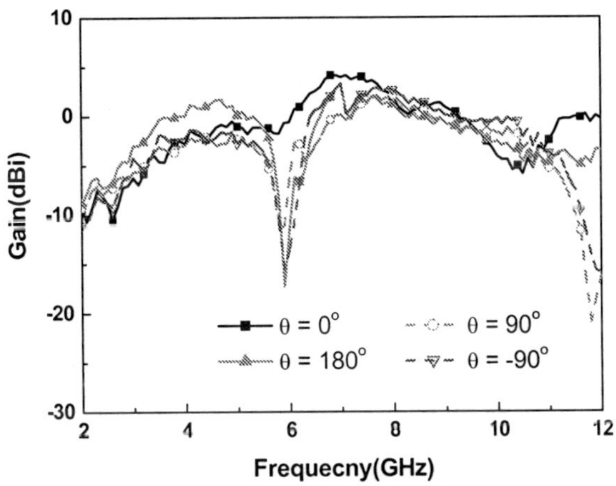

Figure 6 Measure gain response of the propose antenna at θ = 0°, 90°, 180°, -90° in H-plane.

5. CONCLUSIONS

A microstrip-fed folded U-shape planar monopole antenna with band-rejected characteristics has been presented in this paper. It features compact size, ultra-wide impedance bandwidth, almost omni-directional H-plane patterns, and fairly good gain-suppressions at the WLAN band. The antenna was carefully investigated using the dimensionless normalized antenna transfer functions and may found applications in UWB radio integrations by means of multilayer PCB technology.

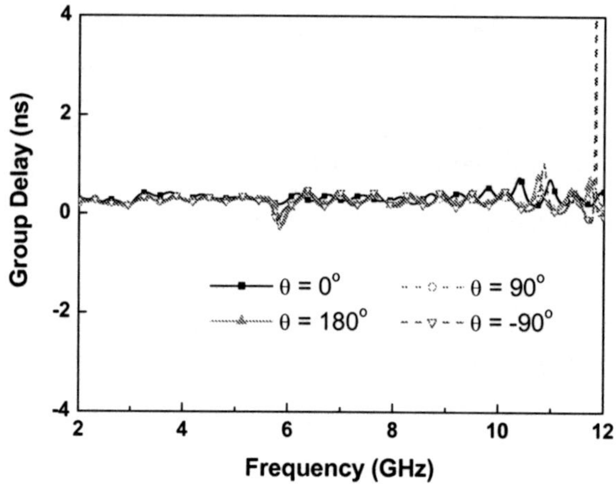

Figure 7 Measured group delay of the propose antenna at $\theta = 0^o$, 90°, 180°, -90° in H-plane.

6. ACKNOWLEDGEMENTS

This work is financially supported by the National Science Council, R.O.C., under Grant 94-2218-E-011-011- and 95-2221-E-011-021-.

7. REFERENCES

1. H. G. Schantz, G. Wolence, and E. M. Myszka, Frequency notched UWB antenna, in *Proc. IEEE UWBSTConf.*, 214-214 (Reston, VA, November 2003)
2. Y. Kim and D.-H. Kwon, CPW-fed planar ultra wideband antenna having a frequency band notch function, *Electron. Lett.* **40**(7), 403-405 (2004).
3. T.-G. Ma and S.-K. Jeng, A planar tapered-slot-fed annular slot antenna with band-notched characteristic for ultra-wideband radios, *IEICE trans. Fundamentals*, **E88-A**(9), 2384-2386 (2005).
4. Z. N. Low, J. H. Cheong, and C. L. Law, Low cost PCB antenna for UWB application, *Antennas Wireless Propaga. Lett.* **4**, 237-239 (2005).
5. D. C. Chang, M. Y. Liu, C. H. Lin, A CPW-fed U type monopole antenna for UWB application,, in *IEEE AP-S Int. Symp. Dig.* **2A**, 512-515 (Washington DC, July 3-8 2005).
6. J. Jung, W. Choi, and J. Choi, A small wideband microstrip-fed monopole antenna, *IEEE Microwave Wireless Compon. Lett.* **15**(10), 703-705 (2005).
7. J. Liang, C. C. Chiau, X. Chen, and C. G. Parini, Study of a printed circular disc monopole antenna for UWB systems, *IEEE Trans. Antennas Propagat.* **53**(11), 3500-3504 (2005).
8. T. G. Ma and S. K. Jeng, Planar miniature tapered-slot-fed annular slot antennas for ultra-wideband radios, *IEEE Trans. Antennas Propagat.* **53**(3), 1194-1202 (2005).
9. E. G. Farr and C. E. Baum, Extending the definitions of antenna gain and radiation pattern into time domain," in '*Sensor and simulation note*' **note 350**, Air Force Research Laboratory (1992).
10. C. E. Baum, E. G. Farr and C. A. Frost, Transient gain of antennas related to the traditional continuous-wave (CW) definition of gain, *Ultra-Wideband Short-Pulse Electromagnetics* **4**, 109-118, (1998).
11. B. Scheers, M. Acheroy and A. Vander Vorst, Time-domain simulation and characterisation of TEM horns using a normalised impulse response, *IEE Proc.-Microw. Antennas Propag.* **147**(12), 463-468 (2000).

COMPARISON OF UWB ANTENNAS CONSIDERING PATTERN VARIATION WITH FREQUENCY

Tharaka Dissanayake, and Karu P. Esselle *

1. INTRODUCTION

The radiated waveform *shape* of a good UWB antenna should be the same in all directions of interest. This property is termed "waveform omni-directionality"[1]. Such antennas can be characterized by a single normalized transfer function or impulse response, making system optimization, equivalent circuit modeling, propagation analysis [2] and pulse shaping filter design[3] much easier. Waveform omni-directionality is directly related to the antenna pattern stability with respect to frequency. It may not be feasible to realize UWB antennas with absolutely stable patterns; however, a quantitative figure for pattern stability enables one to compare UWB radiation from various antennas in a range of directions of interest. In this paper we use a newly proposed Pattern Stability Factor (PSF) to compare several probe-fed monopole UWB antennas and microstrip-fed printed planar monopole UWB antennas. The PSF is also used to determine the pattern stability bandwidth of those antennas. The PSF is defined in the frequency domain because UWB regulations (e.g. on bandwidth, mask level etc.) are specified in this domain but it also represents waveform fidelity and omni-directionality in the time domain. It is important to note that the PSF is independent of the gain of the antenna. For a good PSF, a flat gain response (versus frequency) is not necessary but the gain frequency response curves should be similar (apart for a constant) in all directions of interest. A pattern stable antenna fulfills this condition because, although its absolute gain pattern may change with frequency, the fractional change is the same in all directions of interest.

* Tharaka Dissanayake and Karu P. Esselle, Department of Electronics, School of Information and Sciences, Macquarie University, NSW2109, Sydney, Australia.

2. PATTERN STABILITY FACTOR

Correlation between radiated pulses in two directions $\vec{R}$ and $\vec{r}$ can be written as:

$$f^2(\vec{R}, \vec{r}) = \frac{\left[\int_{-\infty}^{\infty} e(\vec{r}, t)e(\vec{R}, t)dt\right]^2}{\int_{-\infty}^{\infty} \left|e(\vec{R}, t)\right|^2 dt \int_{-\infty}^{\infty} \left|e(\vec{r}, t)\right|^2 dt} \quad (1)$$

where $e(\,.\,, t)$ denotes time domain electric field in the far field. This $f^2(\vec{R}, \vec{r})$ is independent of the amplitude of the far field. A similar expression can be written in the frequency domain with the help of Fourier transform and Parceval's power theorem:

$$F^2(\vec{R}, \vec{r}) = \frac{\left|\left[\int_{BW} E(\vec{r}, f)E^*(\vec{R}, f)df\right]\right|^2}{\int_{BW} \left|E(\vec{r}, f)\right|^2 df \int_{BW} \left|E(\vec{R}, f)\right|^2 df} \quad (2)$$

where $E(\,.\,, f)$ is the Fourier transform of $e(\,.\,, t)$ and absolute value of the numerator is taken to make the result a real value, and BW denotes the frequency band of interest. The average correlation between a reference direction $\vec{R}$ and all other directions of interest is then obtained as

$$C(\vec{R}) = \left(\int_{\Omega} F^2(\vec{R}, \vec{r})d\Omega\right) \Big/ \left(\int_{\Omega} d\Omega\right)$$

where the integration is w.r.t $\vec{r}$ over the directions of interest, denoted by Ω. The PSF is defined as the average of C:

$$PSF = \left(\int_{\Omega} C(\vec{R})d\Omega\right) \Big/ \left(\int_{\Omega} d\Omega\right) \quad (3)$$

where the integration is w.r.t the reference directions R. It indicates how radiation is correlated in the range of directions of interest, within the frequency band considered. For an antenna with ideal pattern stability, PSF=1 because,

$$E(\vec{r}, f) = k(\vec{r})E(\vec{R}, f)$$

where $k(\,.\,)$ is a constant with respect to frequency, or

$$e(\vec{r}, t) = \alpha(\vec{r})e(\vec{R}, t)$$

where α is a constant w.r.t. time.

3. RESULTS

UWB antennas shown in Fig. 1 have been simulated using CST Microwave Studio time domain solver. Probes are placed on the H-plane to sense E-field. (For probe-fed antennas H-plane is the XY plane; for printed antennas it is the XZ plane. An ideal monopole has an omni-directional co-polar radiation pattern on this plane). Each antenna is excited by a modulated Gaussian pulse, band limited to the 10dB return loss (RL) bandwidth of the antenna. For the calculation of PSF, the far-field transfer function $E(\vec{r}, f) / V(f)$, where V(f) is the source voltage, is used so the results are independent of the excitation pulse. PSF values on the H-plane over the entire RL bandwidth are given in Table 1. PSF values over a narrower frequency range are usually higher then these values. (Note that over a very narrow frequency range, PSF is almost 1 for any antenna.) We somewhat arbitrarily defined the UWB antenna PSF bandwidth as the frequency range within which PSF = 0.95, and the UWB antenna overall bandwidth as the range in which PSF = 0.95 *and* RL >10 dB. The overall bandwidths of the 9 antennas are plotted in Fig. 2, considering the lower limit of the band as the independent variable. Since the overall bandwidth is always bounded by the upper limit of the RL bandwidth, a straight line means 1:1 RL bandwidth to PSF bandwidth ratio.

Conical monopole and cross-planar monopole have the highest stability due to their improved rotational symmetry, which leads to nearly or exactly omni-directional patterns on the H-plane at all frequencies. Their PSF is > 0.95 over the entire RL bandwidth. For all other antennas the overall bandwidth is less than the RL bandwidth, as shown in fig. 2. The shape of the ground plane is also important. The edge diffraction from a circular ground plane is more symmetrical with respect to the angle of observation compared to a square ground. Therefore, for the same radiating element a circular ground plane performs better than a relatively larger square ground plane. Even though the PSF improvement is small, the maximum overall bandwidth has improved by about 2GHz. Also note that creating a ground plane window near the square monopole does not affect the impedance bandwidth, but it brings down the PSF bandwidth by about 1GHz.

PSF of printed monopoles are > 0.9, apart for the circular antenna and windowed square monopole. Note that square and triangular printed antennas are electrically small compared to the probe-fed antennas considered here. Usually the largest dimension determines the lower limit of RL bandwidth. Therefore, RL bandwidths of the printed antennas are relatively smaller than those of probe-fed monopoles. The printed circular monopole has dimensions comparable to the circular planar monopole on circular ground. Good overall performance is recorded by triangular shaped antennas[4] due to their small size.

Table 1. PSF on the H-plane over the entire 10dB return loss bandwidth.

UWB Antenna	10dB RL Bandwidth (GHz)	PSF on H-plane
Triangular Ring (TRA)	5.0-10.0	0.936
Triangular (TA)	5.0-10.0	0.947
Square (SA)	6.5-10.6	0.941
Square with Window (SAW)	6.5-10.6	0.891
Printed Circular (PCA)	3.1-9.0	0.847
Circular planar monopole on square ground (CSG)	3.1-10.6	0.873
Cross planar monopole (CPM)	3.1-10.6	0.983
Circular planar monopole on circular ground (CCG)	3.1-10.6	0.878
Conical monopole (CMA)	6.0-10.6	0.998

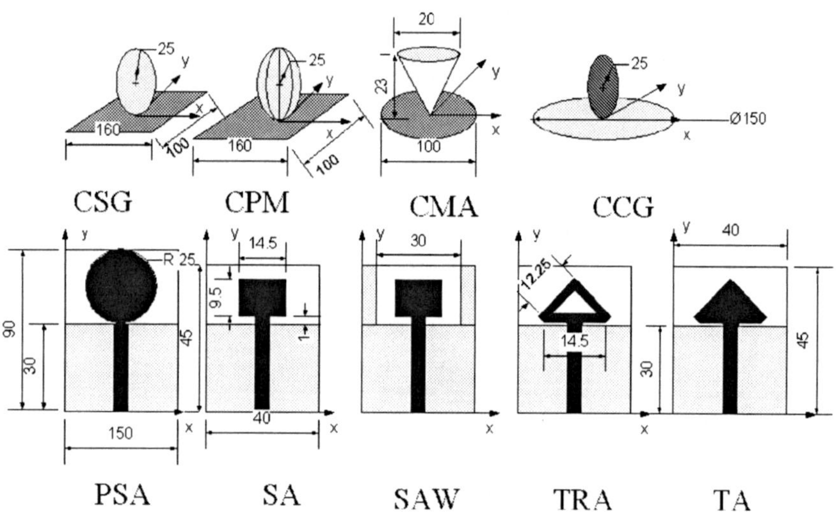

Figure 1. UWB monopole antennas analysed in this paper (see Table 1 for abbreviations)

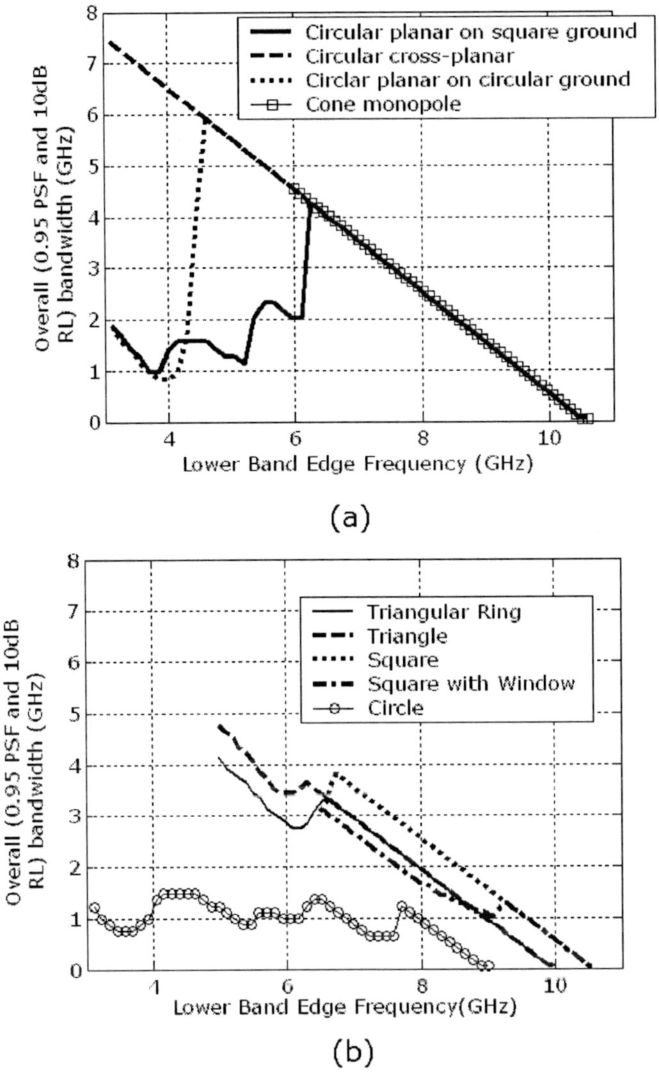

Figure 2. Overall (0.95 PSF & RL > 10 dB) bandwidth for co-polar radiation on the H-plane

4. CONCLUSION

The PSF is useful to quantitatively compare pattern performance of UWB antennas available for a given application. Its calculation is simple. Although the whole H-plane is considered in this example analysis, any range of directions can be used in PSF calculations depending on the application. Furthermore, a frequency-domain weighing function can be included in equation (2), to represent spectral limits such as FCC UWB mask. PSF bandwidth and overall bandwidth calculations, which indicate the best operating frequency range of a given UWB antenna, are also useful when designing and optimizing antennas for UWB applications. Our results indicate that the overall bandwidth is significantly affected by the size and geometry of the antenna, the ground plane as well as any near-field scatterers. It is also possible to trace the pattern stable beam of a UWB antenna using a similar method.

5. REFERENCES

1. Stanley Bo-Ting Wang, Ali M. Niknejad and Robert W. Borderson, "Modeling Omnidirectional Small Antennas for UWB Applications", IEEE Antenna and Propagation International Symposium, June (2004).
2. Florent Munier, Thomas Eriksson, and Arne Svensson. "On The Effect ofAntennas in UWB Systems", In Proceedings of 28th URSI General Assembly in India (2005).
3. Goufeng Lu, and Larry Greenstien, "Antenna and Pulse Designs for Meting UWB Spectrum Density Requirements", IEEE Conference on Ultra-wideband Systems and Technologies, November (2003).
4. T. Dissanayake, and Karu Esselle, " A Printed Triangular-ring Antenna with a 2:1 Bandwidth", Microwave and Optical Technology Letters, **44**(1), (2005)

DISPERSIVE PROPERTIES OF TERMINAL–LOADED DIPOLE ANTENNAS IN UWB LINK

Anatoliy O. Boryssenko and Daniel H. Schaubert*

1. INTRODUCTION

A variety of UWB applications require signals with very wide bandwidths or very short pulses[1-3] that can be severely distorted across channels even for free-space line-of-sight because of antenna dispersion. UWB radiators can play a major role in unavoidable spectral/temporal transformations of signals across channels[3-6]. These signal variations depend implicitly on the antenna shapes, which stimulates R&D work to use narrowband antennas in UWB regime[3-6] and/or develop new types of specific broadband antennas[7-9].

This work supplements typical UWB design and analysis, which often focuses on antenna and signal issues[3-6,10,11], with some ideas on optimal antenna terminations to the transmitter and receiver front-ends. The effect of lumped resistive port terminations on link dispersive properties is explored to meet several design objectives for received signals including peak voltage, energy transmission efficiency, and energy confinement in time. Some illustrative results are derived for canonical flat and solid dipoles like those sketched in Fig. First, a full-wave Method of Moment (MoM) analysis is applied to get the numerical transfer function for link in Fig. 2a. Then, the link transmission features are studied using an equivalent circuit, Fig. 2b, based on the link factored model[12]. Finally, the main results are given in 3:1 band using normalized time and frequency units that make them scalable to any physical band of interest.

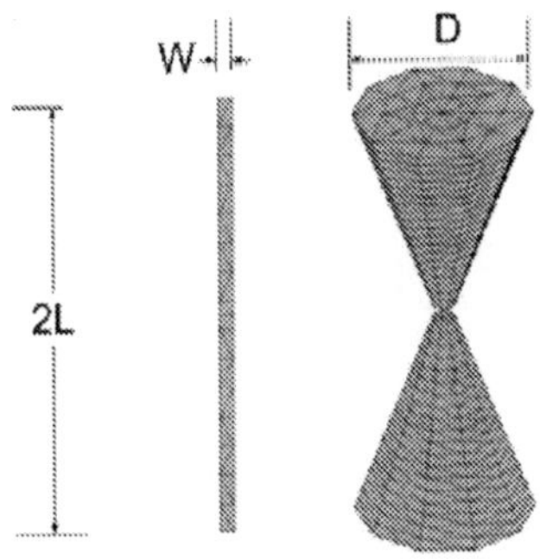

Figure 1. Examples of practical solid and flat dipoles with parameterized length *L*, diameter *D* and width *W*.

This work was supported in part by MURI project DAAD19-01-0477 and contract to US Air Force FA8718-06-C-0047.

* Center for Advanced Sensor and Communication Antennas, University of Massachusetts, Amherst, USA

2. FULL-WAVE AND CIRCUIT TWO-ANTENNA LINK-MODEL

A frequency domain (FD) link model, Fig. 2a, includes two antennas where the transmitter is represented by the voltage source $V_G(\omega)$ with the internal impedance $Z_G(\omega)$ and the receiver by just the load impedance $Z_L(\omega)$ with the received voltage $V_L(\omega)$. For the present study, the link antennas operate in the far field with respect to each other and are ideally aligned in terms of polarizations and maximum gain directions to provide a useful reference for real-world channels. The link FD transfer function, which can be numerically predicted using a suitable EM solver, e.g. MoM, is numerically defined as

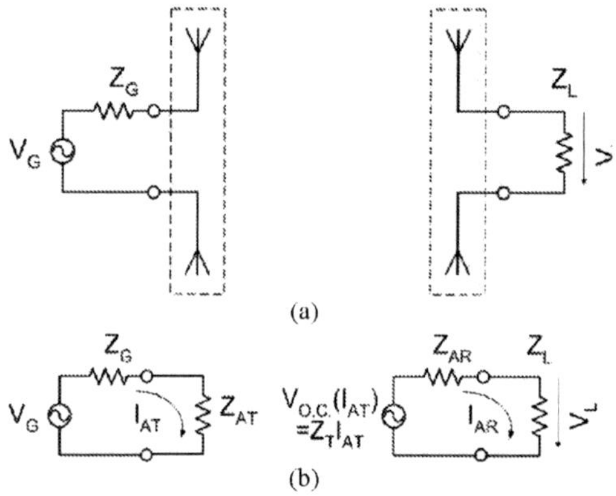

Figure 2. (a) Two-antenna link model with lumped transmitter generator and receiver load connected to antenna ports. (b) Two-port link equivalent circuit.

$$H_{LG}(\omega) = \frac{V_L(\omega)}{V_G(\omega)} \frac{1}{H_{F.S.}(\omega, R)} \quad (1)$$

where angular and distance dependencies are suppressed, and the free-space propagation factor

$$H_{F.S.}(\omega, R) = \frac{\exp(-jkR)}{R} \quad (2)$$

is removed to ensure distant-independent accuracy. Scaling to any band of interest is possible by using the normalized frequency and time:

$$f^* = f \cdot \tau_0, \; \omega^* = 2\pi \cdot f^* \quad (3)$$

$$t^* = t / \tau_0 \quad (4)$$

where the characteristic time is defined as $\tau_0 = L/c$ and c is the free-space speed of light. The quantity, Eq. (1), can be normalized with the help of Eq. (4) and the antenna geometrical parameter L[12], Fig. 1

$$H_{LG}(\omega^*) = H_{LG}^0(\omega^*) L \quad (5)$$

Any link such as in Fig. 2a can be to converted to an equivalent two-port circuit like that in Fig. 2b. The circuit parameters are defined using the EM-computed transfer function, Eq. (5), and input impedances $Z_{AT}(\omega)$ and $Z_{AR}(\omega)$ of the link antennas. The link transfer function in terms of the circuit elements, Fig. 2b, can be factored[12] as

$$H_{LG}^0(\omega^*) = \frac{1}{Z_{AT}(\omega^*) + Z_G(\omega^*)} Z_T(\omega^*) \frac{Z_L(\omega^*)}{Z_{AR}(\omega^*) + Z_L(\omega^*)} \quad (6)$$

where the transmission impedance term $Z_T(\omega)$ collects together all wave-related link properties. It is proved[12] using the generic EM reciprocity and linearity that the link transmission impedance in Eq. (6) does not depend on antenna port loading. Thus, the terminal and wave-related link features can be separated, which gives a number of advantages, e.g., each link of a fixed geometry can be simulated just one time for some reference port loads. Then, arbitrary port loads can be analyzed at the much lower computational cost by manipulating Eq. (6) or modeling the link with the help of a circuit simulator based on the equivalent circuit, Fig. 2b.

3. FREQUENCY-DOMAIN LINK PERFORMANCE MERITS

The normalized magnitude (gain) and group delay of the link transfer function, Eq. (6), can be used to characterize links in FD. These quantities are obtained from the exponential presentation as follows

$$H_{LG}^{0}(\omega) = A_{LG}^{0}(\omega)\exp\{j\theta(\omega)\} \tag{7}$$

$$\tau_{LG}(\omega) = -d\theta / d\omega\,. \tag{8}$$

Intuitively, in FD signal transmission with minimal distortions can be formulated as supporting "flat" gain and "linear" phase responses (constant group delay) within the band. These link merits are affected by both the antenna geometry and port loads as illustrated in Figs. 3 & 4 for flat strip and bicone dipole antennas, accordingly. In particular, link gain is always higher for the bicone, Fig. 4a, compared to the flat narrow strip, Fig. 3a. Also, the arbitrary loaded bicone has typically better group delay properties compared to arbitrary loaded strips but the optimally loaded strip ($R_{G(L)}$=600Ω), demonstrates a bit steady group delay in f*=0.1:0.8, Fig. 3b. Therefore, proper design choice for the antenna shapes and port loads is very important because of their collective impact on the gain and group delay of the transfer function, Eq. (6), across the band.

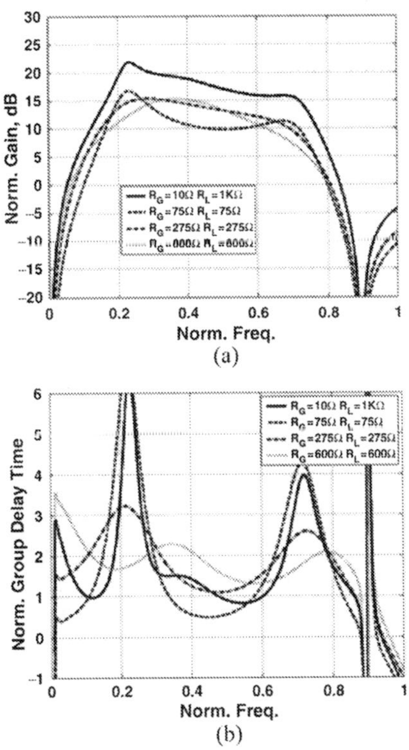

Figure 3. Normalized link FD transfer function for the flat strip dipole (W/L=1/15) with several port loads: (a) normalized magnitude of link transfer function; (b) normalized group delay.

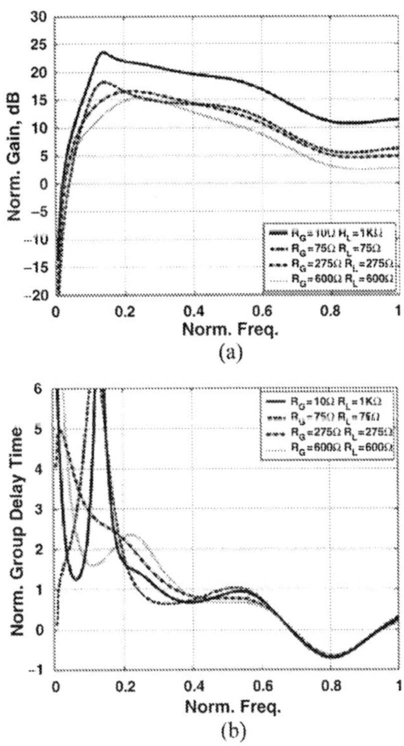

Figure 4. Normalized link FD transfer function for the bicone (D/L=6/15) with several port loads: (a) normalized magnitude of link transfer function; (b) normalized group delay.

4. TIME-DOMAIN LINK PERFORMANCE MERITS

The received time-domain (TD) signal waveform at the receiver load can be derived from the original numerical FD data through Fourier transform as

$$S(t) = \frac{1}{2\pi} \int_{BW} H_{LG}^{0}(\omega) \cdot V_G(\omega) \cdot \exp(j\omega t) d\omega \qquad (9)$$

A set of time-harmonic signals of unit amplitude in the given band BW of interest

$$V_G(\omega) = \begin{cases} 1, & \omega \in \mathrm{BW} \\ 0, & \omega \notin \mathrm{BW} \end{cases} \qquad (10)$$

can be used to drive the transmitting antenna that forms a bandpass channel impulse response. The received TD waveforms are shown in Fig. 5a & 6a using Eqs. (9)-(10) for the flat narrow strip dipole and solid bicone, accordingly. These numerical data are computed within the normalized operational band BW^* = 0.2…0.6 with 3:1 frequency coverage where the upper frequency limit is set as $L=0.6\lambda_{min}$.

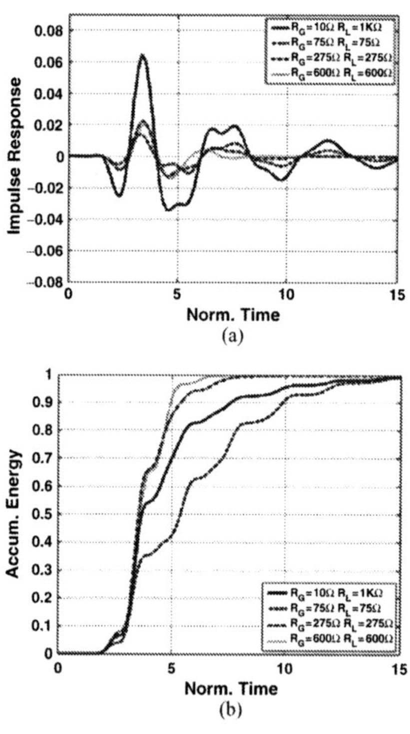

Figure 5. Normalized TD link transfer features for the flat narrow strip dipole (W/L=1/15) with several port loads: (a) received waveform; (b) accumulated energy of received signal.

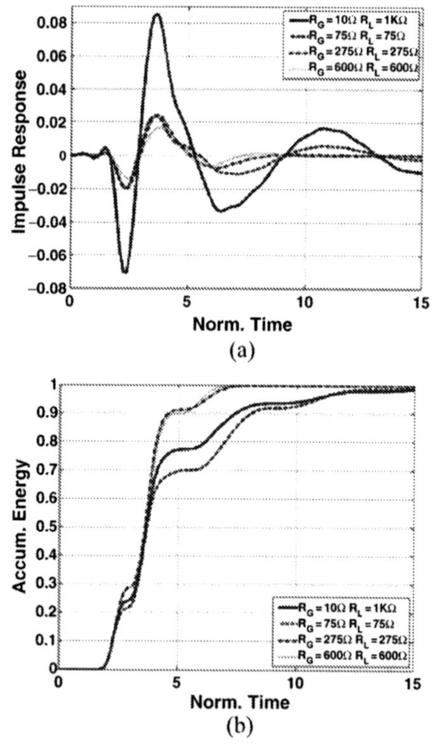

Figure 6. Normalized TD link transfer features for the bicone (D/L=6/15) with several port loads: (a) received waveform; (b) accumulated energy of received signal.

The signal accumulated energy at the receiver load R_L vs. time, Figs. 5b & 6b, is:

$$\mu(t) = (W_L)^{-1} \int_0^t S_L^2(\tau)/R_L^2 \, d\tau \tag{11}$$

where normalization to the total received signal energy W_L is applied:

$$W_L = \frac{1}{2\pi} \int_{BW} \frac{|V_L(\omega)|^2}{R_L} \cdot d\omega \tag{12}$$

This property characterizes spreading of signal energy in time because of the collective effect of assigned bandwidth, antenna dispersion and its port loads. The given TD properties of both simulated antennas are pretty comparable in terms of signal shapes and localization in time except the difference in amplitudes for some load case. E.g., in Figs. 5a & 6a, the peak received voltage is larger for the bicone dipole than for the strip one that are loaded for providing the maximized peak voltage, viz. R_G = 10 Ω and R_L = 1 kΩ.

4. PORT-LOAD EFFECT ON RECEIVED SIGNALS

The contour plots in Figs. 7 & 8 are computed in the band BW^* = 0.2…0.6 for a number of different resistive lumped terminations at the transmitter, R_G, and receiver, R_L, link ends in the range 10…1000 Ω to study particular design objectives. The regions of optimum performance in Figs. 7 & 8 are given in lighter color. For example, the link transmission efficiency or energy gain G, Figs. 7a & 8a, is defined as ratio of the energy delivered to the receiving antenna load to that available from the generator:

$$G = W_L / W_G \tag{13}$$

where the energy available from the generator W_G is[10] :

$$W_G = 1/2\pi \int_{BW} |V_G(\omega)|^2 / [4R_G] d\omega \tag{14}$$

The equivalent energy rise time t_0 bounds a given amount of the signal accumulated energy, Eq. (11), at a specified level, e.g., from 0.1% (-30 dB) up 95% (-0.2 dB) [12]:

$$\mu(t_0) \cong 0.95 \tag{15}$$

The normalized energy rise time plotted in Figs. 7b & 8b is derived as

$$t_0^* = t_0 / \tau_0 \tag{16}$$

In particular, maximizing the energy transmission efficiency is provided by symmetrical loading at both the link ends that depends on a specific antenna structure, viz. R_G=R_L=300 Ω for the flat strip dipole, Fig. 7a, and R_G=R_L=200 Ω for the bicone, Fig. 8a. The optimized energy rise time can be nearly the same for the flat strip and solid bicone when they are properly terminated, Figs. 7b & 8b. However, the flat strip shows a little improved energy confinement compared to the bicone, which correlates with the group delay properties in Figs. 3b & 4b. As well, the loads required for optimal merit values, Eqs. (13) & (16), are in better mutual correspondence for the strip dipole, Fig. 7, than for the bicone, Fig. 8. Therefore, some design objectives can be mutually exclusive with respect to the specified antenna shapes and port resistive terminations, requiring some trading in practical multi-objective UWB link design with real, dispersive, and port-loaded antennas.

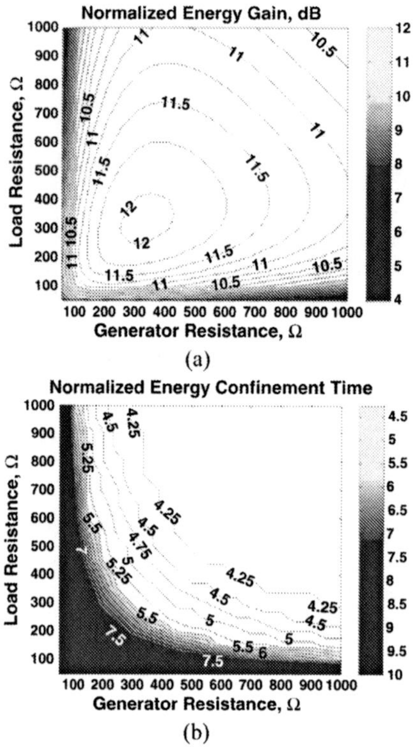

Figure 7. Contour plots for norm. (a) energy transmission efficiency; (b) energy confinement time of flat strip dipole (W/L=1/30) link.

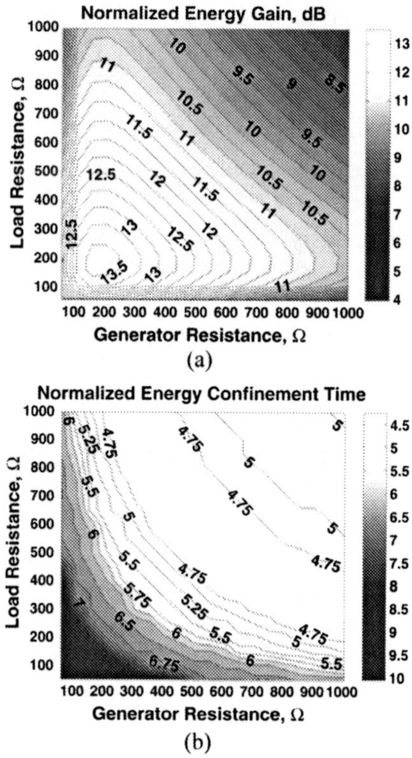

Figure 8. Contour plots for norm. (a) energy transmission efficiency; (b) energy confinement time of bicone (D/L=6/15) link.

5. REFERENCES

1. J. D. Taylor, Ed, *Ultra-Wideband Radar Technology*, (CRC Press, 2001).
2. K. Siwiak, D. McKeown, *Ultra-Wideband Radio Technology*, (Wiley, 2004).
3. R. A. Scholtz, D. M. Pozar, W. Namgoong, "Ultra-Wideband Radio", *EURASIP J. Applied Signal Proc*, 3, pp. 252-272 (2005).
4. Z.H. Chen, X. H. Wu, H. F. Li, N. Yang, M.Y.W. Chia, "Consideration for Some Pulses and Antennas in UWB Radio Systems", *IEEE Trans. Ant. & Propag.*, **52**(7), pp. 1739-1748 (2004).
5. W. Sorgel, W. Wiesbeck, "Influence of Antennas on the Ultra-Wideband Transmission", *EURASIP J. Applied Signal Proc.*, 3, pp. 296-305 (2005).
6. G. Lu, P. Spasojevic, L. Greenstein, "Antenna and Pulse Design for Meeting UWB Spectrum Density Requirements", *IEEE Conf. Ultra Wideband Systems and Technologies*, pp. 162-166 (2003).
7. C. E. Baum, E.G. Farr, "Impulsive Radiating Antennas", *Ultra-Wideband Short-Pulse Electromagnetics*, pp. 139-147 (Plenum Publisher, 1993).
8. H. G. Schantz, *The Art and Science of Ultrawideband Antennas*, (Artech House, 2005).
9. H. F. Harmuth, *Antennas and Waveguides for Nonsinusoidal Waves*, (Academic Press, 1984).
10. D. M. Pozar, "Waveform Optimization for Ultra-Wideband Radio System", *IEEE Trans. Ant. & Propag.*, **51**(9), pp. 2335-2345 (2003).
11. A.O. Boryssenko, D.H. Schaubert, "Electromagnetic-Related Aspects of Signaling and Signal Processing for UWB Short Range Radios", *J. VLSI Signal Proc.*, **43**(1), pp. 89-104 (2006).
12. A. O. Boryssenko, D. H. Schaubert, "Antenna Link Transfer Function Factorization Applied to Optimized Channel Design", *IEEE Trans. Ant. & Propag.*, **54**(10), pp. 2878-2889 (2006).

AN APPROACH TO THE DETERMINATION OF THE PHASE CENTER OF VIVALDI-BASED UWB ANTENNA

Bo-shi Jin, Qun Wu, Yu-ming Wu, Li Bian, and Le-Wei Li

1. INTRODUCTION

The phase center of ultra-wide band (UWB) antenna acts a lot to the performance in the time domain [1]. Usually, the width of transmitted impulse is about several hundreds of nanoseconds, so even fine phase center drift will influence the phase of far field much and lead to the waveform dispersion in the time domain [2-3].

If we recognize the far field which is radiated by the antenna as a sphere, then the phase center of antenna should be the center of the sphere. Unluckily, the antenna may have the phase center or not because it correlates to the beam widths and the different tangent planes. In other words, an antenna can't only have one phase center, the phase center will distribute over the scope under the different beam widths and tangent planes. The shorter the radius of scope, the more stable the phase center is. But the concept above mentioned is only suitable for the conventional antennas, for most of conventional antennas are narrow band, we can deem the phase center constantly in the operating frequency band. For UWB antenna applications, the phase center is required to be stable as the beam width and frequency vary at the same time. Conventionally, the phase center

Bo-shi Jin, Qun Wu, Yu-ming Wu, Li Bian, Harbin Institute of Technology, Harbin, 150001, China, Email: jinboshi@hit.edu.cn. Le-Wei Li, National University of Singapore, Kent Ridge, 119260, Singapore.

can be deduced according to the directional diagram, but not all of antennas can be implemented because of complicated structures, therefore, numerical methods are effective ways[4]. In this paper, the finite integration method (FIM) is used to calculate the phase center of typical Vivaldi antenna which is used for the ultra-wide band communication system.

2. THE NUMERICAL ANALYSIS OF THE VIVALDI ANTENNA

Vivaldi antenna is a traveling-wave antenna that behaves well in the time domain. In the operating bandwidth, it has the good waveform in the time domain. The model is shown in Fig.1, the working plane is X-Y plane, the slot curve of the antenna is the exponential function, which is expressed as $y = Exp(0.0348x) - 0.48$ and the size is $26 \times 73\text{mm}^2$. Before the theoretical analysis, firstly we can evaluate the position of phase center, for it will contribute a lot to the numerical analysis. For the Vivaldi antenna, it can be approximated to the continuous taper with annular linearly tapered slotline sections. On each section, we considered it as the nonuniform conical transmission line, it can be decomposed into several conical transmission line sections. Slot-sector and flare angle are different for each section since they correspond to different uniform conical slot lines. The phase center can be determined on the symmetry plane and shift along the axis according to the conical transmission-line theory easily [5], because the radiation can't occur in the narrow part of the slot, so the phase center can't be at the original point, it should be at the point which drifts a little from the original point, as shown in Fig.1. The original point sphere misaligns with the phase center sphere, and it should be in the Y-Z plane, the initial analysis will simplify the numerical complexities.

We calculate the phase center of E, H planes and three-dimension, respectively.

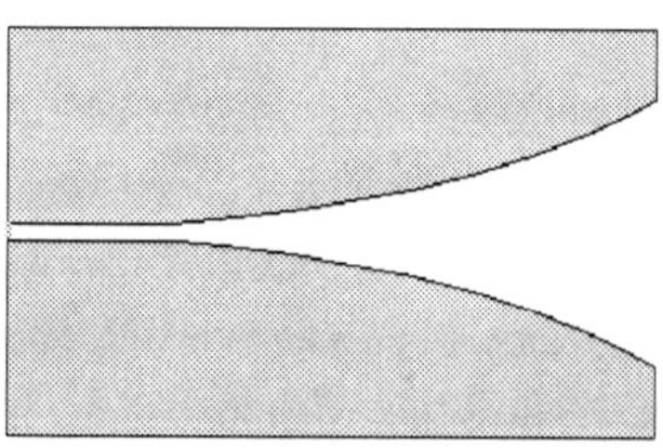

Fig.1 The model of the Vivaldi antenna.

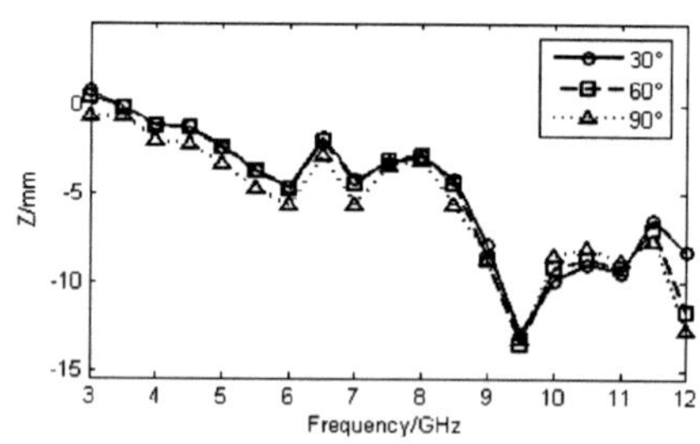

Fig.2 Coordinate Z of the E plane with different beam width.

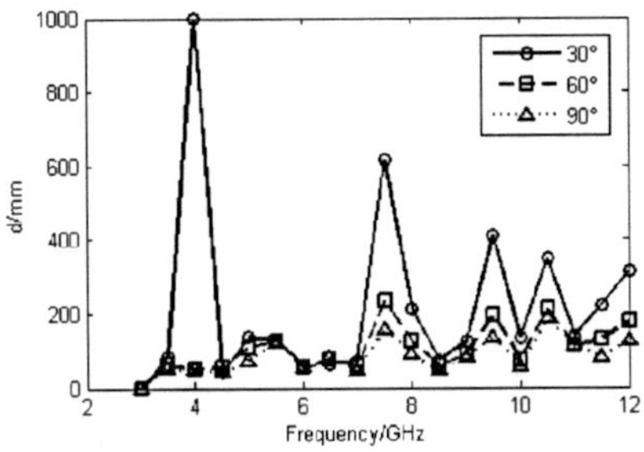

Fig.3 The variation of the H plane with different beam width.

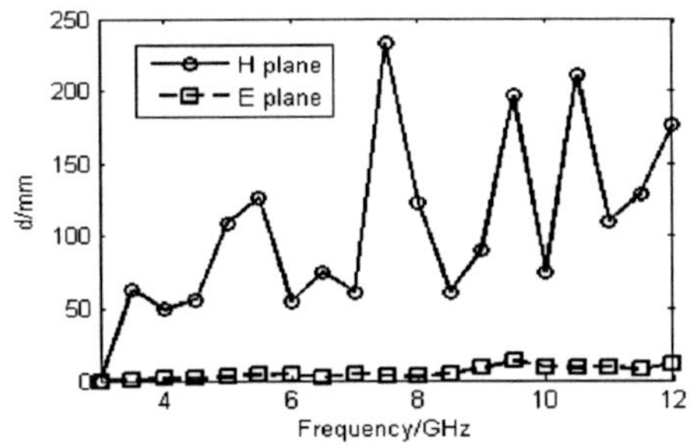

Fig.4 The comparison of phase center between the E and H plane with 60° beam width.

2.1. The Phase Center of E Plane

The E plane of Vivaldi antenna is the θ plane and is the principal polarization plane. According to the results of calculation, we find that the X and Y coordinates of the phase center are fixed at the point (27.5, 0), respectively and shifts along the Z coordinate. The Z coordinate is calculated from 3~12 GHz and the interval is set as 0.5 GHz, shown in Fig.2. The three curves labeled differently are under 30°, 60°and 90° beam width, respectively. From the results above, we find that the Z-coordinate shifts more to the negative direction as the frequency increases and achieves the -13 mm peak value at 9.5 GHz. In the operating bandwidth, the phase center of the E plane varies in the range of 13 mm in the space and the error is within 30 ns in the time domain, correspondingly. Because the interval of most of UWB impulses are about hundreds of nanoseconds, therefore, the Viavldi antenna has the stable phase center in the E plane, and it won't cause the serious effect of dispersion in the time domain.

2.2. The Phase Center of H Plane

With the same method, we calculate the phase center in the H plane (the cross polarization plane), because it varies a lot in the space, so the results of other frequency points are relative distances from the coordinate at 3 GHz which is deemed as the reference point, the results are shown in Fig.3. The compared results of the E and H plane with 60° beam width are shown in Fig.4. From above analysis, we find that the phase centers of the H plane fluctuates much in the space as the frequency increases comparing with the E plane. The narrower the beam width, the more unstable the phase center is,

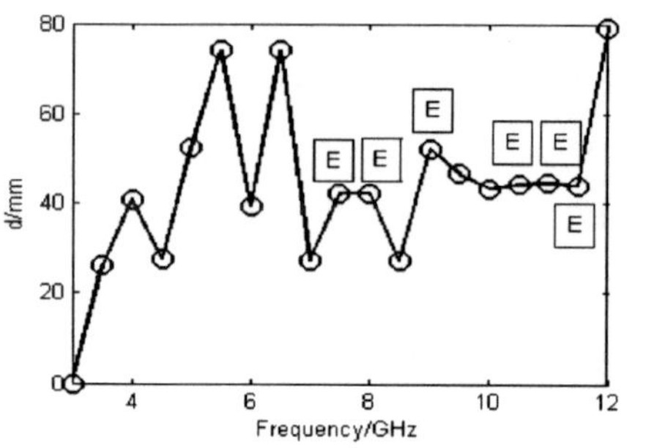

Fig.5 The phase center in the three-dimensional space.

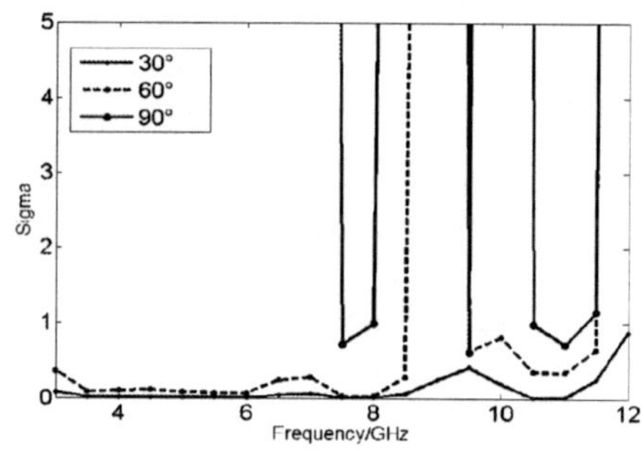

Fig.6 The variation of the H plane with different beam width.

even the phase center can't be found at some frequency points. Therefore, the phase center of the H plane is unstable and it will cause the serious dispersive effect in the time domain.

2.3. The Phase Center of Three-dimension Space

Based on the above methods, we calculate the phase center in the three-dimensional space further, which is shown in Fig.5.When the beam width is narrow, such as 30° and 60°, the phase centers are coincidence with the E plane, with the beam width increasing, the phase centers deviate from the E plane in the low frequency, but in the high frequency, the phase centers are consistence with the E plane again at some points which are labeled with small squares. The reason for the above analysis is that with the frequency increasing, the operating wavelength becomes shorter and then the radiation occurs in the narrow part of the slot, according to the field distribution on the slotline, the transverse electrical field is far beyond the longitude electrical field, therefore, the phase center of the E plane is almost the phase center in the space. In order to prove the good performance of the Vivaldi antenna in the time domain, we also calculate the sigma value of the Vivaldi antenna, which is the maximum standard deviation of the phase center location in the space, the smaller the value, the stable the phase center is, the result is shown in Fig 6. We find that the sigma value is less than 1 in the lower frequency band, and increases sharply at some higher frequency points, and the trend is almost the same as the phase center.

3. CONCLUSION

The determination of phase center of the Vivaldi antenna is presented in this paper. The phase center of the E, H planes and three-dimension are compared, respectively. We can arrive at the conclusion that the phase center of the E plane is more stable in the operating frequency and different beam width and is in agreement with the phase center of three-dimension space in the wide beam width and high operating frequency, besides we also find that the phase center of the H plane is very unstable. Thus we can get the conclusion that we should reduce the cross polarization level in the design. During the analysis of the UWB phase center, we find that the concept of the phase center should be adjusted, for the conventional concept is too strict to be unsuitable to the UWB antenna, many UWB antennas which have good performance in the time domain even don't have the phase center at all if the conventional concept is applied. In my opinion, the stable phase center for the UWB antennas should be defined as the stability of the phase center in the operating frequency band within the 3 dB beam width.

References:

[1] Schantz, H.G, Dispersion and UWB antennas, Joint with Conference on Ultra Wideband Systems and Technologies, Joint UWBST and IWUWBS, , 2004, pp. 161-165.

[2] Zhao, C.-D., Analysis on the properties of a coupled planar dipole UWB Antenna, *IEEE Antennas and Wireless Propagation Letters*, vol.3 (1), pp. 317 320 (2004)

[3] Schantz, H.G., A brief history of UWB antennas, *IEEE Aerospace and Electronic Systems Magazine*, vol .19(4), pp.22-26 (2004).

[4] Moheb, H., A. Sebak, and L. Shafai, Phase centre analysis of array antennas and its significance for microwave landing system. IEE Conference Publication, 1999, pp. 213-216,

[5] Stockbroeckx, B. and A.V. Vorst, Electromagnetic modes in conical transmission lines with application to the linearly tapered slot antenna, *IEEE Transactions on Antennas and Propagation*, vol.48 (3), pp. 447-455. (2000).

APPLICATION OF UWB ANTENNA DESCRIPTORS TO ANTENNA PERFORMANCE ASSESSMENT

Pavel Miškovský, José M. González-Arbesú, J. Romeu*

1. INTRODUCTION

Ultra-wideband systems (UWB) are exploiting a large frequency band. Ranging from 3.1 to 10.6 GHz, they have a fractional spectrum of more than 20%, as defined by Federal Communications Commission[1] (FCC). Several antenna performance descriptors, such as input impedance bandwidth, gain, beamwidth or effective antenna length, can face strong frequency dependence. In this context, the traditional narrowband approach to antenna design is no more appropriate for UWB antennas. Moreover, the frequency dependence of the radiation pattern produces that antennas behave as temporal and spatial filters, radiating differently distorted time-domain waveforms in different directions. For the same excitation, the comparison of two antennas radiating two different waveforms, is almost impossible. For all these reasons, a more *compact* time domain antenna descriptor without frequency and direction dependence is needed. In this sense, the compactness of the descriptor means that it should summarize the performance of the antenna (or antenna system) in a single parameter, in order to simplify its design using optimization techniques.

Lamensdorf[2] used a time domain approach, defining signal fidelity as a measure of distortion, based on cross-correlation of compared signals. McLean[3] also defined new pattern descriptors for UWB antennas (correlated energy pattern and correlation coefficient pattern), useful in systems with correlation detection. These descriptors consider a single antenna and indicate how the radiated UWB signals will change with the time and as a function of direction.

Our proposal is to characterize antennas for UWB applications, using figures of merit that do not depend on direction. Figures of merit applied to a single antenna in free space, as well as to a complete transmitter-receiver system will be presented and applied to actual antennas through electric field measurements.

* Pavel Miškovský, Telecommunications Technological Center of Catalonia (CTTC), Mediterranean Technology Park (PMT), Av. Canal Olímpic s/n, 08860, Castelldefels, Barcelona, Spain.
José M. González-Arbesú, Jordi Romeu, Technical University of Catalonia (UPC), Campus Nord, Ed. D3 c\Jordi Girona 1-3, 08034, Barcelona, Spain.

2. BASIC FORMULATION FOR SINGLE ANTENNA

In UWB systems using correlation detection, the received pulse is being correlated with a template pulse form. In this case the received pulse shape is one of the most important parameters, on which depends the overall system performance.

2.1. Spatially Averaged Fidelity

For the basic formulation, a single antenna is considered as depicted in Fig.1, driven by a voltage source $V_G(f)$ having an internal impedance $Z_G(f)$. To evaluate the performance of a single antenna we propose using the Spatially Averaged Fidelity (SAF_1), defined by Eq. (1). The definition of SAF_1 takes an average over a hypothetical sphere that surrounds the antenna. Nevertheless, Eq. (1) could be slightly modified to average only a region in the surrounding space where the pulse radiated ought to match a given template.

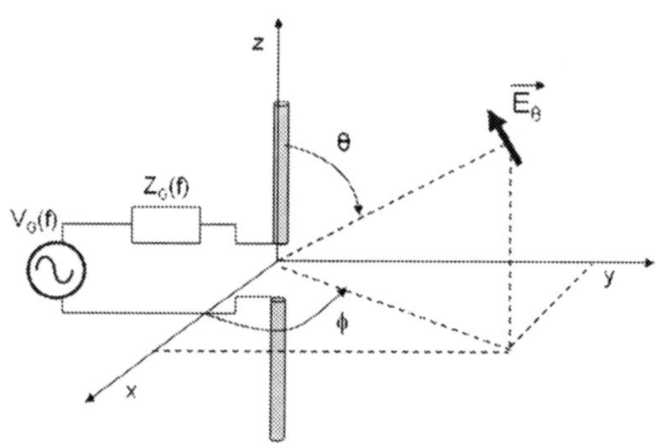

Figure 1. Single antenna configuration

The fidelity SAF_1 correlates the total radiated field **E** at a distance R with a conveniently selected template T in all directions. Then a spatial averaging is realized in order to resume the antenna performance direction-dependence, in a single number.

$$SAF_1 = \frac{\int_0^{2\pi}\int_0^{\pi}\left|\int_{-\infty}^{+\infty}\vec{E}(t,\theta,\phi)\cdot T(t)\cdot\vec{a}(\theta,\phi)\,dt\right|^2 R^2\sin\theta\,d\theta\,d\phi}{\int_{-\infty}^{\infty}|T(t)|^2\,dt\cdot\int_0^{2\pi}\int_0^{\pi}\int_{-\infty}^{\infty}\left|\vec{E}(t,\theta,\phi)\right|^2 dt\,R^2\sin\theta\,d\theta\,d\phi} \qquad (1)$$

In the numerator of equation 1, only the part of the total radiated energy of the field **E** correlated with the template T, is considered. This quantity is averaged over all angular directions (θ,ϕ). As stated by McLean[3], the template can be formulated as a product of a time varying scalar function T(t) and unit vector **a** expressing the polarization of the receiving antenna. The denominator of Eq. (1) normalizes SAF_1 by the respective energies of the template and the total radiated field. Proceeding this way, the maximum value of SAF_l could be 1.

The shape of a template is arbitrary, but a reasonable choice could improve the angular performance of the antenna[3], also in terms of SAF_l.

2.2. Application of SAF_l

The performance of an antenna can dramatically change with a given template. In order to assess this fact and to verify the applicability of SAF_l, three antennas have been compared: a quarter-wavelength monopole at 6.5 GHz, an UWB conical monopole (about the same height), and an UWB ridge horn. The UWB conical monopole, as well as the ridge horn antenna, has a very good matching within the frequency band defined by FCC. The electric fields radiated by the three antennas were measured in an anechoic chamber, in a range from 2.5 to 10.5 GHz with 10 MHz frequency step. The symmetry of the radiation patterns was taken into consideration in order to reduce the computation load. To avoid the perturbing influence of the measurement system on the performance of the ridge antenna, only the broadside hemisphere was measured. As a feeding pulse, an ideal wideband pulse having a bandwidth from 2.5 to 10.5 GHz was used. To show the variation of SAF_l with the template an illustrative example is given in Fig. 2.

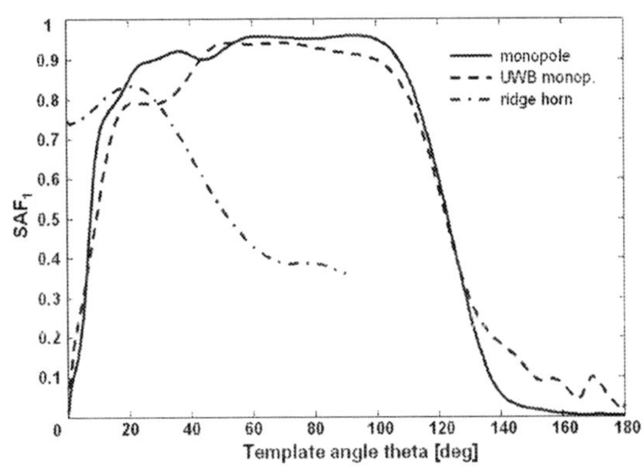

Figure 2. SAF_l as function of different templates. The template is the shape of the pulse radiated in the direction given by the angle theta.

In this case the template was the electromagnetic pulse radiated by each antenna in the direction defined by the abscissa angle θ. From the point of view of SAF_l, over a large range of templates, the quarter-wavelength monopole (solid line) has a similar performance as the UWB monopole (dashed line) and more stable than UWB ridge horn antenna (dash-dotted). This result does not mean that the antenna is a good UWB antenna, but that the shapes of the radiated pulses are quite similar in all the directions. So, for this particular type of template, the quarter-wavelength monopole and UWB conical monopole are less sensitive on template variation, and they distort less the radiated pulse than the ridge horn.

Another three UWB template types were used to compare the antennas performance in terms of SAF_l. The first template is an ideal pulse filling the UWB spectrum according to FCC. In time domain this pulse is a sinc function. The second and third templates are the first and the second derivative of the sinc function. The computed SAF_l for the three antennas under consideration and the mentioned pulses are presented in Table 1.

Table 1. Computed SAF_l using the feeding pulse and its derivative as templates.

Template	UWB monopole	λ/4 monopole	Ridge horn
Sinc	0.63	0.83	0.27
Sinc derivative	0.11	0.05	0.33
Sinc 2nd derivative	0.60	0.79	0.20

In the case of both monopoles, the performance is similar when the sinc or its second derivative is used as a template. This is due to the fact that both template shapes are almost the same. In the case of a template equal to the first sinc derivative, the values of SAF_l are very low for both monopoles, though slightly better for the UWB monopole. The variation of the current distribution on the monopole changes significantly in this wide frequency band. Consequently, the radiated pulses can not be considered as the perfect derivative of the input pulse, as commonly understood for narrowband monopoles. The ridge horn radiates a derivative of the feeding pulse but only within a narrow angular range, explaining the low figures of SAF_l for this antenna.

3. FORMULATION FOR A TRANSMIT-RECEIVE SYSTEM

In a real UWB system, where transmitting and receiving antennas are used, not only the radiated waveform respecting the FCC mask[1] is of significant interest, but especially the pulse waveform delivered to the receiver load.

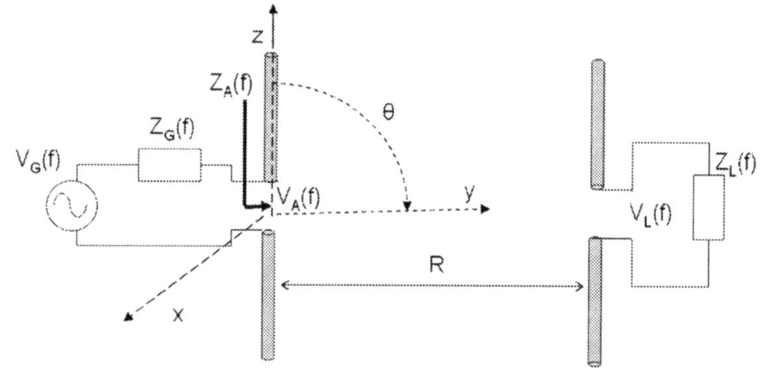

Figure 3. UWB system configuration: transmitting subsystem at left and receiving subsystem at right.

An UWB system, as depicted in Fig. 3, is considered. The transmitter antenna is driven by a voltage source $V_G(f)$, having an internal impedance $Z_G(f) = R_G(f) + jX_G(f)$. The receiver antenna is loaded by impedance $Z_L(f) = R_L(f) + jX_L(f)$ and has a terminal voltage $V_L(f)$. The input impedance of the transmitter antenna is $Z_A(f) = R_A(f) + jX_A(f)$. The antennas are separated by a distance R, not necessarily being a far-field distance.

For the evaluation of antenna performance within a system like described in Fig.3, we propose two figures of merit: Spatially Averaged Fidelity (SAF_2) and Averaged Energy Efficiency (AE). The procedure of evaluating SAF_2 and AE is the same as in the case of SAF_1, placing the receiver subsystem in all angular positions (θ,ϕ) around the transmitting subsystem.

3.1. Spatially Averaged Fidelity

In the same way as for SAF_1, a Spatially Averaged Fidelity (SAF_2) of the received pulse at receiver load can be defined by Eq. (2) for the complete system of Fig.3. In this case the correlation is done between the voltage on the load and the template.

$$SAF_2 = \frac{\int_0^{2\pi}\int_0^{\pi}\left[\int_{-\infty}^{\infty} V_L(t,\theta,\phi)\cdot T(t)\ dt\right]^2 R^2 \sin\theta\ d\theta\ d\phi}{\int_{-\infty}^{\infty}|T(t)|^2\,dt\cdot\int_0^{2\pi}\int_0^{\pi}\int_{-\infty}^{\infty}[V_L(t,\theta,\phi)]^2\,dt\ R^2\sin\theta\ d\theta\ d\phi} \tag{2}$$

This equation is more realistic than SAF_1. Voltage at load terminal $V_L(f)$ implicitly accounts for the frequency variations of the system due to the mismatch between the voltage source and the transmitting antenna, the mismatch between the receiving antenna and the load, and the radiation patterns of both antennas. The value of the parameter SAF_2 also depends on the coupling between the transmitting and receiving antennas, and on the values of source and load impedances. A high value of SAF_2 would mean a high degree of similitude between the template and the received pulse at the load (at all directions of space with respect to the transmitting antenna). Unfortunately, SAF_2 accounts for the shape of the received pulses but it does not account for the efficiency of the system.

3.2. Averaged Energy Efficiency

For a single antenna the radiation efficiency is defined as the ratio of the radiated energy (at a distance R from the antenna) to the energy of the pulse delivered from the generator to the transmitting antenna Eq. (3). Z_0 represents the impedance of free space.

$$\eta = \frac{\int_0^{2\pi}\int_0^{\pi}\int_{-\infty}^{\infty}\left|\vec{E}(t,\theta,\phi)\right|^2 dt\, R^2\sin\theta\, d\theta\, d\phi}{Z_0\int_{-\infty}^{\infty}\frac{|V_A(t)|^2}{R_A}dt} \tag{3}$$

Similarly to this definition, an Averaged Energy Efficiency (AE) is defined by Eq. (4). The numerator of Eq. (4) takes into consideration the spatially averaged energy delivered to the receiver load and the denominator considers the spatially averaged energy of the pulse delivered from the generator to the transmitting antenna. In both expressions, (Eq. 3 and Eq. 4), the mismatch between the transmitting antenna and the generator is considered. The mismatch between the receiving antenna and the load is only present in Eq. (4).

$$AE = \frac{\int_0^{2\pi}\int_0^{\pi}\int_{-\infty}^{\infty} \frac{|V_L(t,\theta,\phi)|^2}{R_L} dt \sin\theta \, d\theta \, d\phi}{\int_0^{2\pi}\int_0^{\pi}\int_{-\infty}^{\infty} \frac{|V_A(t,\theta,\phi)|^2}{R_A} dt \sin\theta \, d\theta \, d\phi} \quad (4)$$

3.3. Application of SAF_2 and AE

The figures of merit, described by Eqs. (2) and (4), have been applied to compare lossy dipole antennas. The UWB system configuration from Fig. 3 has been simulated using Numerical Electromagnetics Code (NEC)[4]. Two dipoles of 10 cm length were placed at distance R = 5.46 m. Generator internal impedance and receiver antenna loading were assumed constant $R_G = R_L = 50$ ohms. A Rayleigh pulse of natural width of 100 ps is used as a generator pulse and its derivative, the Gaussian monocycle, as a template.

In Fig. 4 are shown the simulation results of SAF_2 and normalized AE for two cases. First a constant conductivity was applied along the whole dipole (Fig. 4 left). Second, a linearly variable conductivity dipole (Fig.4 right) was analyzed. The value of conductivity on the horizontal axis (Fig.4 right), represents the maximum value of conductivity applied in the dipole center. The optimum value of conductivity can be found at the intersection point of SAF_2 and normalized AE.

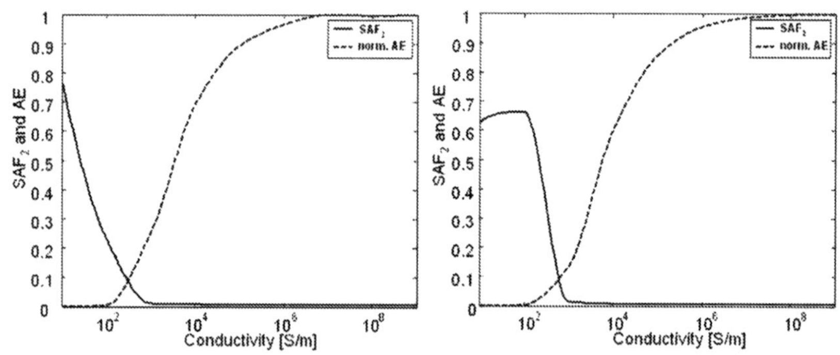

Figure 4. Simulated SAF_2 (continuous line) and AE (dashed line) for constant (left) and variable loading (right)

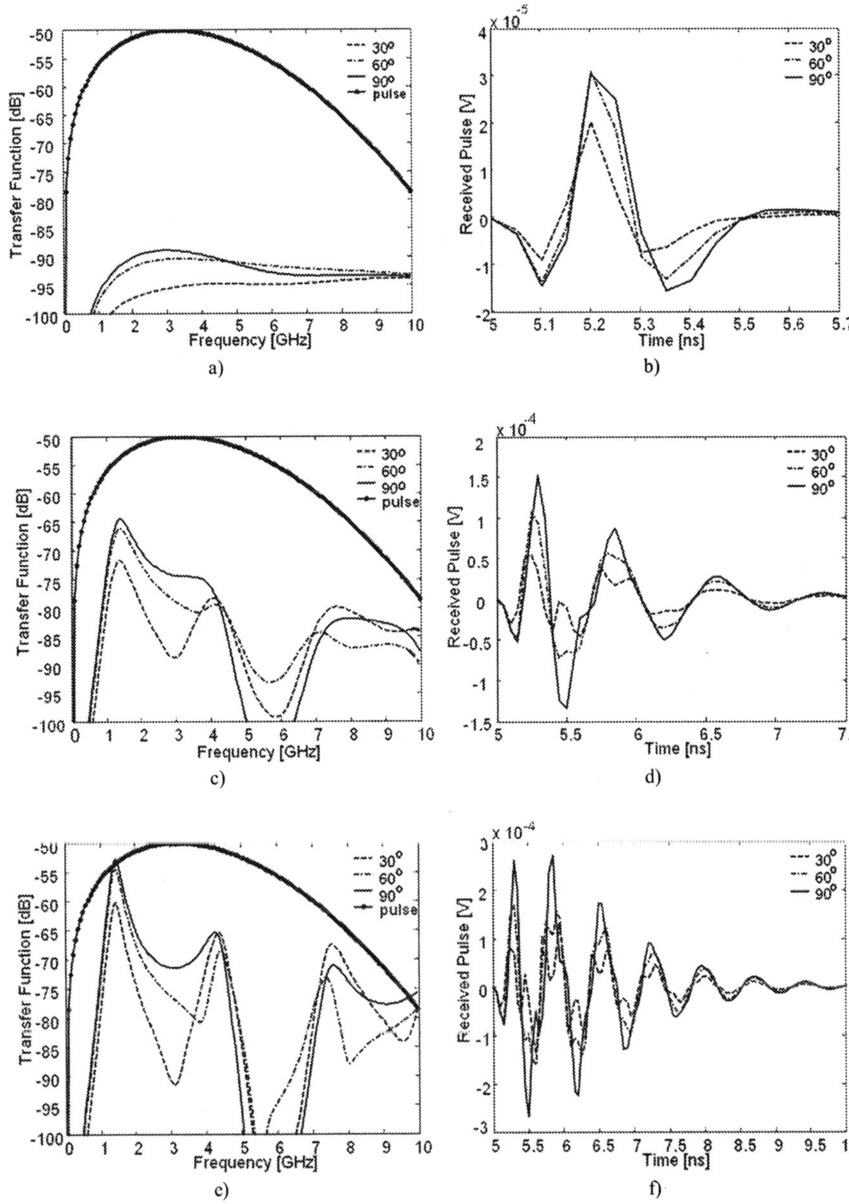

Figure 5. Transfer functions $|V_L(f)/V_G(f)|$, and received pulses at the load, of the system from Fig.3, for 3 different conductivities: 10^2 S/m (a,b), 10^3 S/m (c,d) and 10^7 S/m (e,f). Different relative positions of the receiver antenna, with respect to the transmitter, are presented in each graph: $\theta = 90°$ (continuous line), $\theta = 60°$ (dot-dashed line) and $\theta = 30°$ (dashed line). Dotted line represents the spectrum of the feeding pulse.

In the following section, only a variable conductivity dipole is analyzed. Transfer functions and received pulse shapes, respectively for three values of conductivity (10^2, 10^3, and 10^7 S/m), are presented in Fig. 5. For low conductivity, where SAF_2 is high (Fig. 4 right), the received pulse is very similar to the template because the transfer function is highly wideband for all directions. The efficiency AE is poor, because of the high losses attenuating the transmitted pulse. For high conductivity, it is exactly the opposite. The narrower transfer function filters the transmitted pulse. For the optimum conductivity (10^3 S/m) the pulse suffers some distortion but is not excessively attenuated and still matches the template shape quite well. For this value of conductivity, the efficiency starts to increase and the fidelity is still quite high (Fig. 4 right).

4. CONCLUSION

Two figures of merit evaluating the antenna performance within an UWB system have been presented: Spatially Averaged Fidelity for a single antenna (SAF_1), Spatially Averaged Fidelity for a transmit-receive system (SAF_2), and Averaged Energy Efficiency (AE). First two parameters, SAF_1 and SAF_2, summarize the resemblance between received pulses in free space and at the load respectively, and a given template in all directions of space.

By using SAF_1 we have shown how the antenna performance depends on the chosen template and that a good practice should be specifying this template for each particular antenna and application. Using the figures of merit SAF_2 and AE it has been demonstrated how antenna losses could improve the bandwidth of an UWB system.

5. ACKNOWLEDGEMENTS

This work have been supported by the Generalitat de Catalunya (Grup de Recerca Singular, grant number SGR2005-0069), and Ministerio de Educación y Ciencia (grants TEC 2004/04866-C04-02, TIC2003-09317-C03-03, and the Ramon y Cajal Programme).

Authors would like to thank Beatriz Monsalve Carcelen from AntennaLab - UPC, for designing, fabricating and lending the UWB monopole antenna.

6. REFERENCES

1. Federal Communications Commission (FCC), Revision of Part 15 of the Commission's Rules Regarding Ultra-Wideband Transmission Systems, April, 2002; http://www.wireless.fcc.gov/rules.html
2. D. Lamensdorf, L. Susman, Baseband - Pulse antenna techniques, *IEEE Ant. & Prop. Mag.*, vol. 36, no. 1, pp. 20-30, Feb.1994.
3. J.S. McLean, H. Foltz, R. Sutton, Pattern descriptors for UWB antennas, *IEEE Trans. Ant.& Prop.*, vol. 53, no. 1, pp. 553-559, Jan. 2005.
4. G.J. Burke and A.J. Poggio, Numerical Electromagnetics Code (NEC) - Method of Moments, Tech. Rep. UCID18834, Lawrence Livermore Lab. Livermore, CA, 1981.

BROADBAND TUNING LIMITS ON UWB ANTENNAS BASED ON FANO'S FORMULATION

M. C. Villalobos, H. D. Foltz, J. S. McLean, I. Sen Gupta*

ABSTRACT

The well-known limits on the radiation Q of electrically-small antennas do not lead to a straightforward, rigorous computation of wideband impedance matching limits, especially when higher-order spherical modes are used and higher-order tuning networks are permitted. Fano's formulation of the wideband matching of arbitrary impedances does provide a rigorous solution, and has been previously applied to the lowest-order (TM_{01} or TE_{01}) mode. In this paper we apply Fano's theory to higher-order spherical modes. Graphs of numerical limits on high-pass and band-pass tuning versus size are presented.

1. INTRODUCTION

The well-known Chu-Wheeler limits[1,2] on the radiation Q of small antennas are rigorous; however, there are difficulties in applying them to UWB antennas. The relationships between Q and achievable impedance bandwidth that are commonly used in the literature are accurate only when applied to second-order systems (in terms of spherical harmonics, the TM_{01} or TE_{01} modes) or to narrow band antennas. A UWB antenna will usually not be electrically small at the upper band edge, and thus we cannot make many of the assumptions that are common to the theory of small antennas. In particular, the radiation Q and its associated resonant circuit model are not accurate in describing the behavior. A more rigorous approach is that of Fano[3], who gave very general limits on broadband impedance matching which

*M. C. Villalobos, H. D. Foltz, and I. Sen Gupta are with the University of Texas - Pan American. J. S. McLean is with TDK R&D Corporation. This work was supported in part by the U.S. Army Research Office, Grant DAAD19-02-1-0008, NSF Award 0421352, and the UTPA Computing and Information Technology Center (CITeC).

apply to any load impedance that can be represented by a finite number of linear, lumped, passive network/circuit elements. The Fano approach and related methods have been previously applied to small antennas by Gustafsson and Norbedo[4], Hansen[5,6], and Hujanen *et al*[7].

The approach we take here is based on circuit models for the wave impedance of spherical modes in free space. As shown by Chu[1], the wave impedance of spherical modes as seen at a spherical boundary can be exactly represented by a finite LC high-pass (series-C shunt-L) ladder network terminated in a fixed resistance. It has been shown[8] that computation of radiation Q based on this circuit model yields results equivalent to computations based directly on the field expressions.

2. BRIEF DISCUSSION OF FANO'S INTEGRAL RELATIONS

Fano's analysis begins by classifying networks according to the placement of their transmission zeros in the complex plane. Based on this classification, the reflection coefficient, including the effect of an arbitrary passive matching network, is subject to a series of integral relationships that effectively places limits on how well the network can be matched.

The series-C shunt-L ladder networks have all their transmission zeros at the origin. In this case, Fano's integral relations are

$$\int_o^{\infty} \omega^{-2(k+1)} \ln(\frac{1}{|\Gamma|}) d\omega = (-1)^k \frac{\pi}{2} F^0_{2k+1} \tag{1}$$

for $0 < k < N$, where N is the multiplicity of the transmission zero at zero. The coefficient in the right hand side is given by

$$F^0_{2k+1} = A^0_{2k+1} - \frac{2}{2k+1} \sum_i \lambda_{ri}^{-(2k+1)}, \tag{2}$$

where

$$A^0_{2k+1} = \frac{1}{2k+1} \left(\sum_i \lambda_{oi}^{-(2k+1)} - \sum_i \lambda_{pi}^{-(2k+1)} \right) \tag{3}$$

and the λ_{oi} are the zeros of the reflection coefficient of the network formed by the spherical modes, the antenna, and any matching network combined; the λ_{pi} are the poles, and the λ_{ri} are the zeros that lie in the right half plane. For a physically realizable network, the poles and zeros will either be real or in complex conjugate pairs. Since the λ_{ri} by definition have positive real parts, and if complex come in complex conjugate pairs, the summations in (2) and (3) are real and positive, and we have $F^0_{2k+1} \leq A^0_{2k+1}$. Also, as pointed out in Fano[3], although the summations in (2) and (3) are over all the poles and zeros of the load *and* the matching network, the coefficients A^0_{2k+1} for $0 < k < N$ will depend only on the load network, which in this case is the spherical mode or modes.

In this paper, we consider the antenna and any added matching network combined to form a single matching network between the source and the modes of free space. Thus the limits apply to an arbitrary lossless antenna, with an arbitrary lossless matching network of unlimited complexity. It should be noted that a lossy antenna or matching network can improve the reflection coefficient seen by the source, but that the power *radiated* will not exceed that predicted for the lossless case.

3. APPLICATION TO SPHERICAL MODES

For a single spherical mode of any order, each shunt-L or series-C in the equivalent ladder network introduces a transmission zero at DC and N is equal to the order of the equivalent LC network, which in turn is one more than the Legendre function order n in a spherical mode TM_{mn} or TE_{mn}. For example, for a TM_{m3} wave, $N = 4$ and $k = 0, 1, 2, 3$.

As a first example, consider the lowest order tranverse magnetic spherical mode, TM_{m1}. In terms of complex frequency s, the wave impedance (normalized to free space) from Chu's model is:

$$Z = \frac{c_0}{sa} + \frac{1}{1 + \frac{c_0}{sa}} = \frac{s^2a^2 + sac_0 + c_0^2}{s^2a^2 + sac_0}. \tag{4}$$

The reflection coefficient is then:

$$\Gamma = \frac{Z-1}{Z+1} = \frac{c_0^2}{2s^2a^2 + 2sac_0 + c_0^2} \tag{5}$$

which has no zeros, and two poles located at $s = \frac{-c_0 \pm jc_0}{2a}$. From (3) the A coefficients are $A_1^0 = \frac{2a}{c_0}$ and $A_3^0 = -\frac{4a^3}{3c_0^3}$. The integral relationships governing this mode are then

$$\int_0^\infty \frac{1}{\omega^2} \ln\left|\frac{1}{\Gamma}\right| d\omega = \frac{\pi}{2}\left(\frac{2a}{c_0} - 2\sum \lambda_{ri}^{-1}\right) \tag{6}$$

and

$$\int_0^\infty \frac{1}{\omega^4} \ln\left|\frac{1}{\Gamma}\right| d\omega = \frac{\pi}{2}\left(\frac{4a^3}{3c_0^3} + \frac{2}{3}\sum \lambda_{ri}^{-3}\right). \tag{7}$$

For the TM_{m2} modes, it can be shown that $A_1^0 = \frac{2a}{c_0}$, $A_3^0 = 0$, and $A_5^0 = \frac{a^5}{15c_0^5}$, leading to integral relationships:

$$\int_0^\infty \frac{1}{\omega^2} \ln\left|\frac{1}{\Gamma}\right| d\omega = \frac{\pi}{2}\left(\frac{2a}{c_0} - 2\sum \lambda_{ri}^{-1}\right), \tag{8}$$

$$\int_0^\infty \frac{1}{\omega^4} \ln\left|\frac{1}{\Gamma}\right| d\omega = \frac{\pi}{2}\left(\frac{2}{3}\sum \lambda_{ri}^{-3}\right), \tag{9}$$

and

$$\int_0^\infty \frac{1}{\omega^6} \ln\left|\frac{1}{\Gamma}\right| d\omega = \frac{\pi}{2}\left(\frac{a^5}{15c_0^5} - \frac{2}{5}\sum \lambda_{ri}^{-5}\right). \tag{10}$$

The integrals are analogous to the "matching area" defined in Wheeler[2]. Again, the λ_{ri} are set by the design of the antenna and the matching network, and the summations are real and positive. To optimize use of this "matching area", Γ should have the maximum possible magnitude of 1 outside the desired band, and be constant with $|\Gamma| = \Gamma_0$ within the band. When this behavior is assumed, the integration on the left-hand sides of the above equations can be performed analytically. It can then be seen that when the requisite bandwidth is very large, as in the case of a UWB antenna, little is lost by imposing a high-pass tuning.

4. NUMERICAL RESULTS FOR TM_{m2} MODES

For high-pass matching of the TM_{m2} modes, the computed in-band reflection coefficient tolerance Γ_0 as a function of $k_L a = \frac{\omega_L a}{c_0}$ is shown in Figure 1 graphically, along with the computed values. For each value of $k_L a$, N varied from $2, 3, \ldots, 10$. The value of $\Gamma_0 = 1/e^K$ was obtained through the following optimization problem:

$$\text{maximize} \quad K \quad \text{subject to} \quad \begin{aligned} K - \tfrac{\omega_L \pi}{2}\left(\tfrac{2a}{c_0} - 2\textstyle\sum_{i=1}^N \lambda_{ri}^{-1}\right) &= 0 \\ K - \tfrac{3\omega_L^3 \pi}{2}\left(\tfrac{2}{3}\textstyle\sum_{i=1}^N \lambda_{ri}^{-3}\right) &= 0 \\ K - \tfrac{5\omega_L^5 \pi}{2}\left(\tfrac{a^5}{15c_0^5} - \tfrac{2}{5}\textstyle\sum_{i=1}^N \lambda_{ri}^{-5}\right) &= 0. \end{aligned}$$

Letting $S_{ri} = \alpha_i + j\beta_i$, where $j = \sqrt{-1}$, be the reciprocal of λ_{ri} and since λ_{ri} has positive real parts and if complex, occur in conjugate pairs, then the above optimization problem can be remodeled in its real form as:

$$\begin{aligned} &\text{maximize} && K \\ &\text{subject to} && K - \tfrac{\omega_L \pi}{2}\left(\tfrac{2a}{c_0} - 2\textstyle\sum_{i=1}^N \alpha_i\right) = 0 \\ &&& K - \tfrac{3\omega_L^3 \pi}{2}\left(\tfrac{2}{3}\textstyle\sum_{i=1}^N \alpha_i^3 - 3\alpha_i\beta_i^2\right) = 0 \\ &&& K - \tfrac{5\omega_L^5 \pi}{2}\left(\tfrac{a^5}{15c_0^5} - \tfrac{2}{5}\textstyle\sum_{i=1}^N \alpha_i^5 - 10\alpha_i^3\beta_i^2 + 5\alpha_i\beta_i^4\right) = 0 \end{aligned}$$

$$\begin{aligned} \beta_i + \beta_{i+1} &= 0 \quad i = 1, 3, \ldots, \tilde{N} \\ \beta_i(\alpha_i - \alpha_{i+1}) &= 0 \quad i = 1, 3, \ldots, \tilde{N} \\ \alpha_i &\geq 0 \quad i = 1, 2, \ldots, N \end{aligned}$$

where $\tilde{N} = N$ for N odd, and $\tilde{N} = N - 1$ for N even. The optimization problem was solved using the software CONOPT in conjunction with the NEOS server[9].

5. REFERENCES

1. L.J.Chu, "Physical Limitations of Omnidirectional Antennas," *J. Applied Physics*, vol. 19, pp.1163-1176, Dec. 1948.
2. H.A. Wheeler, "The wide-band matching area for a small antenna", *IEEE Trans. Antennas Propagat.*, vol. 31, pp.364 - 367, March 1983.
3. R.M. Fano, "Theoretical Limitations on the Broadband Matching of Arbitrary Impedances," *J. Franklin Institute*, vol.249, pp.57-83, Jan. 1950, and pp.139-155, Feb. 1960.
4. M. Gustafsson and S. Nordebo, "Bandwidth, Q factor, and resonance models of antennas" *Progress in Electromagnetics Research*, 62, 1-20, 2006.
5. R.C. Hansen, "Fano limits on matching bandwidth", *IEEE Antennas and Propagation Magazine*, vol. 47, no. 3, pp. 89 - 90, June 2005.

6. A.R. Lopez, "Rebuttal to 'Fano limits on matching bandwidth' ", *IEEE Antennas and Propagation Magazine*, vol. 47, no. 5, pp. 128-129, October 2005.
7. Hujanen, A., Holmberg J. , and Sten., J. C.-E., "Bandwidth Limitations of Impedance Matched Ideal Dipoles," *IEEE Trans. Antennas Propagat.*, pp. 3236-3239, Vol. 53, No. 10, Oct. 2005.
8. J.S. McLean, "A Re-examination of the Fundamentals Limits on the Radiation Q of Electrically Small Antennas," *IEEE Trans. Antennas and Propagation*, vol. 44, no.5, pp.672-676, May 1996.
9. E.D. Dolan, R. Fourer, J. Moré, and T.S. Munson, "Optimization on the NEOS Server," *SIAM News*, vol. 35, pp. 5, 8–9, 2002.

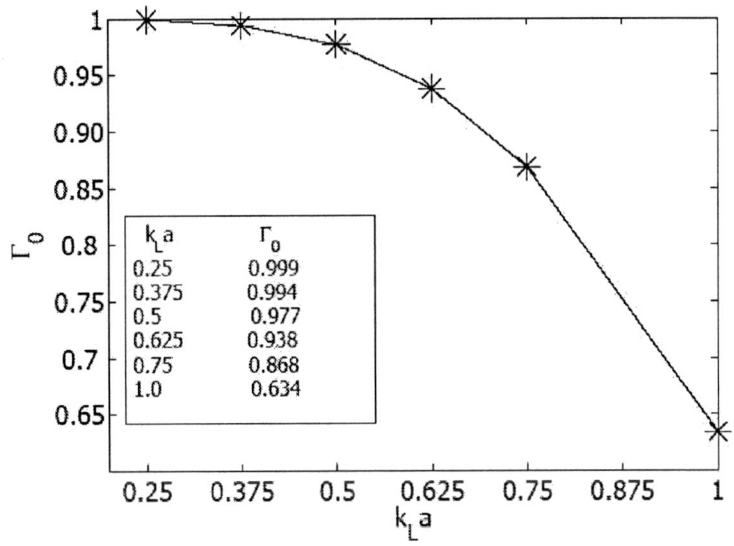

Figure 1. In-band reflection coefficient Γ_0 vs. $k_L a$, electrical size at lower band edge

ELECTROMAGNETIC LENS DESIGN AND GENERALIZED E AND H MODES

A. P. Stone and C. E. Baum

1. INTRODUCTION

Among several possible approaches to the design of lens waveguide transitions is a differential-geometric method. In general one starts with Maxwell's equations together with boundary conditions and general theorems such as conservation of energy and reciprocity and looks for various mathematical concepts for representing the solution of an EM problem. One may start with inhomogeneous TEM plane waves which propagate on ideal transmission lines with two or more independent perfectly conducting boundaries. These types of inhomogeneous media can be used to define lenses for TEM waves without reflection or distortion between conical and cylindrical transmission lines. While there may be practical limitations (i.e., the properties of materials used to obtain the desired permittivity and permeability of the inhomogeneous medium) perfect characteristics are not really necessary. This approach to EM lens design was initiated by C. E. Baum and has been applied successfully in many applications. Specifically the scaling method creates a class of equivalent electromagnetic problems each having a complicated geometry and medium from an electromagnetic problem having a simple (Cartesian) geometry and medium. Thus the scaling method transforms an EM problem by a change of coordinates, and is a method that is well known in fluid dynamics and mechanics. In earlier work transient lens for propagating TEM modes with dispersion have been considered. We may also consider the properties of E and H modes in such lenses. The presence of longitudinal field components brings in additional constraints on the coordinate systems that are allowable. As a consequence the cases of transient lenses supporting E and H modes is limited to a subset of those supporting TEM modes.

The technique developed by Baum[1] for the design of EM lenses utilizes the expression of the constitutive parameters $\overleftrightarrow{\epsilon}$ and $\overleftrightarrow{\mu}$ and Maxwell's equations in a general orthogonal curvilinear coordinate system, yielding what we will call the formal quantities. These are customarily denoted by affixing primes as superscripts.

The line element is

$$(d\ell)^2 = h_1^2(du_1)^2 + h_2^2(du_2)^2 + h_3^2(du_3)^2 \tag{1}$$

and the coordinates are (u_1, u_2, u_3). The scale factors, h_i, relate the formal parameters ϵ_i' and μ_i' to the real world parameters ϵ_i and μ_i in the diagonal case via the equations

$$(\epsilon_{ij}') = \begin{pmatrix} \frac{h_2h_3}{h_1}\epsilon_1 & 0 & 0 \\ 0 & \frac{h_1h_3}{h_2}\epsilon_2 & 0 \\ 0 & 0 & \frac{h_1h_2}{h_3}\epsilon_3 \end{pmatrix} \tag{2}$$

$$(\mu_{ij}') = \begin{pmatrix} \frac{h_2h_3}{h_1}\mu_1 & 0 & 0 \\ 0 & \frac{h_1h_3}{h_2}\mu_2 & 0 \\ 0 & 0 & \frac{h_1h_2}{h_3}\mu_3 \end{pmatrix} \tag{3}$$

The scale factors also relate the formal fields E_i' and H_i' to the real fields E_i and H_i via the equations

$$E_i' = h_iE_i, \quad H_i' = h_iH_i \tag{4}$$

for $i = 1, 2, 3$. Maxwell's equations for the formal fields are

$$\begin{aligned} \nabla' \times \vec{E}' &= -s\vec{B}' = -s\overleftrightarrow{\mu}' \cdot \vec{H}' \\ \nabla' \times \vec{H}' &= s\vec{D}' = s\overleftrightarrow{\epsilon}' \cdot \vec{E}' \end{aligned} \tag{5}$$

Here we use $s = \Omega + j\omega$, the two-sided Laplace-transform variable or complex frequency. This suppresses the time-derivatives for our convenience in notation, and furthermore allows the constitutive parameters to be frequency dependent if desired. This last point is significant only in the case of dispersive media, which need not concern us here. Since we are not going back and forth between the time and frequency domains, we do not need to indicate the fact that all fields are Laplace transforms (functions of complex frequency).

Thus if we assume diagonal forms for the tensors $\overleftrightarrow{\epsilon}$ and $\overleftrightarrow{\mu}$, Maxwell's equations in expanded form become

$$\begin{aligned} \frac{\partial E_3'}{\partial u_2} - \frac{\partial E_2'}{\partial u_3} &= -s\mu_1'H_1' \\ \frac{\partial E_1'}{\partial u_3} - \frac{\partial E_3'}{\partial u_1} &= -s\mu_2'H_2' \\ \frac{\partial E_2'}{\partial u_1} - \frac{\partial E_1'}{\partial u_2} &= -s\mu_3'H_3' \end{aligned} \tag{6}$$

and

$$\begin{aligned}\frac{\partial H_3'}{\partial u_2} - \frac{\partial H_2'}{\partial u_3} &= s\epsilon_1' E_1' \\ \frac{\partial H_1'}{\partial u_3} - \frac{\partial H_3'}{\partial u_1} &= s\epsilon_2' E_2' \\ \frac{\partial H_2'}{\partial u_1} - \frac{\partial H_1'}{\partial u_2} &= s\epsilon_3' E_3'\end{aligned} \tag{7}$$

These equations will be the starting point in our search for conditions on the parameters $\overleftrightarrow{\mu}$ and $\overleftrightarrow{\epsilon}$ in the case of E (or TM) and H (or TE) modes.

2. E-WAVE (FORMAL FIELDS)

We begin our investigation with the E-wave case (transverse magnetic field), and so we will take $H_3' = 0$ and seek conditions on the formal parameters ϵ_i' and μ_i' which lead to solutions of the formal Maxwell equations. As usual in the case of waveguides we seek solutions for the formal-field components in terms of some operator on E_3' which we will later take as some mode function of u_1 and u_2 (transverse coordinates) times some propagation function of u_3. Our starting point once again will be Maxwell's equations as they appear in (6) and (7), which come from $\nabla' \times \vec{E}' = -s\overleftrightarrow{\mu}' \cdot \vec{H}'$ and $\nabla \times \vec{H}' = s\overleftrightarrow{\epsilon}' \cdot \vec{E}'$. If we put $H_3' = 0$ in these equations we obtain

$$\begin{aligned}\frac{\partial E_3'}{\partial u_2} - \frac{\partial E_2'}{\partial u_3} &= -s\mu_1' H_1' \quad , & \frac{\partial H_2'}{\partial u_3} &= -s\epsilon_1' E_1' \\ \frac{\partial E_1'}{\partial u_3} - \frac{\partial E_3'}{\partial u_1} &= -s\mu_2' H_2' \quad , & \frac{\partial H_1'}{\partial u_3} &= s\epsilon_2' E_2' \\ \frac{\partial E_2'}{\partial u_1} - \frac{\partial E_1'}{\partial u_2} &= 0 \quad , & \frac{\partial H_2'}{\partial u_1} - \frac{\partial H_1'}{\partial u_2} &= s\epsilon_3' E_3' \, .\end{aligned} \tag{8}$$

The consequences of the condition that $H_3' = 0$ will lead to restrictions on the formal parameters ϵ_i' and μ_i' (except for μ_3' since $H_3' = 0$) as well as solutions for the formal fields E_i' and H_i'.

We assume that the formal constitutive parameters have the forms similar to those specified in the TEM case, namely

$$\begin{aligned}\mu_n' &= \mu_n^{(0)'}(u_1, u_2) g_\mu(u_3) \\ \epsilon_n' &= \epsilon_n^{(0)'}(u_1, u_2) g_\epsilon(u_3)\end{aligned} \tag{9}$$

for $n = 1, 2$, with μ_3' irrelevant since u_3 is the assumed propagation direction. We take

$$\epsilon_3' = \epsilon_3^{(0)'}(u_1, u_2) g_{\epsilon 3}(u_3) \tag{10}$$

and then try solutions of the form

$$\begin{aligned}\vec{E}'_t &= \vec{E}_t^{(0)'}(u_1,u_2)g_e(u_3)\\ \vec{H}'_t &= \vec{H}_t^{(0)'}(u_1,u_2)g_h(u_3)\\ E'_3 &= E_3^{(0)'}(u_1,u_2)g_{e3}(u_3),\ H'_3 = 0\\ \overleftrightarrow{D}' &= \overleftrightarrow{\epsilon}'\cdot\vec{E}',\ \overleftrightarrow{B}' = \overleftrightarrow{\mu}'\cdot\vec{H}'.\end{aligned} \tag{11}$$

Now for physical realizability g_μ and g_ϵ are real, nonzero and frequency independent. The propagation functions g_e, g_h and g_{e3} are in general complex functions of the complex frequency s (exponential like) and all have u_3 derivatives nonzero except possibly at special frequencies like $s = 0$, or degenerate cases like propagation perpendicular to u_3 (waveguide cutoff). The functions of u_1, u_2 are taken as independent of the complex frequency s.

Thus we obtain

$$\begin{aligned} g_{e3}\frac{\partial E_3^{(0)'}}{\partial u_2} - \frac{dg_e}{du_3}E_2^{(0)'} &= -s\mu_1^{(0)'}g_\mu g_h H_1^{(0)'}\\ E_1^{(0)'}\frac{dg_e}{du_3} - g_{e3}\frac{\partial E_3^{(0)'}}{\partial u_1} &= -s\mu_2^{(0)'}g_\mu g_h H_2^{(0)'}\\ \frac{\partial E_2^{(0)'}}{\partial u_1} - \frac{\partial E_1^{(0)'}}{\partial u_2} &= 0\end{aligned} \tag{12}$$

and

$$\begin{aligned} H_2^{(0)'}\frac{dg_h}{du_3} &= -s\epsilon_1^{(0)'}g_\epsilon g_e E_1^{(0)'}\\ H_1^{(0)'}\frac{dg_h}{du_3} &= s\epsilon_2^{(0)'}g_\epsilon g_e E_2^{(0)'}\\ g_h\left[\frac{\partial H_2^{(0)'}}{\partial u_1} - \frac{\partial H_1^{(0)'}}{\partial u_2}\right] &= s\epsilon_3^{(0)'}g_{\epsilon 3}g_{e3}E_3^{(0)'}\end{aligned} \tag{13}$$

Moreover, since $\nabla'\cdot\left[\overleftrightarrow{\epsilon}'\cdot\vec{E}'\right] = 0$, we have

$$\frac{\partial[\epsilon_1^{(0)'}E_1^{(0)'}]}{\partial u_1} + \frac{\partial[\epsilon_2^{(0)'}E_2^{(0)'}]}{\partial u_2} = -\frac{1}{g_\epsilon g_e}\frac{\partial[g_{\epsilon 3}g_{e3}]}{\partial u_3}\epsilon_3^{(0)'}E_3^{(0)'} \tag{14}$$

and also, from $\nabla'\cdot\left[\overleftrightarrow{\mu}'\cdot\vec{H}'\right] = 0$, we have

$$\nabla'_t\cdot\left[\overleftrightarrow{\mu}_t^{(0)'}\cdot\vec{H}_t^{(0)'}\right] = 0. \tag{15}$$

We may now define some constants since we can separate variables in (13) and (14). Thus we can set

$$K_h \equiv -sg_\epsilon g_e\left[\frac{dg_h}{du_3}\right]^{-1} = \frac{H_2^{(0)'}}{\epsilon_1^{(0)'}E_1^{(0)'}} = -\frac{H_1^{(0)'}}{\epsilon_2^{(0)'}E_2^{(0)'}} \tag{16}$$

$$v_0\gamma_{e3} \equiv -s g_{\epsilon 3} g_{e3} g_h^{-1} = \frac{1}{\epsilon_3^{(0)'} E_3^{(0)'}} \left[\frac{\partial H_1^{(0)'}}{\partial u_2} - \frac{\partial H_2^{(0)'}}{\partial u_1} \right] \tag{17}$$

$$\gamma_d \equiv \frac{1}{g_\epsilon g_e} \frac{d[g_{\epsilon 3} g_{e3}]}{du_3} = -\frac{1}{\epsilon_3^{(0)'} E_3^{(0)'}} \left[\frac{\partial[\epsilon_1^{(0)'} E_1^{(0)'}]}{\partial u_1} + \frac{\partial[\epsilon_2^{(0)'} E_2^{(0)'}]}{\partial u_2} \right]. \tag{18}$$

The constant γ_d can be expressed in terms of K_h and $\gamma_{\epsilon 3}$ and are in general nonzero and bounded except perhaps for special values of s. They also have the same frequency dependence.

We would like the E modes to propagate in the same medium as the TEM modes which are limiting cases of an E mode. Once the propagation functions for the fields are known, the transverse field components are also known and related to $E_3^{(0)}$. The factors which occur in the constitutive parameters are also known. Since the Maxwell equations have been satisfied, the E modes are determined.

Thus we can have E modes in the same media as the TEM modes discussed previously, provided we have an additional constraint on the u_3 part of the permittivity. Specifically $g_{\epsilon 3}$ varies reciprocally with respect to g_c (previously specified) for the special case of K_0 independent of u_3. However, the medium can now be both inhomogeneous and anisotropic.

3. H-WAVE (FORMAL FIELDS)

The H-wave case (transverse electric field) can be studied in a similar manner. We may impose the condition that the formal field component, E_3', vanishes and then look for conditions on the formal parameters, ϵ_i' and μ_i', which lead to solutions of the formal Maxwell equations. The analysis is dual to the E-wave case, and the results will be dual. (Duality is the symmetry on interchange of electric and magnetic parameters.) Thus solutions can be sought for the formal field components in terms of an operator on H_3' which will eventually be taken as some mode function of the transverse coordinates, u_1 and u_2, multiplied by a propagation function of u_3. There is a differential equation for $H_3^{(0)'}$ as

$$\Phi_h' = \frac{H_3^{(0)'}}{\gamma_g} = \frac{1}{\gamma_d \gamma_g} \frac{1}{\mu_3^{(0)'}} \nabla_t' \cdot \left[\overleftrightarrow{\mu}_t^{(0)'} \cdot \nabla_t' \Phi_h' \right] \tag{19}$$

For appropriate boundary conditions for the magnetic field (normal derivative of Φ_h' zero on some closed contour (perfectly conducting boundary) in the u_1, u_2 plane) we have a waveguide problem for Φ_h' with $\gamma_d\gamma_g$ assuming the role of an eigenvalue (a transverse wave number or propagation constant). Again we find real-valued solutions for Φ_h' with negative eigenvalues as

$$\begin{aligned} \gamma_d \gamma_g &= -K_0^2 = \frac{s}{v_0'} \gamma_g \\ K_0^2 &= \text{positive numbers of same dimension as } [s/v^{(0)'}]^2. \end{aligned} \tag{20}$$

We find that H modes can propagate in the same media as the TEM modes. Thus we obtain the result $g_{\mu 3}(u_3)$ is constrained to vary inversely with respect to a previously specified $g_c(u_3)$. That is,

$$g_{\mu 3} g_c = 1. \tag{21}$$

A complete discussion of these results may be found in.[2]

4. SCALING TO REAL MEDIUM FOR E-MODES

We may now consider the scaling of the u_n coordinates to something other than Cartesian coordinates. The results for the TEM case are directly applicable. If the real medium is constrained to be isotropic we have

$$\begin{aligned} \epsilon &= \epsilon_1 = \epsilon_2 = \epsilon_3 \\ \mu &= \mu_1 = \mu_2 (= \mu_3) \end{aligned} \tag{22}$$

so that we have five relevant constitutive-parameter components to consider, one more than in the TEM case. In particular we have

$$\begin{aligned} \epsilon_1^{(0)'} &= \epsilon_2^{(0)'} \equiv \epsilon_T^{(0)'} \quad , \quad \mu_1^{(0)'} = \mu_2^{(0)'} \equiv \mu^{(0)'} \\ h_1 &= h_2 \equiv h_t \end{aligned} \tag{23}$$

with surfaces of constant u_3 limited to spheres and planes (See[3] and[4]). We obtain

$$\begin{aligned} \epsilon_3^{(0)'}(u_1, u_2) g_c^{-1}(u_3) &= \frac{h_t^2}{h_3} \epsilon \\ \epsilon_t^{(0)'}(u_1, u_2) \epsilon_3^{(0)'}(u_1, u_2) &= h_t^2 \epsilon^2 \end{aligned} \tag{24}$$

giving

$$\epsilon_t^{(0)'}(u_1, u_2) g_c(u_3) = h_3 \, \epsilon. \tag{25}$$

If the requirement that $\mu = \mu_0$ (uniform and isotropic) is added with ϵ isotropic, we obtain

$$\begin{aligned} Z &= \left[\frac{\mu_0}{\epsilon}\right]^{\frac{1}{2}} \\ \epsilon &= \epsilon(u_1, u_2) \\ v &= [\mu_0 \epsilon]^{-\frac{1}{2}} = v(u_1, u_2) \\ h_3 &= \frac{v(u_1, u_2)}{v^{(0)'}} g_c(u_3) \end{aligned} \tag{26}$$

Hence we then have

$$h_t = \frac{\left[\epsilon_t^{(0)'}(u_1, u_2)\, \epsilon_3^{(0)'}(u_1, u_2)\right]^{\frac{1}{2}}}{\epsilon(u_1, u_2)} = \text{function of } u_1, u_2 \text{ only} \tag{27}$$

which implies that surfaces of constant u_3 can only be planes. It is interesting to note that such a case of a bending lens with constant-ϕ surfaces being planes is considered in Baum.[5]

At this point we can note that it is possible to have a simpler form for h_3 as

$$h_3 = h_3(u_1, u_2) = \frac{v(u_1, u_2)}{v^{(0)'}} = \text{function of } u_1, u_2 \text{ only} \tag{28}$$

by setting

$$g_c \equiv 1 = \frac{1}{g_{\epsilon 3}}. \tag{29}$$

This corresponds to choosing the formal medium to be uniform with respect to the u_3 coordinate, a somewhat simpler form for the formal medium. If one in addition were to force the formal permittivity to be isotropic, then

$$\begin{aligned} \epsilon_3' &= \epsilon_3^{(0)'} g_c^{-1} = \epsilon_t^{(0)'} g_c = \epsilon_t' \equiv \epsilon' \\ g_c^2 &= \frac{\epsilon_3^{(0)'}}{\epsilon_t^{(0)'}} \neq \text{function of } u_3. \end{aligned} \tag{30}$$

Then if we set, as a matter of convention,

$$\begin{aligned} g_c &= 1 \\ h_t &= h_3 \equiv h \\ \epsilon' &= h\epsilon \ , \quad \mu' = h\mu. \end{aligned} \tag{31}$$

This gives the case in Baum and Stone [4, (Appendix C)] which admits of only two types of solutions: Cartesian coordinates and the inversion of Cartesian coordinates.

This is a very restrictive case, so the constraint is not very significant, and we can allow the formal permittivity to be anisotropic.

5. CONCLUDING REMARKS

We now have a significant set of results for E and H modes. The basic form for those is found by separating out the u_3 (propagation) coordinate from the u_1, u_2 (transverse) coordinates, and requiring propagation in one direction without reflection. Various assumptions (constraints) on the constitutive parameters lead to constraints on the allowable coordinate systems. We can note that these results apply only to such modes, and not to all possible solutions of the Maxwell equations which may include additional contributions (e.g., hybrid HE modes).

The present results also allow for more general anisotropic real and/or formal media to be considered, including the case of TEM modes. This may lead to other interesting cases for transient lens design.

It should be noted that in a previous paper (Baum & Stone[6]), it was shown that the generalized form of a TEM plane wave involving inhomogeneous constitutive parameters could be used for the formal fields for generating lens designs based on differential geometric scaling. This method, which is described in earlier works such as Mo, Papas[6] and Baum and Stone,[8] has led to the specification of lens materials with inhomogeneous permittivity. Practical considerations will of course influence the construction of these lenses. Bends in waveguides pose special problems for high-voltage ultra-wideband systems as only straight sections of conventional transmission lines can support the pure TEM mode that is necessary to preserve the rise time of a transmitted pulse. It has been shown that a waveguide bend can be embedded within a purely dielectric lens. Moreover, Bigelow and Farr[9] have introduced the concept of a graded dielectric lens. Most recently, Bigelow, Farr, and Prather[10] described the design of a layered dielectric lens in a 90° bend in an electrically large coaxial transmission line.

Finally we note the fundamental assumption of $\vec{E}$ and $\vec{H}$ each having both u_1 and u_2 components. This could be relaxed by allowing the fields to have only one transverse component (e.g., E_1 and H_2). So there are various possible other cases to consider.

REFERENCES

1. Baum, C. E. 1968. A scaling technique for the design of idealized electromagnetic lenses. *Sensor and Simulation Note* 64.
2. Baum, C. E., & A. P. Stone. 2001. Generalized TEM, E, and H modes, *Sensor and Simulation Note* 453.
3. Eisenhart, L. P. 1960. *A Treatise on the differential geometry of curves and surfaces.* New York: Dover.
4. Baum, C. E. & A. P. Stone. 1991. *Transient lens synthesis: Differential geometry in electromagnetic theory,* New York, Taylor & Francis.
5. Baum, C. E.. 1998. Use of generalized inhomogeneous TEM plane waves in differential geometric lens synthesis, *Sensor and Simulation Note* 405, December 1996; Proceedings of the URSI International Symposium on Electromagnetic Theory, Thessaloniki, Greece, May, pp. 636-638.
6. Mo, T. C., C. H. Papas, & C. E. Baum. 1973. Differential-geometry scaling method for electromagnetic field and its applications to coaxial waveguide junctions. *Sensor and Simulation Note* 169, March, available from C. E. Baum, Air Force Research Laboratory/DEHP, 3550 Aberdeen Ave. SE, Kirtland AFB, NM 87117, USA. Also in a shorter version, General scaling method for electromagnetic field with application to a matching problem. *J. Math. Phys.* 14:479-483.
7. Baum, C. E., & A. P. Stone. 2000. Synthesis of inhomogeneous dielectric, dispersionless TEM lenses for high-power application. *Electromagnetics* 20:17-28.
8. Baum, C. E., & A. P. Stone. 1999. Unipolarized generalized TEM plane waves in differential geometric lens synthesis. *Sensor and Similation Note* 433.
9. Bigelow, W. S., & E. G. Farr, 1998. Minimizing dispersion in a TEM waveguide bend by a layered approximation of a graded dielectric material. *Sensor and Simulation Note* 416. Available from C. E. Baum, Air Force Research Laboratory/DEHP, 3550 Aberdeen Ave. SE, Kirtland AFB, NM 87117, USA.
10. Bigelow, W. S., E. G. Farr, & W. D. Prather. 2000. Compensation of an electrically large coaxial transmission line bend by a layered dielectric lens. *Sensor and Simulation Note* 445.

PRODUCING LARGE TRANSIENT ELECTROMAGNETIC FIELDS IN A SMALL REGION: AN ELECTROMAGNETIC IMPLOSION

Carl E. Baum*

1. INTRODUCTION

One approach to a transient electromagnetic radiator is an impulse radiating antenna (IRA) [1]. In this case an antenna aperture is focused at infinity. While there are several ways to realize an IRA (reflector, lens, array), the reflector type is very practical, especially for large ones, if electronic beam steering is not required. Basic physical considerations are discussed in [2].

What if, instead, one has a target which is relatively close to the antenna so that focusing at infinity may not be appropriate. Focusing at the target can then be used to increase the field strengths there [3, 4]. Such is the subject of the present paper.

2. INCOMING SPHERICAL WAVES

Instead of thinking of a wave radiating away from a point source, let us reverse this wave (time reversal) so as to converge on some position (eventually a target). There are various ways to look at this. One way considers some surface S surrounding some volume V containing a point of interest, say $\vec{r} = \vec{r}_0$.

Imagine some transient wave propagating inward toward $\vec{r}_0$, but which has not yet reached this point. On S, there will be tangential electric $\vec{E}_s(\vec{r}_0,t)$ and magnetic $\vec{H}_s(\vec{r}_s,t)$ fields $(\vec{r}_s \in S)$. If one then applies such fields all over S, this will produce the same inward propagating wave in V with no fields outside of V (equivalence principle). These sources can also be considered as magnetic and electric surface current densities as well. This can form the basis for an array of sources on S to approximately realize the desired fields. One can also use just one type of source, say for tangential electric field [5], but this produces waves both inside and outside of S.

*University of New Mexico, Dept. of Electrical and Computer Engineering, MSC01 1100, Albuquerque, NM, 87131-0001

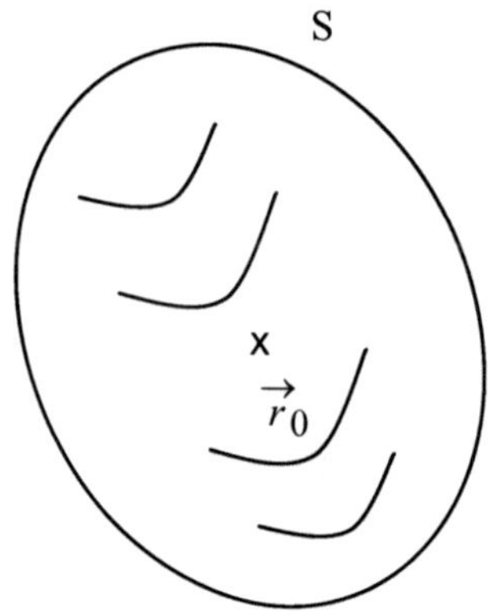

Fig. 2.1. Surface Surrounding Target Position.

As time goes along the wave reaches $\vec{r}_0$ where peculiar things may happen with the wave emerging from this region and becoming an outgoing wave propagating toward S. The details depend on the presence or absence of a target centered on $\vec{r}_0$. In the absence of a target the fields can get very large, depending on the rate of rise of the incoming wave. Now $\vec{r}_0$ is a focus of the wave which needs detailed consideration for transient pulses.

Various types of incoming waves can be considered. If S is a sphere centered on $\vec{r} = \vec{0}$, one can consider particular vector spherical harmonics for source distributions. These are boundary conditions for the spherical vector wave functions, appropriately transformed into time domain.

For IRA-like waves we think in terms of TEM waves (dispersionless) which are guided on conductors. Of course, if inside S there are no conductors, the assumed TEM fields on S will produce some other kind of wave, but still focused on $\vec{r}_0$.

3. PROLATE-SPHEROIDAL REFLECTOR

Now, instead of a paraboloidal reflector, consider a prolate-spheroidal reflector. This is based on the two foci of an ellipse. A prolate sphere S_p is a body of revolution with an equation for the surface

$$\left[\frac{\Psi}{b}\right]^2 + \left[\frac{z}{a}\right]^2 = 1 \quad , \quad a \equiv \text{major radius} \quad , \quad b \equiv \text{minor radius} \tag{3.1}$$

This is based on cylindrical (Ψ, ϕ, z) coordinates with

$$x = \Psi\cos(\phi) \quad , \quad y = \Psi\sin(\phi) \quad , \quad \Psi^2 = x^2 + y^2 \tag{3.2}$$

As illustrated in Fig. 3.1, there are two foci at

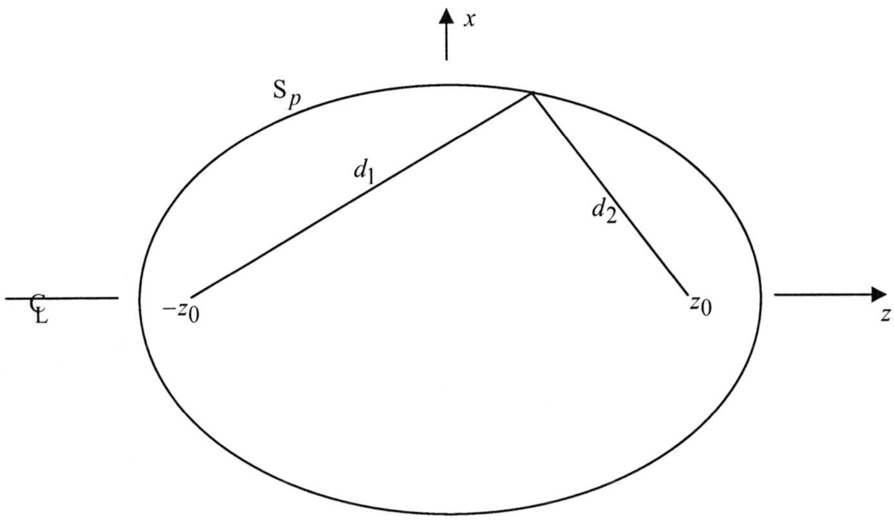

Fig. 3.1 Prolate-Spheroid Cross Section (Ellipse)

$$(\Psi,\phi,z) = (0,\phi,\pm z_0) = \pm \vec{r}_0 \ , \quad z_0 = \left[a^2 - b^2\right]^{1/2} \tag{3.3}$$

The distance traveled by a ray from $-\vec{r}_0$ to $\vec{r}_0$ is

$$d_1 + d_2 = 2a = 2\left[z_0^2 + b^2\right]^{1/2} \tag{3.4}$$

Independent of which direction a ray leaves from $-\vec{r}_0$. At the surface S_p the angle of incidence equals the angle of reflection.

A real reflector would consist of some sector S'_p of S_p as indicated in Fig. 3.2. As with the usual reflector IRA there could be flat-plate conical feed arms. Two are shown here, but the usual 4-arm system (with terminating resistors) would be appropriate. The wave leaving $-\vec{r}_0$ to $\vec{r}_0$ needs to be in uniform, isotropic dielectric medium (whether air or whatever), of permittivity ε (ideally lossless), extending around the target. This includes all the ray paths reflecting from S'_p. Note that one may have some special high-voltage source region around $-\vec{r}_0$ with special equipment and perhaps a lens to shape the spherical wave centered on $-\vec{r}_0$.

At later times (after first signal arrival at $\vec{r}_0$) truncation of the dielectric medium will influence the fields near $\vec{r}_0$. So some consideration of the shaping of such a boundary is appropriate. The edge of S'_p (perhaps a circle, or even a more sophisticated shape) is chosen based on feed-arm locations and the orientation of the fields incident on the reflector [6, 7], so as to give the best results at $\vec{r}_0$.

In earlier papers [8-10], it was shown that the spherical TEM wave launched into paraboloidal and hyperboloidal reflectors was exactly transformed into planar and spherical TEM waves respectively. The reflectors did physically (with a minus sign) what the stereographic transformation did mathematically in transforming one type of TEM wave into another. I suspect that the same is the case with the prolate-spheroidal reflector considered here.

Also illustrated in Fig. 3.2 is an aperture plane S_a. One can take the fields incident on the plane from the left and extrapolate them to the region near $\vec{r}_0$. Some of this appears in [3]. All that is needed is the mathematic form of the early-time TEM wave incident on S_a. This can be compared to the numerical results in [4].

Here we have considered a metal (ideally perfectly conducting) reflector. If, however, $\varepsilon > \varepsilon_0$ one can utilize the positive, instead of negative, reflection with S'_p taken merely as the boundary between the dielectric ε and air ε_0.

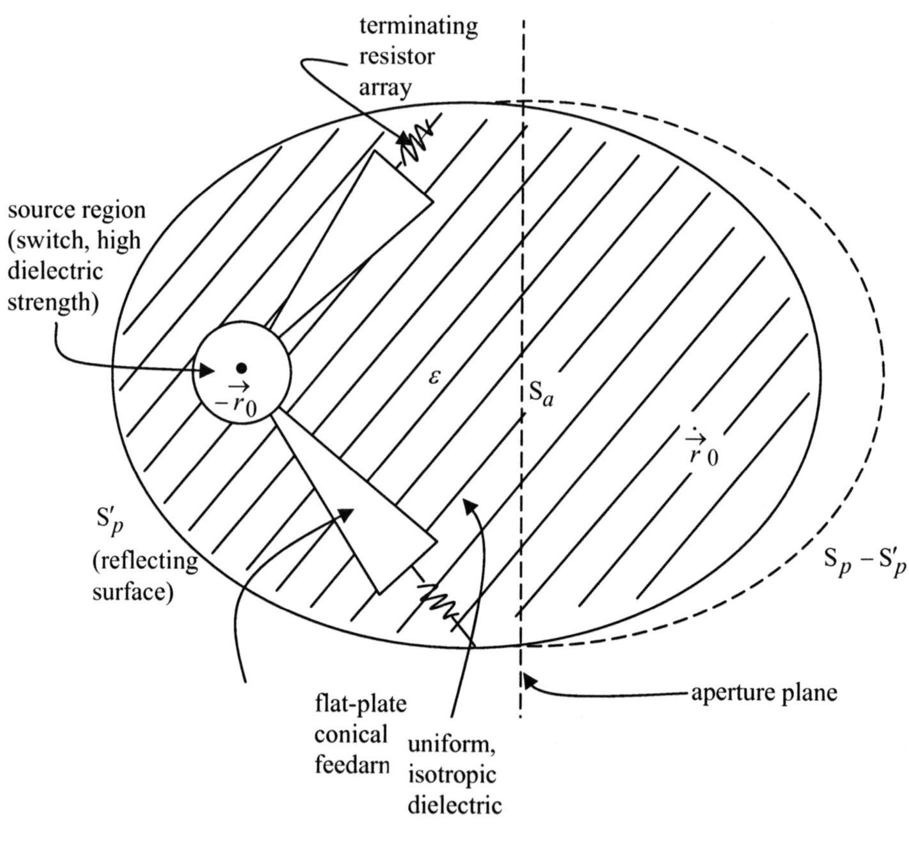

Fig. 3.2. Focused IRA

4. TIMED ARRAY

An alternate approach to focusing a transient wave is with a timed array. One can take a set of N reflector IRAs (of the usual type), point them each at the same target, and trigger them so that the N pulses all arrive at the target at the same time. This has technological problems concerning the timing accuracy with which one can trigger all N IRAs. The timing error needs to be small compared to all times of interest in the waveform from a single IRA at the target.

These two approaches are not incompatible. One can have an array of N "implosion" IRAs (IIRAs, pronounced (aye)(ee)ras) to further increase the field at the target.

Let us then consider an IIRA array as illustrated in Fig. 4.1. In this hexagonal example we have 7 IIRAs, number zero in the center, and numbers 1 through 6 in a hexagonal pattern around this. While the feed arms are indicated at $\phi_0 = 45^\circ$ with respect to a horizontal plane [11], other choices such as the recently popular $\phi_0 = 60^\circ$ are

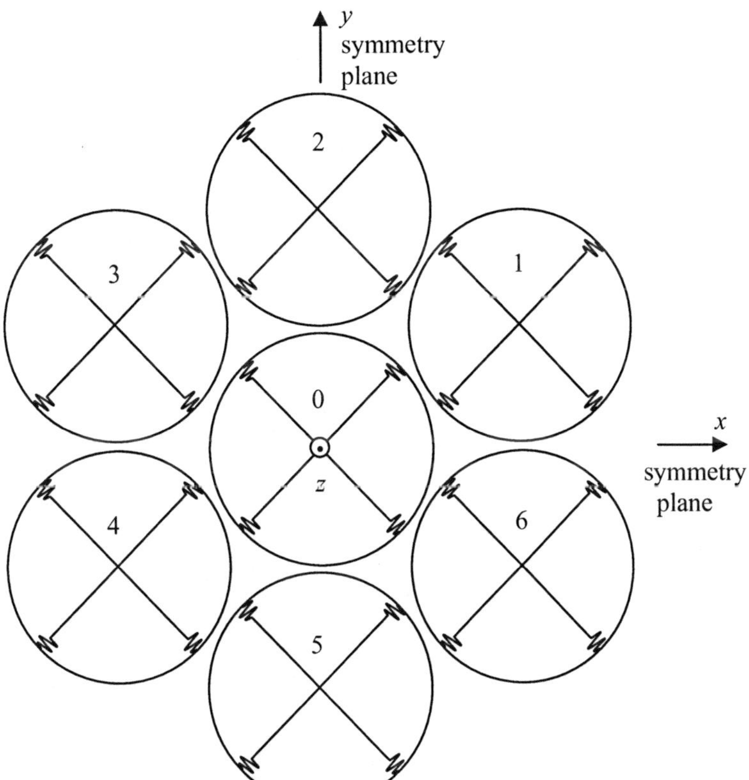

The outer 6 are canted inwards toward the z axis

Fig. 4.1 IIRA Array, Front View

possible. Of course, antennas 1 through 6 are canted inward so that all add at the target focus (say $\vec{r}_0 = z_0 \vec{1}_z$) with each polarization parallel to the *yz* plane Note that the source region, is not depicted here.

Considering the symmetries of such an array, the *xz* and *yz* planes are symmetry planes ($R_x \otimes R_y$ symmetry). The diagram is suggestive of C_{6a} symmetry (6-fold rotation axis with 6 axial symmetry planes) but not exactly. If we choose $\phi_0 = 60^\circ$ and place two more (dummy feed arms on the locally horizontal plane ($\phi_0 = 0^\circ$, 180°) this would complete this type of symmetry. One can consult [12] for discussion of such rotation symmetry in the context of a different kind of high-power source.

The inward canting of the individual antennas is illustrated in Fig. 4.2 which gives a cross section on the *yz* plane. Note that all the IIRAs need not have the same focal distances, as long as they all have the same target focus $\vec{r}_0$. While the feed arms here are shown as "inside" S'_p (the reflector), this need not be the case. There is some optimization to be done in this regard.

5. TIMING ACCURACY

Timing the launches of the N waves, later converging on $\vec{r}_0$, is critical. The key concept is *switch spread* [13]. By this we mean a time window of width t_s, centered (roughly) on t_0, during which some very large fraction (say 90%) of the pulses begin. This is several times the *jitter* (a standard deviation), depending on the actual distribution function of beginning times of the pulses.

For fast-rising, step-like pulses launched into the reflector, the focusing near $\vec{r}_0$ has a time-derivative, giving a narrow pulse with width given by the risetime (t_{mr}) of the original pulse [3]. The amplitude is inversely proportional to t_{mr}. If we have N waves arriving at $\vec{r}_0$, we need all N waves to arrive at the "same" time, i.e., within a spread which is small compared to t_{mr} (in the 100 ps range [14]), if we want the individual field amplitudes to add for maximum field in a short pulse.

There is some information regarding switch spread in the 100s of kV range [15, 16], for a timed-array application (of TEM horns in this case). One might achieve a smaller spread at lower voltages with solid-state technology, but smaller voltages give smaller fields on target. Achieving a sufficiently small spread for the antenna triggering is a significant technological challenge.

6. CONCLUDING REMARKS

The basic concept is fairly simple, the practical implementation, however, involves many details. There are both theoretical and construction questions to be addressed.

Extending our consideration of an array of IIRAs, we can potentially greatly increase the amplitude of a fast pulse incident on a small target in the near field. This however, is accompanied by significant technological problems, specifically switch spread.

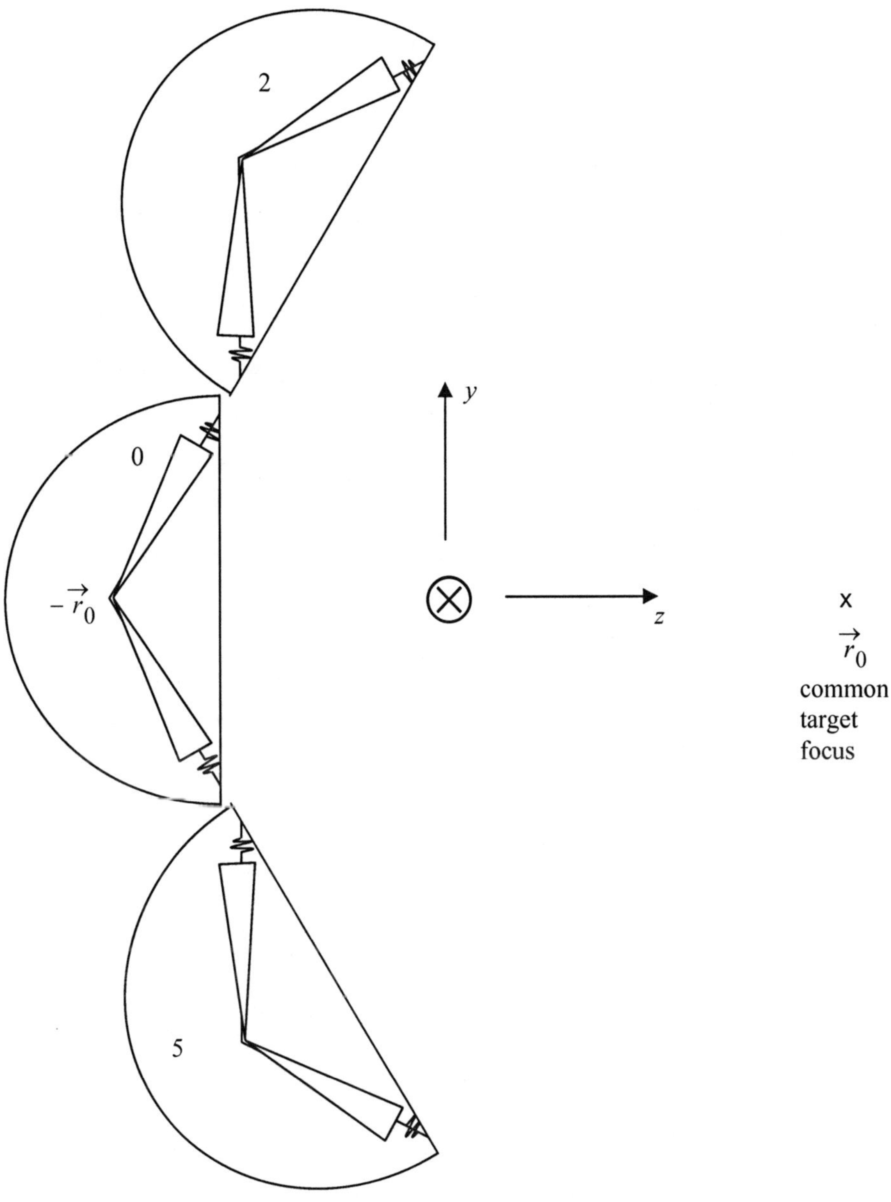

Fig. 4.2 Cross-Section View of Array

REFERENCES

1. C. E. Baum, E. G. Farr, and D. V. Giri, "Review of Impulse-Radiating Antennas", ch. 16, pp. 403-439, in W. R. Stone (ed.), Review of Radio Science 1996-1999, Oxford U. Press, 1999.
2. C. E. Baum, "Radiation of Impulse-Like Transient Fields", Sensor and Simulation Note 321, November 1989.
3. C. E. Baum, "Focused Aperture Antennas", Sensor and Simulation Note 306, May 1987; pp. 40-61, in Proc. 1993 Antenna Applications Symposium, Sept. 1993, U. of Illinois Urbana-Champaign, RL-TR-94-20, February 1994.
4. K. Kim and W. R. Scott, Jr., "Analysis of Impulse-Radiating Antennas with Ellipsoidal Reflectors", Sensor and Simulation Note 481, October 2003.
5. C. E. Baum, "The Distributed Source for Launching Spherical Waves", Sensor and Simulation Note 84, May 1969.
6. C. E. Baum, "Modification of TEM-Fed Reflector for Increased Efficiency", Sensor and Simulation Note 458, July 2001.
7. M. J. Baretela and J. S. Tyo, "Selective Trimming of Impulse Radiating Antenna Apertures to Increase Prompt Radiated Field", Sensor and Simulation Note 461, November 2001.
8. E. G. Farr and C. E. Baum, "Prepulse Associated with the TEM Feed of an Impulse Radiating Antenna", Sensor and Simulation Note 337, Appendix A, March 1992.
9. E. G. Farr and C. E. Baum, "A Canonical Scatterer for Transient Scattering Range Calibration", Sensor and Simulation Note 342, June 1992.
10. C. E. Baum and E. G. Farr, "Hyperboloidal Scatterer for Spherical TEM Waves", Sensor and Simulation Note 343, July 1992.
11. C. E. Bqum, "Selection of Angles Between Planes of TEM Feed Arms of an IRA", Sensor and Simulation Note 425, August 1998.
12. C. E. Baum, "Combining RF Sources Using C_N Symmetry", Circuit and Electromagnetic System Design Note 37, June 1989.
13. C. E. Baum and J. M. Lehr, "Some Considerations for Multichannel Switching", Switching Note 31, January 2002.
14. D. V. Giri et al, "A Reflector Antenna for Radiating Impulse-Like Waveforms", Sensor and Simulation Note 382, July 1995; "Design, Fabrication, and Testing of a Paraboloidal Reflector Antenna and Pulser System for Impulse-Like Waveforms", IEEE Trans. Plasma Science, 1997, pp. 318-326.
15. D. V. Giri et al, "Design, Fabrication, and Testing of a Timed Array of TEM Horns for Beam Steering", Sensor and Simulation Note 469, May 2002.
16. V. Carboni et al, "The Breakdown Fields and Risetimes of Select Gases Under Conditions of Fast Charging (~20 ns and less) and High Pressures (20-100 Atmospheres)", Switching Note 32, May 2002.

COMMUNICATION

OPTIMUM WIRELESS COMMUNICATION THROUGH UNKNOWN OBSCURING ENVIRONMENTS USING THE TIME-REVERSAL PRINCIPLE: THEORY AND EXPERIMENTS

Sermsak Jaruwatanadilok, Akira Ishimaru, and Yasuo Kuga*

1. INTRODUCTION

Wireless communication in unknown and cluttered environments is an important problem which has several practical applications, such as communication in urban areas and disaster areas. This paper presents a method to maximize power transfer efficiency for communication in such environments. The method applies to communication between a transmitting array and a receiving array where the transmitting array sends out a *probe* signal, and at the receiving array, the transfer matrix can be constructed. From this measurement, we perform the time-reversal and eigen analysis. The highest eigenvalue is the best possible transmission efficiency and its corresponding eigenvector represents the transmitting excitation at the transmitting elements to achieve maximum efficiency. The nature of this method makes it possible to operate in unknown, random, and cluttered environments because the maximization is based on the measured signals. Also, this method allows for the adjustment of the system due to the change of the channel, which makes this method adaptive and robust. The time-reversal technique was introduced by Fink[1]. Then, Prada and Fink[2] illustrated the idea of time-reversal imaging to obtain selective focusing. Here, we present the theory of time-reversal communication[3]. We show relevant numerical examples and illustrate experimental verifications in a simple geometry. We further illustrate the effectiveness of this method in laboratory/office environments and in through-the-wall situations. We also investigate the effectiveness of this method in wide-band communication.

* Sermsak Jaruwatanadilok, Box 352500, Department of Electrical Engineering, University of Washington, Seattle, WA 98195, USA.

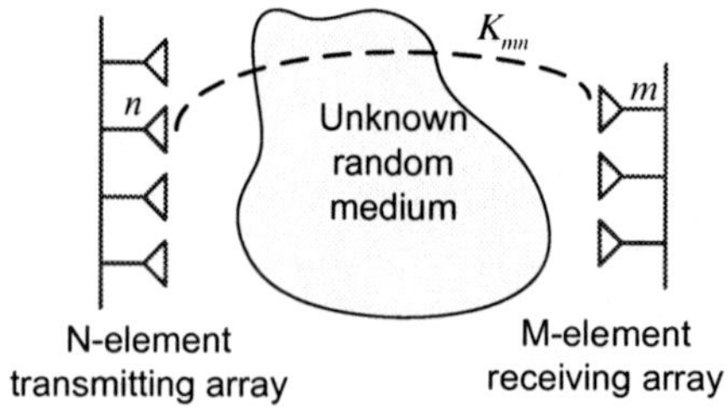

Figure 1. K_{mn} is the transfer function between the mth element of the receiver and the nth element of the transmitter. K_{mn} is measured and therefore known, even though the medium is unknown.

2. THEORY

Let us consider communication between an N-element transmitting array and M-element receiving array in an unknown environment as shown in Fig 1. When a signal is transmitted from the nth element of the transmitter and received by the mth element of the receiver, we can fill in the K_{mn} element of the transfer matrix $\mathbf{K}$. After measuring and constructing the $M \times N$ matrix $\mathbf{K}$, we form an $N \times N$ matrix $\mathbf{T}$ by

$$\mathbf{T} = \tilde{\mathbf{K}}^* \mathbf{K} \tag{1}$$

where $\tilde{\mathbf{K}}^*$ is the conjugate of the transpose of $\mathbf{K}$. Then, we calculate the eigenvectors $\mathbf{V}_i$ and eigenvalues λ_i from

$$\mathbf{T}\mathbf{V}_i = \lambda_i \mathbf{V}_i \tag{2}$$

The largest eigenvalue $\lambda_{\max}$ is equal to the highest transmission efficiency. Its corresponding eigenvector $\mathbf{V}_{\max}$ is to be used as the excitation at the transmitting array.

The physical meaning of $\mathbf{T}$ can be explained as follows. Assume that the transmitting array sends out a signal $\mathbf{V}_t$, then the received signal at the receiving array will be $\mathbf{V}_r = \mathbf{K}\mathbf{V}_t$. If we time-reverse this signal, which is equivalent to the complex conjugate in the frequency domain, we get $\mathbf{V}_r^*$. We then send this signal back to the transmitter. However, the transfer function from the receivers to the transmitters is $\tilde{\mathbf{K}}$. Therefore, we receive $\tilde{\mathbf{K}}\mathbf{V}_r^*$ at the transmitters. If we time-reverse this signal again, we get $\left(\tilde{\mathbf{K}}\mathbf{V}_r^*\right)^* = \tilde{\mathbf{K}}^*\mathbf{V}_r = \tilde{\mathbf{K}}^*\mathbf{K}\mathbf{V}_t = \mathbf{T}\mathbf{V}_t$. Prada and Fink[2] explained this method and applied it to selective focusing for imaging. We applied this to a communication problem. A quantitative measurement of the communication performance is the transmission efficiency η given by

$$\eta = \tilde{\mathbf{V}}_r^*\mathbf{V}_r / \tilde{\mathbf{V}}_t^*\mathbf{V}_t = \tilde{\mathbf{V}}_t^*\tilde{\mathbf{K}}^*\mathbf{K}\mathbf{V}_t / \tilde{\mathbf{V}}_t^*\mathbf{V}_t = \lambda \tag{3}$$

The maximum efficiency can be achieved by the largest eigenvalue $\lambda_{\max}$.

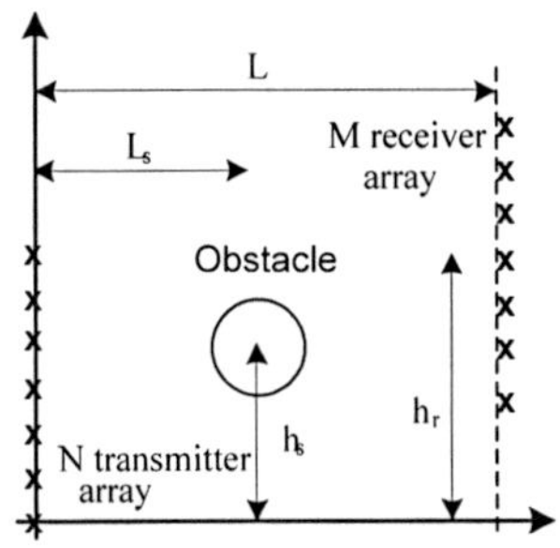

Figure 2. A spherical obstacle with radius a is located between N transmitter array and M receiver array. The spacing between the elements is $\lambda/2$.

3. NUMBERICAL CALCULATIONS AND EXPERIMENTS

Consider communication in the geometry shown in Fig. 2. For this geometry, we can calculate the received signals and estimate the transmission efficiency. The element of the transfer matrix K_{mn} can be calculated by

$$K_{mn} = G_o(m,n) + G_s(m,s,n) \tag{4}$$

where $G_o(m,n) = \exp(-jkl_{mn})/4\pi l_{mn}$, $G_s(m,s,n) = \sum_{n=0}^{\infty} A_n h_n^{(2)}(k,l_{ms}) P_n(\cos\theta)$, $A_n = -(2n+1/4\pi)(-jk) j_n(ka) h_n^{(2)}(kl_{sn})/h_n^{(2)}(ka)$ where $h_v^{(2)}(x)$, $j_n(x)$ are the spherical Bessel functions which relate to the ordinary Bessel functions by $h_v^{(2)}(x) = \sqrt{\pi/2x} H_{v+1/2}^{(2)}(x)$, and $j_n(x) = \sqrt{\pi/2x} J_{n+1/2}(x)$. $G_o(m,n)$ is the Green's function of the free space propagation and $G_s(m,s,n)$ is the result from the scattering from the conducting sphere. Experimental set up for verification is shown in Fig. 3.

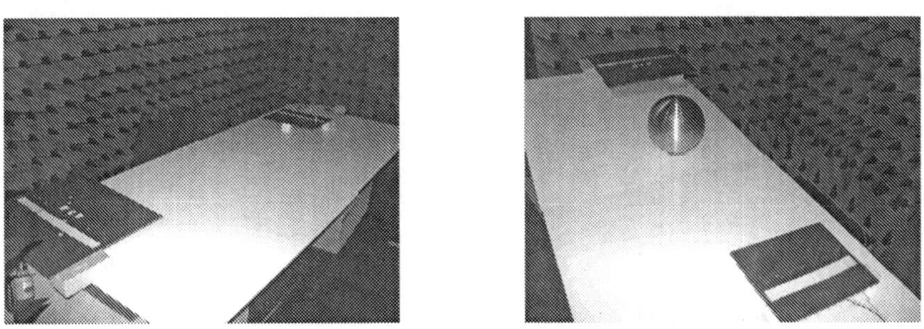

Figure 3. Experimental verifications in the anechoic chamber. Left: free space, Right: around conducting sphere.

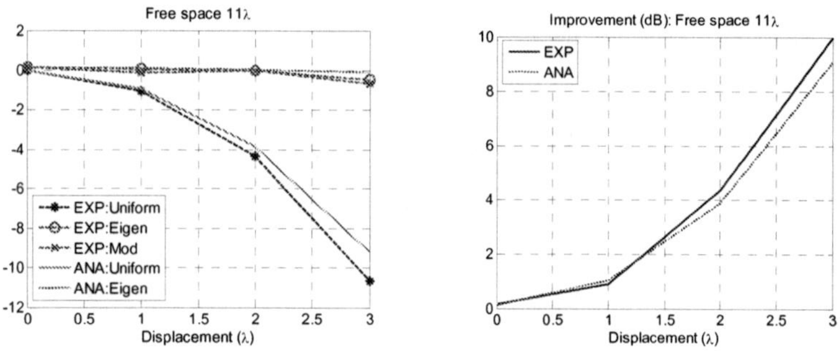

Figure 4. Communication efficiency comparison in the free space case. Left: analytical solution vs experimental data, Right: the improvement of the efficiency comparison.

The communication efficiency is shown as a function of the displacement of the receiving array from the aligned position. The communication efficiency shown is normalized to the free space case in the aligned position. The comparisons show *uniform* as the case where each element of the transmitting antenna radiates the same energy, and *eigen* as the maximum achievable efficiency based on calculations, i.e. λ_{max}. The *modify* represents the communication efficiency when the transmitting antenna is excited according to the time-reversal and eigen analysis ($\mathbf{V}_{max}$).

The results are shown in Fig. 4 for free space and Fig. 5 for the conducting sphere. *ANA* denotes analytical solution and *EXP* indicates experimental data. In both the free space and the conducting sphere cases, excellent agreement between experimental data and the analytical solution is observed. Thus, this shows that the method improves the communication efficiency even in the situation where the line of sight of communication has been obstructed and the improvement is as expected from the analytical solution.

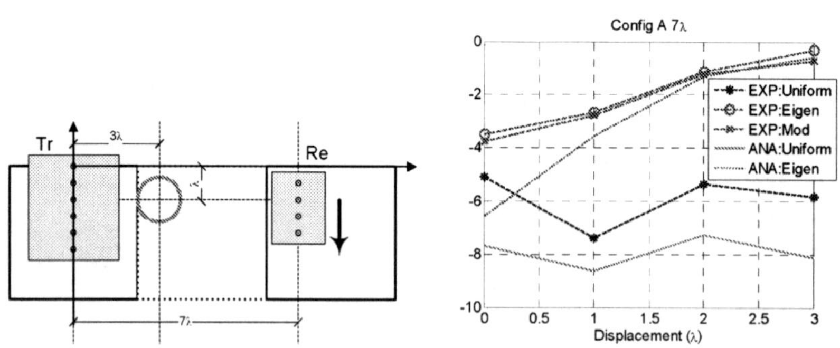

Figure 5. Communication efficiency comparison in conducting sphere cases. Left: schematic of the experimental setup, Right: the efficiency comparison.

Figure 6. Experimental setup in the office/laboratory environment.

We also perform experiments in the laboratory/office environments as shown in Fig. 6. Here, we construct four-element dipole antenna arrays operating at 1 GHz at both the transmitter and the receiver. We test the algorithm in the office/laboratory environments where there are desks, shelves, and experimental apparatus. By moving the transmitter and the receiver in several configurations, we measure the received power when fixing the total transmitting power so that we can quantify the power transmission efficiency. The results in Table 1 show that using the time-reverse eigen value method, we can achieve more efficiency than uniform amplitude scheme.

Table 1. Experimental data for the office/laboratory environment

Configuration	Schematic	Efficiency
Aligned	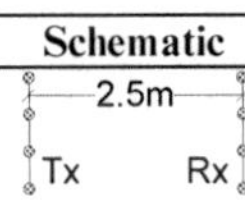	Uniform 9.3155 Eigen: 10.0225 Modify: 9.6140
Off set	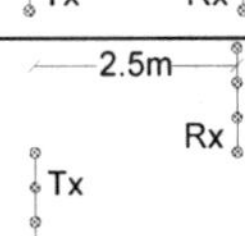	Uniform 1.8277 Eigen: 6.2069 Modify: 6.3309
Off set with angle	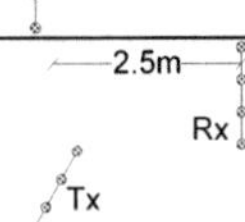	Uniform 0.3483 Eigen: 1.6683 Modify: 1.2130
Through obstacles – offset	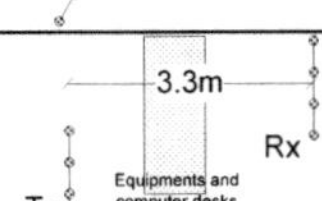	Uniform 0.3217 Eigen: 0.8358 Modify: 0.6413
Through obstacles – offset with angle	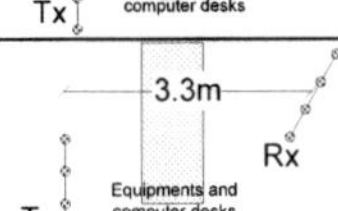	Uniform 0.3831 Eigen: 0.7055 Modify: 0.5114

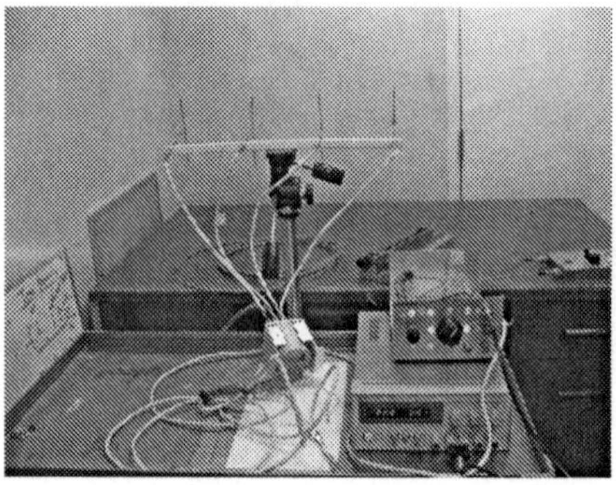
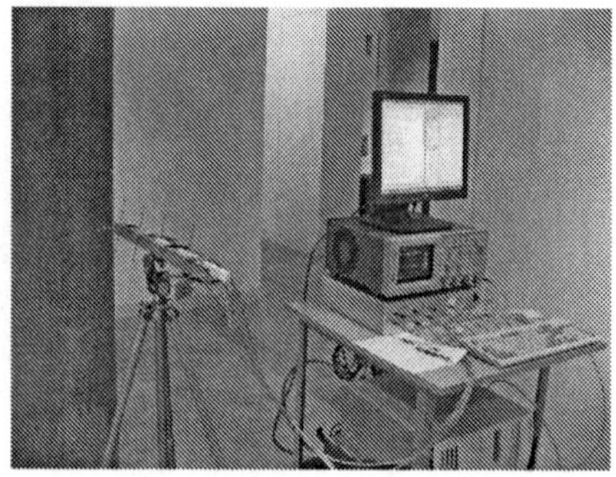

Figure 7. Experimental setup in the through-the-wall situation. Left: transmitting unit, Right: receiving unit.

Similarly, we perform the experiment in the through-the-wall situation where the transmitting array and the receiving array are separated by a dry wall. The experimental set up for this situation is shown in Fig. 7. The results of the experiments are listed in Table 2. The eigen method shows significant improvement in the case where the transmitters and the receivers are off-set with angle. In this particular case, the energy from the transmitters is not focused at the receivers in the uniform case. Therefore, the efficiency decreases dramatically. However, when using the eigen method, we can focus the energy more efficiently which results in substantial improvement shown in efficiency numbers.

Table 2. Experimental data for the through-the-wall situation

Configuration	Schematic	Efficiency
Aligned	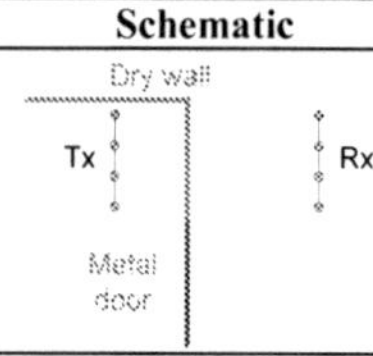	Uniform: 1.2082 Eigen: 14113 Modify: 1.356
Off set	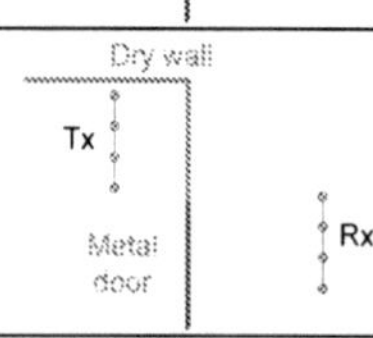	Uniform: 1.3822 Eigen: 1.6459 Modify: 1.4668
Off set with angle	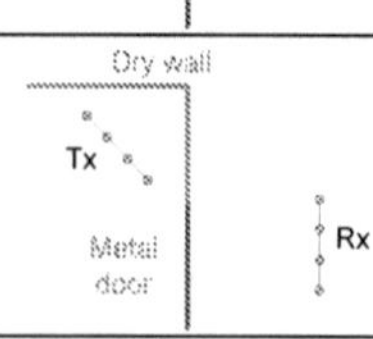	Uniform: 0.1223 Eigen: 0.8095 Modify: 0.8754

We also perform numerical calculations for a wide-band signal with a bandwidth of 20% of the center frequency when the signal is communicated through an obstructing sphere as shown in Fig. 2. The transmission efficiency as a function of frequency is plotted in Fig. 8. Here, *Eigen-cf* denotes the method where the eigen analysis is performed based on the center frequency data and the eigenvector is used in every frequency. *Eigen* is the calculation where the eigen analysis is performed at every frequency. The results show that the transmission efficiency drops off slowly as the frequency is away from the center frequency, indicating that the wide band communication is possible.

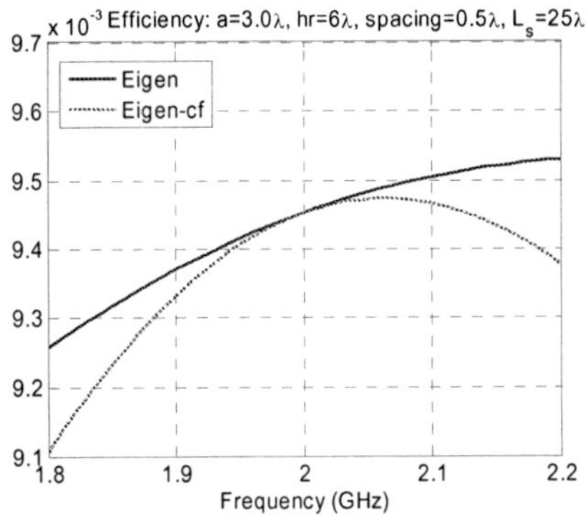

Figure 8. Wide-band calculations.

4. CONCLUSION

In this paper, we present a method to maximize the communication power transfer efficiency through an unknown, random environment. The method is based on time-reversal and eigen analysis. The performance improvement is calculated in a simple case of an obstructing conducting sphere and is verified through experiments. The experiments also include communication in the office/laboratory environments and through-the-wall situations. It is shown that the communication efficiency can be greatly improved using this method. Also, we study the efficiency of this method in the wide-band signals and show that the efficiency slowly drops off away from the center frequency indicating that it is possible to apply this method to a wide-band communication without losing much efficiency. Further studies on signal to noise ratio and dispersion are being conducted for practical applications of this theory.

5. ACKNOWLEDGEMENTS

This work is supported by the Office of Naval Research (Grant # N00014-04-1-0074 and Code 321 Grant # N00014-05-1-0843).

6. REFERENCES

1. M. Fink, "Time reversed acoustics," *Physics Today*, **50**(3), 34-40 (1997).
2. C. Prada and M. Fink, "Eigenmodes of the time reversal operator: A solution to selective focusing in multiple-target media," *Wave Motion*, **20**(2), 151-163, (1994).
3. A. Ishimaru, S. Jaruwatanadilok, and Y. Kuga, "Time-reversal techniques applied to communication through unknown obscuring media," *AMS Meeting*, San Antonio, TX, USA, January 10-15 (2006).

DATA PROCESSING

CORRECTION OF TIME-DOMAIN DATA IN SPECIAL CASES WHERE THE INVERSE TRANSFER FUNCTIONS ARE ANALYTIC TIME DOMAIN OPERATORS

Carl E. Baum*

1. INTRODUCTION

In processing of electromagnetic scattering data associated with an incident broadband pulse, there is the problem of removing the filtering of this data by the radar antenna(s), the characteristics of the incident pulse, and other electronic equipment (such as directional couplers). One approach to this problem is to construct, transfer functions (frequency domain) corresponding to these various factors. Then by use of numerical Fourier transforms one transforms the radar data, divides by the transfer functions and numerically inverse Fourier transforms to obtain the delta-function response of the target.

One of the advantages of looking at the delta-function response is the separation of the target response from that of other scatterers (clutter) by looking only at a time window corresponding to the time of return of the target signal. Such time separation is corrupted by the temporal convolution corresponding to the transfer functions mentioned above. There is also the inevitable problem of noise in the data as well. This can affect the accuracy of the deconvolution required by making the above transfer functions correspond as closely as practical to single delta functions in time (i.e., minimum dispersion).

An alternate approach, applicable in some cases, is to use analytic deconvolution directly in the time domain. This applies to the cases where the inverse of the transfer functions mentioned above can be written analytically in the time domain, and where the form that the deconvolution takes is relatively simple, This paper explores a few such cases.

---

*University of New Mexico, Dept. of Electrical and Computer Engineering, MSC01 1100, Albuquerque, NM, 87131-0001

2. THE TYPICAL PASSIVE INTEGRATOR

As a simple case, let us consider a simple passive integrator (RC integrator) characterized by

$$\begin{array}{l}\tilde{V}_{out}(s) = [1+s\tau]^{-1}\tilde{V}_{in}(s)\,\text{V} \quad , \qquad \tau \equiv \text{ time constant} > 0 \\ \sim \equiv \text{ two-sided Laplace transform over time } t \\ s = \Omega + j\omega = \text{ Laplace-transform variable or complex frequency}\end{array} \tag{2.1}$$

What we wish to obtain is the time integral of $V_{in}(t)$. This is usually expressed as an approximation

$$\tilde{V}_{out}(t) \simeq \frac{1}{\tau}\int_0^t V_{in}(t')dt' \ \text{ for } t << \tau \quad , \quad V_{in}(s) = 0 \ \text{ for } t = 0 \tag{2.2}$$

A more accurate apporach observes that the inverse of the transfer function is a convolution operator

$$1 + s\tau \leftrightarrow \left[\delta(t) + \tau\frac{d}{dt}\delta(t)\right]\circ \quad , \qquad \circ \equiv \text{convolution with respect to time} \tag{2.3}$$

Thus we have

$$\begin{array}{l}\tilde{V}_{in}(s) = [1+s\tau]\tilde{V}_{out}(s) \\ \int_0^t V_{in}(t')dt' = [\tau\delta(t) + u(t)] \circ V_{out}(t) = \tau V_{out}(t) + \int_0^t V_{out}(t')dt'\end{array} \tag{2.4}$$

If we have some data stream for $V_{out}(t)$, then this can be corrected by the addition of a simple integration to give a more accurate version of the time integral of $V_{in}(t)$.

3. TIME-DOMAIN DIRECTIONAL COUPLER

A previous paper [1] considers a special type of time-domain directional coupler. See this paper for the development of a two-conductor (plus reference) ideal transmission line. There we consider only the fully symmetric case for which we have

$$\begin{array}{l}\left(Z_{c_{n,m}}\right) = \begin{pmatrix} Z_{c_{1,1}} & Z_{c_{1,2}} \\ Z_{c_{1,2}} & Z_{c_{1,1}} \end{pmatrix} \equiv \text{ characteristic impedance matrix} \\ \left(X_{n,m}\right) = \frac{1}{R}\left(Z_{c_{n,m}}\right) \quad , \qquad R \equiv \text{ resistance loading all four ports (e.g., 50 } \Omega\text{)} \\ \det\left(\left(Z_{c_{n,m}}\right)\right) = Z_{c_{1,1}}^2 - Z_{c_{1,2}}^2 = R^2 \ , \ \det\left(\left(X_{n,m}\right)\right) = X_{1,1}^2 - X_{1,2}^2 = 1\end{array} \tag{3.1}$$

The port convention has

port 1 = port into which a pulse is launched
port 2 = adjacent port at same end as port 1
port 3 = port out of which the pulse is propagated to a load (e.g., radar antenna, well terminated) (3.2)
port 4 = port adjacent to port 3 which receives the reflected pulse to be recorded (e.g., a radar backscatter) and receives no signal from port 1

In [3] it is observed that for a time T, the round-trip transit time through the coupler, with

$$T = 2\frac{\ell}{\mathrm{v}} \quad , \quad \ell \equiv \text{coupler length} \tag{3.3}$$

v = propagation speed ($\leq c$) in the lossless, dispersionless dielectric medium

the pulse received in port 4 is a faithful replica of the pulse presented back to port 3. The same is true for the pulse from port 1 to port 3. In a radar application, this leaves the antenna transfer function as an additional problem for possible deconvolution.

The scattering matrix elements of interest are

$$\begin{aligned} \tilde{S}_{3,1}(s) &= \left[X_{1,1} \sinh\left(\frac{sT}{2}\right) + \cosh\left(\frac{sT}{2}\right) \right]^{-1} \\ \tilde{S}_{4,3}(s) &= \tilde{S}_{3,1}(s) X_{1,2} \sinh\left(\frac{sT}{2}\right) \quad , \quad \tilde{S}_{4,1}(s) = 0 = \tilde{S}_{1,1}(s) \end{aligned} \tag{3.4}$$

and the product of the first two of these. For convenience we define

$$\xi \equiv \frac{1 - X_{1,1}}{1 + X_{1,1}} < 0 \tag{3.5}$$

Rearranging we have

$$\tilde{S}_{3,1}(s) = 2\left[\left[1 + X_{1,1}\right] e^{\frac{sT}{2}} + \left[1 + X_{1,1}\right] e^{-\frac{sT}{2}} \right]^{-1} = \frac{2e^{-\frac{sT}{2}}}{1 + X_{1,1}} \left[1 + \xi e^{-sT}\right]^{-1} \tag{3.6}$$

Expand the last part as a geometric series

$$\tilde{S}_5(s) \equiv \left[1 + \xi e^{-sT}\right]^{-1} = \sum_{n=0}^{\infty} [-\xi]^n e^{-nsT} \tag{3.7}$$

In time domain this is

$$\tilde{S}_5(s) = \sum_{n=0}^{\infty} [-\xi]^n \delta(t-nT) \tag{3.8}$$

So except for the delay of $T/2$ in (3.6) this is a successive set of delta functions, each $-\xi$ (positive) in decreasing amplitude times the previous and delayed by successive times T.

Note that if we delay S_5 by time T and multiply by ξ (negative) we have

$$\xi S_5(t-T) = -\sum_{n=0}^{\infty} [-\xi]^{n+1} \delta(t-[n+1]T) = -\sum_{n=0}^{\infty} [-\xi]^n \delta(t-nT) \tag{3.9}$$

We then find

$$S_5(t) + \xi S_5(t-T) = \delta(t) \ , \quad \tilde{S}_5(s)\left[1+\xi e^{-sT}\right] = 1 \tag{3.10}$$

So taking a signal coming from the coupler and multiplying by ξ, delaying by T and adding, removes the effect of $S_{3,1}$ except for a constant multiplier and delay by $T/2$, as

$$S_{3,1}(t) + \xi\, S_{3,1}(t-T) = \frac{2}{1+X_{1,1}} \delta\left(t-\frac{T}{2}\right) \tag{3.11}$$

Continuing we have

$$\tilde{S}_{4,3}(s) = \frac{X_{1,2}}{1+X_{1,1}} \tilde{S}_6(s) \quad , \quad \tilde{S}_6(s) = \left[1+\xi e^{-sT}\right]^{-1} \left[1-e^{-sT}\right] \tag{3.12}$$

Apply the previous procedure (as in (3.10))

$$\tilde{S}_7(s) = \tilde{S}_6(s) + \xi e^{-sT} \tilde{S}_6(s) = 1-e^{-sT} \tag{3.13}$$

Thus the previous shift by T, multiplication by ξ, and adding removes the denominator again. In time domain this leaves

$$S_7(t) = \delta(t) - \delta(t-T) \quad , \quad S_{4,3}(t) + \xi S_{4,3}(t-T) = \frac{X_{1,2}}{1+X_{1,1}} S_7(t) \tag{3.14}$$

To deconvolve the remaining $S_7(t)$ one can observe

$$S_8(t) = \delta(t) - \delta(t-Nt) = \sum_{n=0}^{N-1} S_7(t-nT) \tag{3.15}$$

By appropriate choice of N the second delta function can be moved into an appropriate time where there are ideally no clutter signals which could shift into the time window of interest for the target.

Noting that scalar convolution integrals commute, we can remove the effects of $S_{3,1}$ and $S_{4,3}$ by a succession of operations, in any order. There are two operations of multiply by ξ, shift by T, and add. Then there is the succession of shift and adds $N-1$ times in (3.15). Ideally this leaves us with an operator.

$$\frac{2X_{1,2}}{\left[1+X_{1,1}\right]^2}\left[\delta\left(t-\frac{T}{2}\right)-\delta\left(t-\left[N+\frac{1}{2}\right]t\right)\right] \tag{3.16}$$

where the second delta function can be neglected by considerations of the received time-domain waveform. This leaves a signal multiplied by a constant factor and shifted by $T/2$.

Another interpretation of (3.16) concerns what might be called an "equivalent directional coupler", Comparing S_8 to S_7 the double transit time of T has been extended to NT, effectively making the coupler N times as long.

4. IDEAL REFLECTOR IMPULSE RADIATING ANTENNA

A reflector impulse radiating antenna (IRA) has a transmitted waveform when driven by a step function pulse $V_0 u(t)$ as [2]

$$\begin{aligned} V_f(t) &= r\,E_f(t) = V_0 \frac{h_a}{2\pi c f_g} R(t) \qquad \text{(far "voltage")} \\ R(t) &= \frac{1}{T}\left[-u(t)+u(t-T)\right]+\delta_a(t-T) \\ T &= \frac{2F}{c}\ , \qquad F = \text{focal length}\ , \quad c = \left[\mu_0\varepsilon_0\right]^{-1/2} \end{aligned} \tag{4.1}$$

Here only $R(t)$ concerns us. The various factors are explained in the references in [3]. The δ_a is an approximate delta function (a delta function in the limit of $r \to \infty$). $R(t)$ also characterizes such an antenna in reception (an approximate replicator). The step functions form what is called the prepulse. There is also a postpulse [4] (more complicated) which is not considered here.

The problem is to remove the step functions and leave a delta function remaining. Transforming we have

$$\tilde{R}(s) = \frac{1}{sT}\left[-1+e^{-sT}\right] + e^{-sT}$$

$$e^{-sT}\tilde{R}^{-1}(s) = \left[\frac{1}{sT}\left[1-e^{sT}\right]+1\right]^{-1} = \frac{1}{1+\frac{1}{sT}}\left[1-\frac{1}{sT}\left[1+\frac{1}{sT}\right]^{-1}e^{sT}\right]^{-1} \tag{4.2}$$

$$= \frac{sT}{sT+1}\sum_{n=0}^{\infty}\left[sT+1\right]^{-n}e^{nsT}$$

This form has some difficulties. Besides the delay by T, in time domain the leading term is a derivative. Essentially this is keying on the initial step function instead of the later delta function.

An alternate form has

$$e^{sT}\tilde{R}^{-1}(s) = \sum_{n=0}^{\infty}\left[sT\right]^{-n}\left[1-e^{sT}\right]^{n} = 1+\frac{1}{sT}\left[1-sT\right]+\frac{1}{\left[sT\right]^{2}}\left[1-sT\right]^{2}+\cdots \tag{4.3}$$

In this form we have the leading delta function in time domain, but the successive correction terms are rather complicated.

Let us now approach this differently. We can characterize *R(t)* by a recursion beginning with

$$\tilde{V}_{out}(s) = e^{-sT}\tilde{V}_{in}(s) + \frac{1}{sT}\left[-1+e^{-sT}\right]\tilde{V}_{in}(s)$$

$V_{in}(t) \equiv$ signal into "filter" , $V_{out}(t) \equiv$ signal out of "filter"

This is changed to

$$\tilde{V}_{in}(s) \equiv e^{sT}\tilde{V}_{out}(s) + \frac{1}{sT}\left[e^{-sT}-1\right]\tilde{V}_{in}(s)$$

$$\tilde{V}_{in}(t) = \tilde{V}_{out}(t+T) + \frac{1}{T}\left[\int_{-\infty}^{t}\tilde{V}_{in}(t'+T)dt' - \int_{-\infty}^{t}V_{in}(t')dt'\right] \tag{4.5}$$

$$= V_{out}(t+T) + I_1\left(V_{in}(t)\right)$$

$$I_1(V_{in}(t) = \frac{1}{T}\int_{t}^{t+T}V_{in}(t')dt'$$

Define an integral operator

$$I_1(\cdots) = \frac{1}{T}\int_t^{t+T}(\cdots)dt_i$$

$$I_1^n(\cdots) \equiv n = \text{fold multiple integral} = \frac{1}{T^n}\int_t^{t+T}\int_{t_1}^{t_1+T}\cdots\int_{t_{n-1}}^{t_{n-1}+T}(\cdots)dt_n\cdots dt_1 \tag{4.6}$$

Applying this recursively we have

$$\begin{aligned} V_{in}(t) &= V_{out}(t+T) = I_1(V_{in}(t)) \\ &= V_{out}(t+T) = I_1(V_{out}(t+T)) + I_1(V_{in}(t)) \\ &= V_{out}(t+T) + \sum_{n=1}^{\infty} I_1^n(V_{out}(t+T)) \end{aligned} \tag{4.7}$$

So the inverse operator to *R(t)* can be written in the form

$$\begin{aligned} V_{in}(t) &= A(V_{out}(t)) \\ A(\cdots) &= \delta(t+T) \circ + \sum_{n=1}^{\infty} I_2^n(\cdots) \quad , \quad I_2(\cdots) = \frac{1}{T}\int_{t+T}^{t+2T}(\cdots)dt' \end{aligned} \tag{4.8}$$

Alternately we can deal in terms of $V_{out}(t+T)$ without the additional time shift.

This *formal* solution may have some difficulty in implementation due to the infinite series of repeated integrals. This may depend on whether or not a few terms are adequate for implementation in appropriate circumstances. This may also depend on the time duration for which it is to be applied.

5. IDEAL LENS IMPULSE RADIATING ANTENNA

A lens IRA is basically a TEM horn with a lens at the aperture to make the aperture fields a plane wave. The ideal transfer function given in [2, 5] is

$$\begin{aligned} V_f(t) &= r\,E_f(t) = -V_0\frac{h}{4\pi cfg}L(t) \\ L(t) &= \delta(t) + \frac{1}{T}[-u(t) + u(t+T)] \quad , \quad T \equiv \text{round trip transit time} \end{aligned} \tag{5.1}$$

This is basically a time-reversal of *R(t)*.

Following the procedure in the previous section we have

$$V_{out}(t) = V_{in}(t) + \frac{1}{T}\left[-\int_{-\infty}^{t} V_{in}(t')dt' + \int_{-\infty}^{t} V_{in}(t'-T)dt' \right]$$

$$= V_{in}(t) + \frac{1}{T}\int_{t-T}^{t} V_{in}(t')dt'$$

$$V_{in}(t) = V_{out}(t) + I_3(V_{in}(t)) \quad , \quad I_3(\cdots) = \frac{1}{T}\int_{t-T}^{t} (\cdots)dt' \tag{5.2}$$

$$V_{in}(t) = \delta(t) \circ V_{out}(t) + \sum_{n=1}^{\infty} I_3^n(V_{out}(t)) \quad , \quad \tilde{L}^{-1}(s) \leftrightarrow \delta(t) \circ + \sum_{n=1}^{\infty} I_3^n(\cdots)$$

As one can see, this is quite similar to the previous result.

6. CONCLUDING REMARKS

Here we have analytic formulas for deconvolution in the time domain. These have varying degrees of simplicity. The shift-and-add form for the directional coupler is simpler than the recursive integral formulas for the IRAs.

Such analytic procedures will not solve all deconvolution problems. To the degree that they remove some of the dispersion in the data, the corrected data can perhaps be more accurately processed by numerical Fourier transforms since the remaining transfer function to be unfolded then corresponds more closely to a delta function.

REFERENCES

1. C. E. Baum, "Coupled Transmission Lines as a Time-Domain Directional Coupler", Sensor and Simulation Note 451, October 2000.
2. C. E. Baum and E. G. Farr, "Impulse Radiating Antennas", pp. 139-147, in H. L. Bertoni et al (eds.), *Ultra-Wideband, Short-Pulse Electromagnetics*, Plenum Press, 1993.
3. C. E. Baum, E. G. Farr, and D. V. Giri, "Review of Impulse-Radiating Antennas", ch. 16, pp. 403-439, in W. R. Stone (ed.), *Review of Radio Science 1996-1999*, Oxford U. Press, 1999.
4. D. V. Giri and C. E. Baum, "Reflector IRA Design and Boresight Temporal Waveforms", Sensor and Simulation Note 365, February 1994.
5. E. G. Farr and C. E. Baum, "A Simple Model of Small-Angle TEM Horns", Sensor and Simulation Note 340, May 1972.

SECOND TIME INTEGRAL OF THE IMPULSE RESPONSE FOR ENHANCING THE LATE-TIME TARGET RESPONSE FOR TARGET IDENTIFICATION

Carl E. Baum*

1. INTRODUCTION

A general problem, concerning target identification via the late-time aspect-independent complex natural resonant frequencies, is the presence of the large early-time response. It would be advantageous to enhance the late-time response relative to the early-time response. In a general sense this implies some kind of frequency-dependent filter which attenuates the high-frequency parts of the waveform while providing little or no attenuation to the complex resonant frequencies of interest. Such a low-pass filter (passive and/or active) is the subject of this paper.

A desirable feature of such a filter is the temporal separation of early- and late-time responses. Furthermore we would like to be able to still use time gating (windowing) to remove some of the clutter in the signal returned to the radar. So we do not wish to introduce significant dispersion into the scattering data.

There are other problems associated with the signal including the response(s) of the radar antenna(s) and propagation characteristics (e.g., multipath). For present purposes we neglect such problems and assume that appropriate deconvolution, gating, etc., have been applied to give the target delta-function response.

2. MULTIPLE INTEGRATION OF THE TARGET DELTA FUNCTION RESPONSE

Consider first a related problem, low-frequency radiation from an antenna and its implication for pulse radiation. In [1] we have observed that, if a step-like voltage

---

*University of New Mexico, Dept. of Electrical and Computer Engineering, MSC01 1100, Albuquerque, NM, 87131-0001

pulse is applied to an electric dipole, the low frequency content of the radiation is proportional to s (= $\Omega + j\omega$, the Laplace transform variable or complex frequency). This in turn implies that the radiated pulse must have at least one zero crossing. If the late-time voltage pulse is allowed to decay to zero, then the radiated pulse must have at least two zero crossings.

Consider now the far-field scattering problem. Let the incident wave be like a delta-function, i.e., one sided, decaying back to zero with finite, nonzero area (time integral). The low-frequency content of the incident field is then just the nonzero area. The scatterer is characterized at low frequencies by the induced electric and magnetic dipoles ($\vec{p}$ and $\vec{m}$) which are also constant vectors (in the low-frequency limit). The far field scattered by such dipoles is proportional to $s^2 \vec{p}$ and $s^2 \vec{m}$. If we multiply the Laplace transform of the far scattered field by s^{-2}, the resulting low-frequency far scattered field is proportional to $\vec{p}$ and $\vec{m}$. In time domain this is a double time integral which goes to zero at late time since the area (the low-frequency limit of the Fourier transform) must be finite. (Note that the waveform is damped due to the radiation of electromagnetic energy.)

In the singularity-expansion-method (SEM) representation of the scattered field [2] we have the delta-function response

$$\vec{E}_f(t) = \frac{1}{r}\sum_{\alpha} \vec{R}_\alpha \, e^{s_\alpha [t - r/c]} u(t - r/c)$$
$$+ \text{ entire function (temporal form, early time)} \tag{2.1}$$

The second time integral then takes the general form

$$\int_{-\infty}^{t}\int_{-\infty}^{t'} \vec{E}_f(t'')dt''dt' = \frac{1}{r}\sum_{\alpha} \frac{\vec{R}_\alpha}{s_\alpha^2} e^{s_\alpha [t - r/c]} u(t - r/c)$$
$$+ \text{ new entire function} \tag{2.2}$$

Again the damped sinusoids going to zero implies that the entire-function contribution must also be zero at late times. The waveform is also zero before some time, applying as well to the multiple integrals. Here we have taken $t = 0$ as the beginning ot the late-time response (or $t - r/c$ in the far field).

From a frequency-domain point of view we have enhanced the low frequencies by changing the residues $\vec{R}_\alpha$ to $\vec{R}_\alpha s_\alpha^{-2}$. If the delta-function response at early time looks like an impulse, the first integral is a step function and the second integral is a ramp function, thereby suppressing it considerably. (At this point we can note that the early-time ramp response has also been considered by other authors [3].)

Noting that clutter in the radar return is just the scattering from other "targets", the second time integral has the same properties as above. This clutter is then not significantly stretched in time, allowing windowing about the target response to be used as before.

3. ANALOG SECOND TIME INTEGRAL

In analyzing the scattering data one may wish to perform the temporal integration by analog means before digitizing the data. This is because digitizing a small signal (late-time portion) in the presence of a large signal (early-time response) can result in digitization errors due to the large dynamic range to be covered. By the use of passive or active analog integration the early-time signal and high-frequency noise can be suppressed before digitization, allowing one more accurately to represent the late-time data.

As examples let us consider some simple passive-integrator circuits. Fig. 3.1A shows an RC type, assuming the input signal is on a 50 Ω transmission line. The first integrator has a 50 Ω input impedance Z_{in}. The approximate first integral is the voltage across C_1. The integrating resistance is large compared to 50 Ω (say 5kΩ). For the second stage we need a large value resistor to sample the voltage on C_1 and integrate again as the voltage on C_2. In turn this voltage is sampled by a very large input impedance to the recorder.

A more ideal form of integrator is the RLC type, based on what is called an all-pass network. Consider the first stage in Fig. 3.1B and constrain

$$\tau_1 \equiv \frac{L_1}{R} = R\,C_1 \quad , \quad R = 50\,\Omega \tag{3.1}$$

Then if the input impedance to the second stage is 50 Ω, the transfer function of this integrator is

$$\tilde{T}_1(s) = \frac{\tilde{V}_{out}^{(1)}(s)}{\tilde{V}_{in}^{(1)}(s)} = [1 + s\tau_1]^{-1} \tag{3.2}$$

with iinput impedance

$$Z_{in} = 50\,\Omega \tag{3.3}$$

for all frequencies. Similarly for the second stage we require

$$\tau_2 \equiv \frac{L_2}{R} = R\,C_2 \tag{3.4}$$

giving an input impedance of 50 Ω provided it is terminated in 50 Ω (which might be a 50 Ω cable). The transfer function of the second integrator is

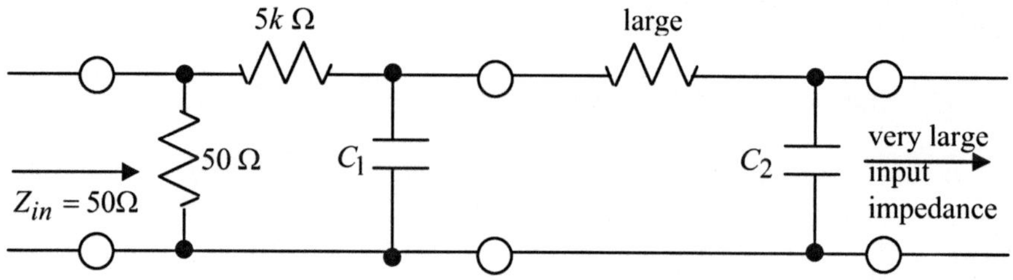

A. RC

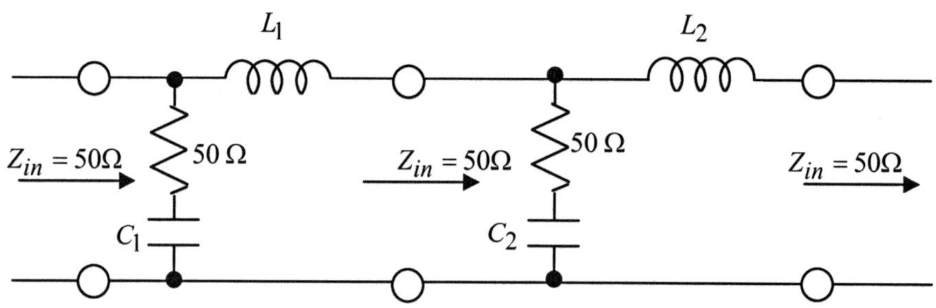

B. LRC

Fig. 3.1 Cascaded Passive Integrators

$$\tilde{T}_2(s) = \frac{\tilde{V}_{out}^{(2)}(s)}{\tilde{V}_{in}^{(2)}(s)} = [1 + s\tau_2]^{-1} \tag{3.5}$$

where

$$\tilde{V}_{in}^{(2)}(s) = \tilde{V}_{out}^{(1)}(s) \tag{3.6}$$

The transfer function of the double integrator is then

$$\begin{aligned}\tilde{T}(s) &= \tilde{T}_1(s)\tilde{T}_2(s) = [1 + s\tau_1]^{-1}[1 + s\tau_2]^{-1} \\ &= \frac{\tilde{V}_{out}^{(2)}(s)}{\tilde{V}_{in}^{(1)}(s)}\end{aligned} \tag{3.7}$$

Rearranging we have

$$s^{-2}\tilde{V}_{in}(s) = \left[\tau_1 + s^{-1}\right]\left[\tau_2 + s^{-1}\right]\tilde{V}_{out}(s) \tag{3.8}$$

which in time domain is

$$\begin{aligned}
&\int_{-\infty}^{t}\int_{-\infty}^{t'} V_{in}(t'')\,dt''dt' \\
&= \left[\tau_1\delta(t) + u(t)\right] \circ \left[\tau_2\delta(t) + u(t)\right] \circ V_{out}(t) \\
&= \left[\tau_1\,\tau_2\delta(t) + \left[\tau_1+\tau_2\right]u(t) + tu(t)\right] \circ V_{out}(t) \\
&= \tau_1\,\tau_2\,V_{out}(t) + \left[\tau_1+\tau_2\right]\int_{-\infty}^{t} V_{out}(t')\,dt' + \int_{-\infty}^{t}\int_{-\infty}^{t'} V_{out}(t'')\,dt''dt'
\end{aligned} \tag{3.9}$$

$\circ \equiv$ convolution with respect to time.

This is another example of "analytic deconvolution" to add to those in [4].

This gives a simple correction technique to obtain the second time integral of $\tilde{V}_{in}^{(1)}(t)$. This removes the requirement that times of interest be short compared to τ_1 and τ_2. So we can choose τ_1 and τ_2 such that the complex resonances of interest are not significantly attenuated before digitizing the signal, while significantly attenuating the higher frequencies.

4. CONCLUDING REMARKS

There are various possible filters one may apply to the scattering data to emphasize the important information for target identification. Here we have explored the second time integral of the delta-function response because of its special properties in emphasizing the late-time response without introducing dispersion of the early-time data into the late-time regime. Having applied such a filter, the various natural-frequency identification techniques such as E-pulse, matrix pencil, etc., are still applicable.

In implementing the second time integral there is some advantage in using analog filters before digitizing the data. This can reduce the impact of digitization errors in the late-time data, effectively increasing the dynamic range.

REFERENCES

1. C. E. Baum, "Some Limiting Low-Frequency Characteristics of a Pulse-Radiating Antenna", Sensor and Simulation Note 65, October 1968.
2. C. E. Baum, "Representation of Surface Current Density and Far Scattering in EEM and SEM With Entire Functions", Interaction Note 486, February 1992; ch. 13, pp. 273-316, in P. P. Delsnato and A. W. Saenz (Eds.), *New Perspectives on Problems in Classical and Quantum Physics, Part II, Acoustic Propagation and Scattering, Electromagnetic Scattering*, Gordon and Breach, 1998.
3. E. M. Kennaugh and D. L. Moffatt, "Transient and Impulse Response Approximations", Proc. IEEE, 1965, pp. 893-901.
4. C. E. Baum, "Correction of Time-Domain Data in Special Cases Where the Inverse Transfer Functions are Analytic Time-Domain Operators", Mathematics Note 95, November 2003.

BLIND SOURCE SEPARATION FOR EXTRACTION OF TARGET SCATTERING CENTERS

Ismail Jouny*

1. ABSTRACT

This chapter focuses on using blind source separation of complex signals to extract scattering centers of radar targets that can then be used for target identification. The concept introduced in this paper assumes that the signal returned (or scattered) from a target is linear or convolutive mixture of several scattering centers along the body of an unknown aircraft. This information can then be used in a pattern recognition scheme for the purpose of identifying the non-cooperative unknown target. The results presented prove the concept using synthetic radar data. Work is ongoing to apply this concept to real radar data as received through an array of sensors (antennas).

2. INTRODUCTION

The problem of extracting the target scattering centers as seen by ultra wideband radar (UWB) has received considerable attention over the last few decades [2-4]. Renewed interest in UWB has motivated further investigation into identifying key scattering centers of a radar target. The wider the bandwidth, the clearer the distinction between scattering centers. Earlier work has focused on using autoregressive (AR), or moving average (MA) models to represent target scattering. These models (with parameters estimated using variable techniques) were successful in identifying two to three major scattering centers, particularly near the engine inlet of an aircraft. Model mismatch, extraneous scatterers, and noise have contributed to the limited performance of these algorithms.

The subject of extracting scattering centers and identifying the scattering mechanisms along the profile of an unknown target has major applications in electromagnetics research. These applications are motivated by the need to validate the geometric theory of diffraction and to use the extracted centers in a radar target identification scheme.

---

*Lafayette College, Easton, PA, 18042

This paper introduces a novel solution to this problem by using antenna arrays and employing specific blind source separation techniques that are best suited for this problem. The main assumption is that the returned radar signal that interrogated the target is a mixture of scattered signals the emanated from distinct scattering centers along the aircraft. Figure 1 shows the concept being examined in this paper.

Figure 1: Signal scattered from different scattering centers along the aircraft are received by different ultra-wideband radar sensors. Blind Source Separation is used to identify scattering centers.

3. PROBLEM FORMULATION

The geometric theory of diffraction has been used to derive scattering in the optical scattering region of a target. This is the frequency band often associated with ultra wideband radar scattering. The theory allows targets to be represented by a number of major scatterers along the aircraft. Scattering from each of these centers depends the shape of the illuminated object. A wedge would have a different signature from that of a point scatterer, and that of a flat plate type object. These distinctions become more pronounced as the bandwidth of the illuminating radar gets wider. In narrower band radars, the target scattering is assumed to be in the resonance region and the whole target may have one or two scattering centers depending on its size.

Recent interest in UWB has sparked research in identifying scattering mechanism, and has led to the application of numerous signal processing techniques including wavelet decomposition, bispectral analysis, and model based techniques. In this paper, a different approach that relies on antenna arrays and blind source separation is being proposed, and the concept is tested using synthetically generated radar data.

The geometric theory of diffraction permits modeling a radar target as a lumped set of scatterers. Electromagnetic plane waves are scattered from objects in a transient phenomenon that depends on the geometry of the object. Point scatterers (theoretical scatterers) have an impulsive response represented by an impulse with delay proportional to travel to and back from the target. Any scatterers that have more defined geometric shape such as a sphere, flat plate, dihedral, corner, curved edge, etc will have a response that is more dispersive in frequency and is often modeled as fractional integrals or derivatives of an impulse. The impulse response of a target composed of N scatterers is thus modeled as

$$h(t) = \sum_{k=1}^{N} a_k \delta^{(n_k)}(t - t_k)$$

where a_k is the magnitude of the response of the kth scatterer and $\delta^{(n)}$ denotes the nth fractional derivative or integral of the impulse function. Previous studies have used the following models for specific scatterers [2]:
n=1 for a flat plate, dihedral
n=0.5 for a curved surface
n=0 for a point scatterer, straight edge, or a doubly curved surface
n=-0.5 for a curved edge diffraction, and
n=-1 for a corner diffraction.
The delay to and from the kth scatterer normalized with respect to a reference point on the target is denoted by t_k. The frequency response of the target is thus modeled as

$$H(\omega) = \sum_{k=1}^{N} a_k (j\omega)^{n_k} \exp(-j\omega t_k)$$

In a real scenario, the number of scatterers is unknown, and so is the location of each scatterer and the level of differentiation or integration.

The above model is similar to a mixture of sources with different temporal behavior, statistical properties, and spectral features. Therefore the above mixture is suited for a number of source separation techniques. The lack of knowledge of the coefficients and the angle of each scatterer with respect to the radar dictate the use of blind source separation methods. The nature of the signal is compatible with techniques that exploit temporal differences between the signatures of various scatterers. The signal is also compatible with BSS techniques that exploit statistical difference (skewness and Kurtosis), and finally the difference in the spectral response permits the use of BSS techniques that exploit spectral differences between scatterers. Furthermore, the time-domain model allows BSS techniques that separate real signals such as SOBI [4], and the frequency based model permits the use of BSS techniques for complex signals such as EASI or JADE [5,6]. The channel is noisy but the delay is insignificant and therefore an instantaneous mixture assumption is appropriate. However, convolutive mixture techniques can also be used assuming that an appropriate channel model can be devised.

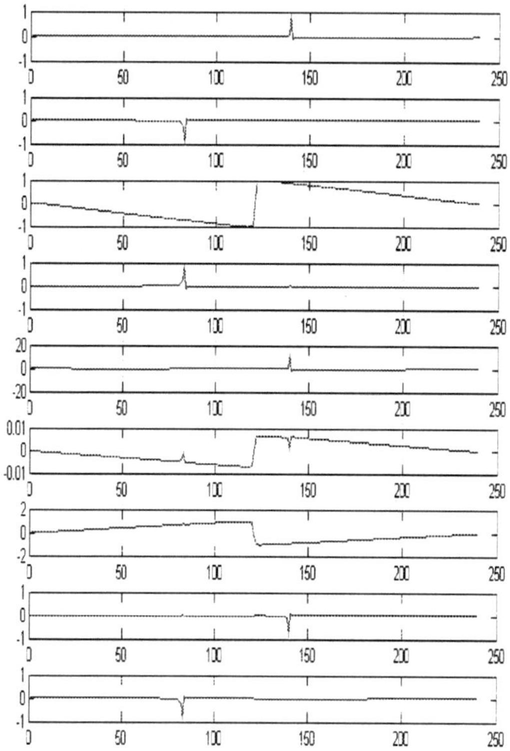

Figure 2: Original three different scatterers (top three graphs). As recovered using Jade (next three) and as recovered using Sobi (bottom three). X-axis represents return time in nanoseconds relative to the target reference point. Y-axis represents scatterer strength.

4. RESULTS AND DISCUSSION

Figure 2 shows the extracted radar scattering center for a synthetic target composed of three scatterers (top three graphs) with distinctly different geometrical shapes. The first BSS technique used is Jade applied to the frequency-domain scattering model (separated scatterers are the middle three graphs) and SOBI is used (bottom three graphs) in the time domain. Clearly, based on this example, the time domain approach yielded better definition of the three distinct scatterers.

The target simulated has a point scatterer, a flat plate type scatterer, and a corner type scatterer. The frequency of operation is from one to twelve GHz, with steps of 50 MHz. The mixing matrix is generated using an array manifold that assumes less than three degrees difference in scattering angle of two consecutive scatterers. This is in accordance with a scenario where the target is within the power range of the radar. The array simulated has 6 ultra-wideband sensors.

It should be noted that radars with large bandwidth to central frequency ratio are often considered ultra wideband, and radars with large antennas relative to range resolution are also considered ultra wideband. Radars that operate using a large (bandwidth) x (signal duration) product can also be classified as UWB [7]. This work focuses on radars operating in frequencies where the target dimension is significantly larger than the range resolution

$$\Delta R = \frac{c}{2B}$$

where c is the speed of light, and B is the bandwidth. In this study ΔR is about 1.25 meters. Figures 3 and 4 show the extracted scattering centers of a target with three scatterers of the same nature. These figures show that certain scatterers are more difficult to discern than point scatterers. They also show that BSS can separate responses due to individual scatterers regardless of their nature or geometrical attributes.

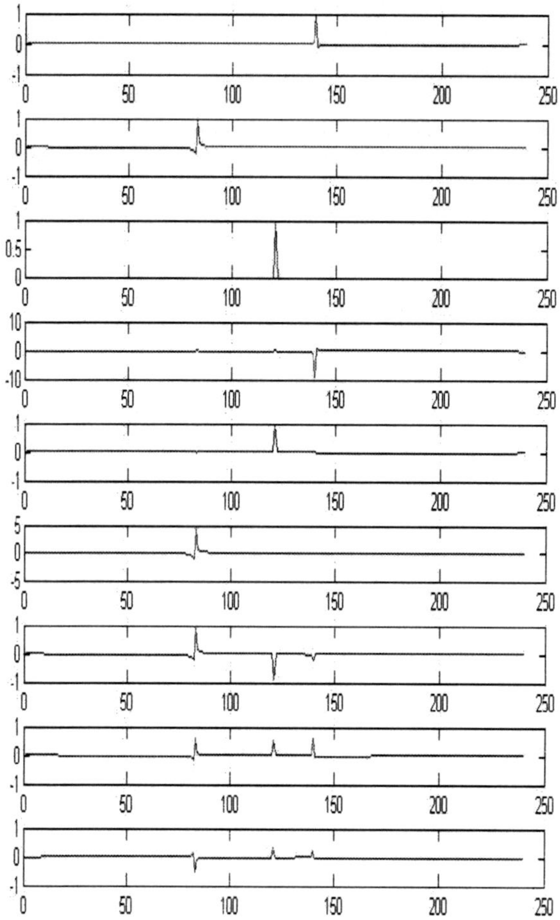

Figure 3. Original three similar scatterers (top three graphs). As recovered using Jade (next three) and as recovered using Sobi (bottom three).

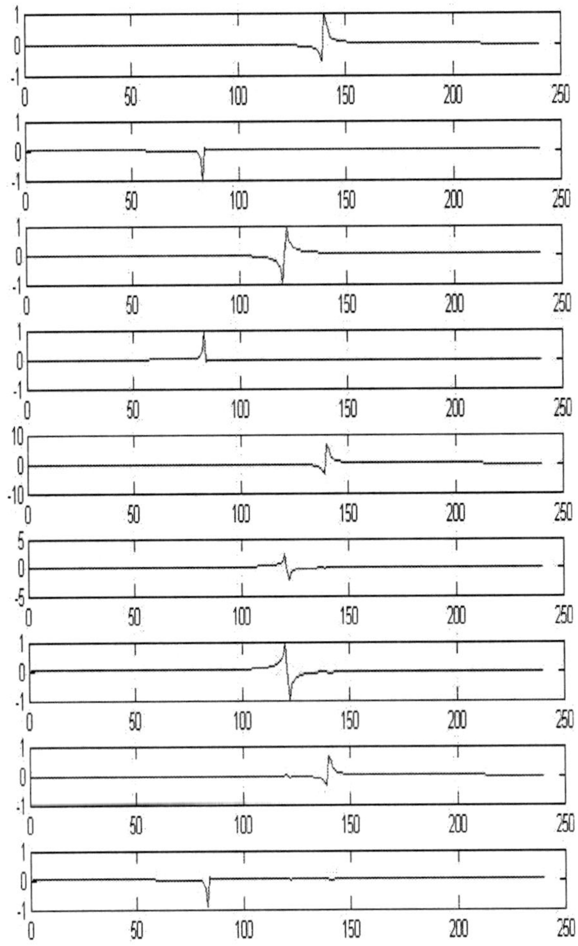

Figure 4: Original three similar scatterers (top three graphs). As recovered using Jade (next three) and as recovered using Sobi (bottom three).

5. CONCLUSIONS

This chapter introduces a new method for extracting scattering centers from a radar target (such as aircraft) using blind source separation. This technique is particularly suited for ultra wideband radars. The results obtained using synthetically generated radar data prove that target scattering centers can be separated and identified using an array of sensors and blind source separation. Although this paper focused on using two common BSS techniques, the nature of the returned signal is fitting to several separation methods. Work continues on extracting scattering centers of real radar targets.

6. REFERENCES

1. Belouchrani, A., Abed-Meraim K., Cardoso J. F., and Moulines, E. "A blind source separation technique using second order statistics", IEEE Transactions on Signal Processing, 45, No.2, 434-444, Feb. 1997.
2. Potter and Moses, Attributed scattering centers for SAR ATR, IEEE Transactions on Image Processing, 6, No.1, 79-91, January 1997.
3. Strifors and Guillermo, Scattering of electromagnetic pulses by simple-shaped targets with radar cross section modified by a dielectric coating, IEEE Transactions on Antennas and Propagation, 46, No.9, 1252-1262, September 1998.
4. Kritikos and Teti, A Time-Frequency analysis method for radar scattering, IEEE Transactions on Microwave Theory and Techniques, 46, No. 3, 257-360, March 1998.
5. Cardoso, J.F., Souloumiac, A., Blind beamforming for nonGaussian signals", IEE Proceedings-F, 140, No. 6, 362-370, Dec. 1993.
6. Cardoso J.F., and Laheld P., Equivariant adaptive source separation, IEEE Transactions on Signal Processing, 45, No. 2, 434-444, Dec. 1996.
7. Shirman, Y.D., Leshchenko, S. P., and Orlenko, V. M., Wideband Radar (Advantages and Problems), Proceedings of IEEE-Ultrawideband and Ultrashort Impulse Signals Conference, Sep. 2004, Ukraine.

EXTENDED APPROACHES FOR INTEGRATED M-SEQUENCE BASED UWB SENSORS

Martin Kmec, Ralf Herrmann, Jürgen Sachs, Peter Peyerl, and Peter Rauschenbach*

1. INTRODUCTION

During recent years, low power ultra wideband (UWB) sensors have become more and more interesting for a great deal of civil, industrial and military applications. UWB sensors may be used for imaging and localization systems, ground–penetrating radars, non-destructive testing and radio channel investigation (e.g. channel sounding) and many more. This has resulted in a diversity of requirements concerning the measurement system – touching both its electronics and signal processing. The article deals with flexible ultra-wideband electronics whose behaviour can be adapted to different parameters. This concerns for example the bandwidth, the operational frequency band, the measurement speed, the system costs, the number of channels and others.

The working principle is based on a special type of very wideband Pseudo-Random Code – the M-Sequence. The mother device which is working in the baseband provides a bandwidth up to 7 GHz (typically 5 GHz), this is explained in detail in[1]. Here, only a short summary will be given as a starting point for some modified approaches which are aimed to allow very wideband operation at an arbitrary frequency band or to increase the system flexibility by replacing RF-filters by digital ones.

2. DESIGN CHALLENGES

In designing ultra-wideband sensor systems, different aspects must be considered. In first point is the choice of a suitable stimulus signal. The classical and as yet mostly used UWB-approaches are based on short pulses or on swept sine waves (as long as one refrains from an instantaneous spectrum). As an alternative to these techniques, we developed an M-Sequence based system, i.e. a method which deals with random noise to

* Martin Kmec, Ralf Herrmann, Jürgen Sachs, Technische Universitaet Ilmenau, Institute for Information Technology, Electronic Measurement Research Laboratory, P.O.B. 100565, 98684 Ilmenau, Germany, martin.kmec@tu-ilmenau.de, (tel.) +49 3677 69 1160, (fax.) +49 3677 1113 http://www-emt.tu-ilmenau.de. Peter Peyerl, Peter Rauschenbach, MEODAT GmbH, Werner-von-Siemens-Str. 3, 98693 Ilmenau, Germany

stimulate the test objects. In the main, the advantage of such a system is that the signal energy is homogenously distributed over the time avoiding high signal peaks. This allows monolithic circuit integration of the critical RF-components.

A second point is to have a fast semi-conductor technology available providing a reasonable noise performance. Traditionally, wideband high frequency circuits are manufactured from technologies based on III-V compounds. However, both silicon based CMOS and bipolar devices have made great progress in their high-frequency performance. The state-of-the art n-MOSFETs and SiGe-HBTs can achieve cut-off frequency above 100 and 200 GHz, respectively[2, 3]. This makes the technology very attractive for ultra-wideband sensors. In order to effectively utilize both types of devices, one popular approach is to develop BiCMOS processes that allow high design flexibility and the integration of mixed signal high speed circuits on the same chip.
Fundamental requirements that must be met by the circuit design are:

- A reasonable matching of the input and output ports i.e. typically 50 Ω or 100 Ω (e.g. for differential ports). (In the case where the electronics is placed directly in the feed point of an antenna other values may be of advantage.)
- Low power consumption despite a short switching time, high bandwidth and reasonable IP3.
- A flat gain curve over the entire bandwidth, a good linearity (i.e. a high IP3) and low noise contributions by input amplifiers.
- The on-chip cross-talk suppression by differential circuiting and shielding measures.

3. BASIC M-SEQUENCE CONCEPT

Figure 1 shows the basic structure of the ultra wideband M-sequence concept. A single tone RF generator (RF-clock) pushes a digital shift register. Supposing an appropriate internal feedback, it provides the M-sequence (Maximum length binary sequence). This signal serves as a test signal. It is a periodic stimulus with a bandwidth determined by the toggle frequency of the shift register. Most of the signal power of an M-Sequence is concentrated below $f_c/2$ where f_c is the master clock rate. One period of the MLBS contains 2^n-1 pseudo randomly distributed chips in which n is the number of stages of the shift register.

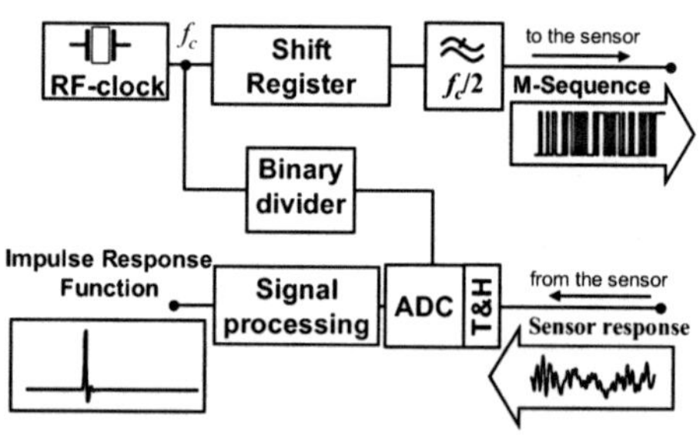

Figure 1 Basic M-sequence Concept

On the receiver side, the signal of interest is captured by a track and hold (T&H) circuit operating in an undersampling mode so that low cost ADCs and acquisition memories can be applied. By limiting the bandwidth to $f_c/2$, for every chip of the MLBS only one sample of the measurement signal is needed in order to respect the sampling theorem. Under these conditions, the sampling clock may be directly derived from the RF master clock by a binary divider.

In order to gain the impulse response function of the scenario under test in the easiest way, a digital impulse compression is performed with the knowledge of the transmit wave form. The corresponding algorithm is called Fast Hadamard Transform.

The RF key-components of the system, the M-Sequence generator, the binary divider and the T&H-circuit are manufactured in 0.25μm SiGe:C BiCMOS-Semiconductor technology[2, 3, 4]. Measurements on current chips (the shift register is based on a Galois feedback structure; semi-conductor technology SGC25A of IHP[2]) gave a maxim toggle rate of about 15 GHz for both M-Sequence generator and binary divider. Thus, the maximum usable bandwidth is from dc to 7.5 GHz. New simulations for redesigning the chips in an improved technology (SGC25B[2, 3]) resulted in a maximum system clock of 30GHz with reasonable power dissipation. The available shift register length is either 9 or 12 and the dividing factor of the binary divider can be selected between 2^4 till 2^9 in steps of a power of two.

The T&H employs a novel design topology which allows a significant improvement in the trade-off between bandwidth, linearity, hold-mode feed-through and power consumption. The achieved input signal bandwidth of the current T&H generation is about 12 GHz. Its feed-through is less than – 40 dB, measured with a 12 GHz input sine wave at a sampling rate of f_s = 20 MHz. The droop rate is about 20 % per ms, which allows sampling rates down to the KHz range. The maximum sampling rate is about 2 GHz. The T&H chip consumes 350 mW at 3.3V. Figure 2 shows its layout.

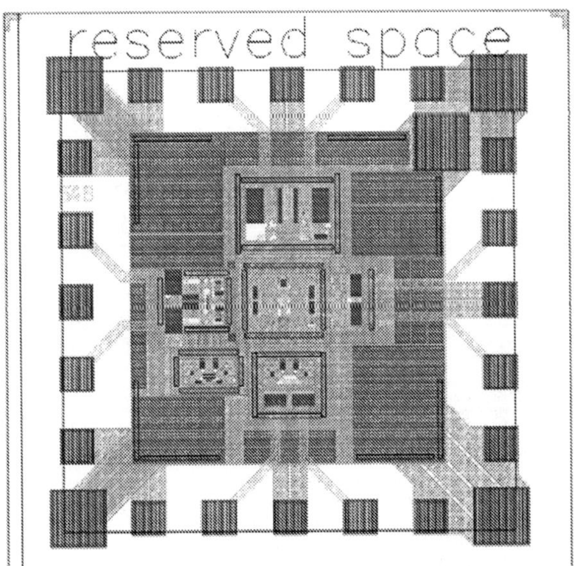

Figure 2 Layout of a Track&Hold circuit. The chip area is 1mm x 1mm.

4. M-SEQUENCE UP-DOWN CONVERSION

An M-Sequence device working on the basic concept operates in the base band. This is of advantage for many surface penetrating radar applications but it is in conflict to generic UWB applications[5] which need to respect the FCC radiation rules[6]

By these rules, the UWB radiation is restricted to the band from 3.1 GHz to 10.6 GHz. In order to match the M-Sequence concept to that frequency band, it has been extended by a direct frequency up-down conversion[7, 8] in which the carrier frequency is provided by the RF-master clock (see Figure 3). This results in the operational band 0,5 … 1,5 f_c. By selecting $f_c \approx 7$ GHz, the occupied spectrum is within the FCC frequency limits.

The direct conversion approach requires a very wideband IQ-down converter in the receive channel. In order to evade IQ-mismatch, the indicated down conversion operates successively on the same transmission channel by phase-switching (0° - 90°) the carrier signal[8, 9]. In that way, the transmission behavior of the I- and Q-channel is identical which largely simplifies the sensor calibration.

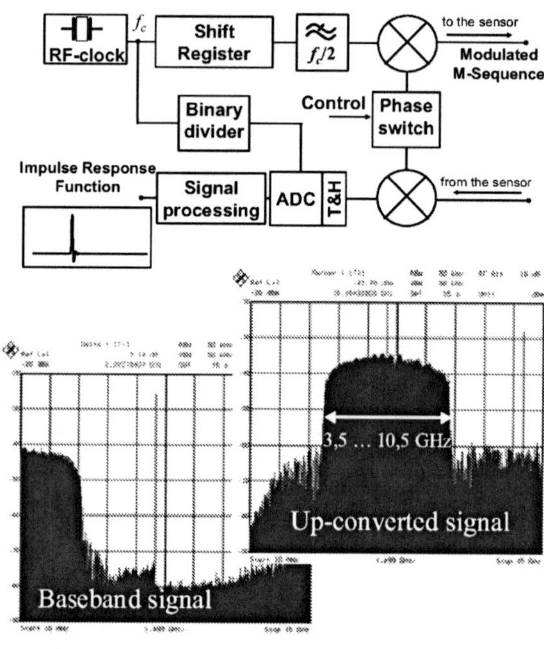

Figure 3 Schematics and typical spectra of the direct up-down conversion approach by a switched IQ demodulator.

The switching sequence is synchronous to the T&H and the ADC clock. Therefore, a serial I and Q data stream is produced, which is reordered to a complex impulse response function at a later stage of signal processing. The overall measurement time is twice that of the basic system conception by keeping all other parameters constant. An implemented device is based on a 12th order M-sequence resulting in an ambiguity range of 585 ns (at 7 GHz master clock) and a maxim measurement speed of 24 complex impulse response function per second for hard disc storage and 1660 for storage on a PCI board.

Since there is also an increasing interest [10, 11, 12, 13] in very wideband real-time measurements in the mm-Wave range e.g. 60 GHz, the direct conversion system (Figure 3) was extended by a further conversion stage (see Figure 4) provided from the Heinrich Hertz Institute Berlin (Germany) and respectively from the Helsinki University of Technology (Finland) for initial experiments[10]. The resulting measurement speed remains the same as mentioned above.

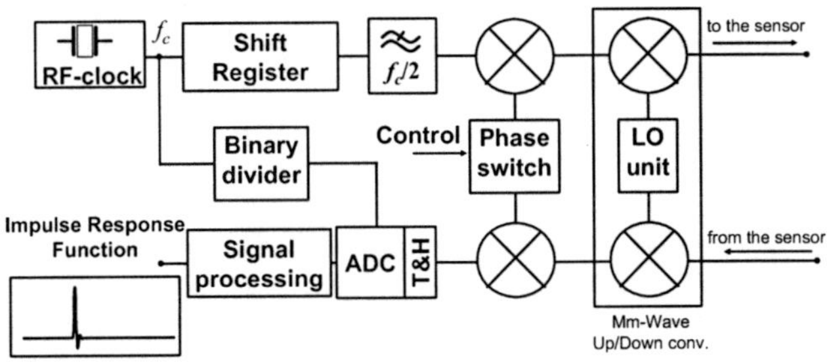

Figure 4 Two stage up-down conversion for mm-Wave applications

The system was used to investigate mm-Wave indoor propagation and the influence of the human body activity on the channel characteristics. Figure 5 shows the record of the time dependent loss of the total received energy with three people walking around and arbitrary shadowing the line of sight connection. The depth of the notches depends upon the mutual constellation of the persons. If only one person is involved, the notch depth is always the same.

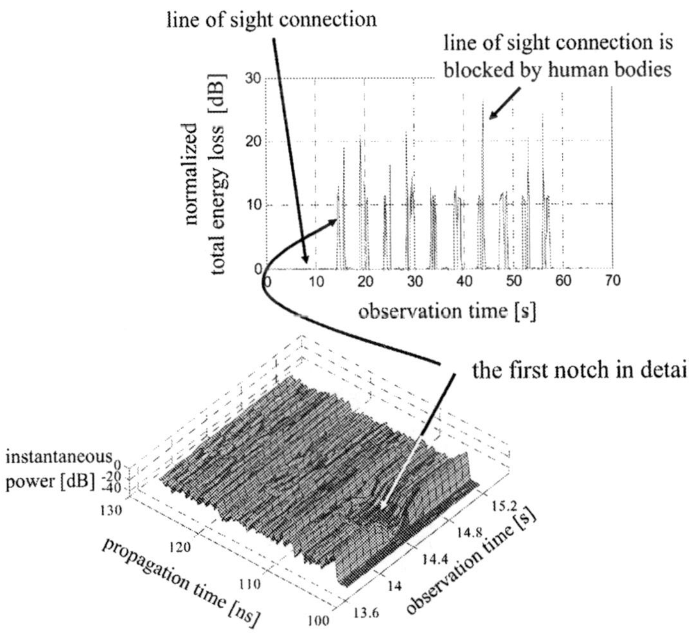

Figure 5 Temporal variations of the transmission loss in the 60 GHz band between two stationary antennas caused by three walking people. The 3D-graph shows the power profile with details of the shadowing.

5. DENSE SAMPLED M-SEQUENCE AND BANDWIDTH EXTENSIONS

The M-Sequence approaches presented above require a less integration friendly anti-aliasing filter which moreover needs to be replaced if the RF-master clock (i.e. the bandwidth) needs to been changed. In order to evade this filter, the equivalent sampling rate needs to be increased[14] i.e. one has to take a number of different probes from every chip of the test signal (note the previous approaches suppose only one probe per chip, hence the maximum bandwidth of $f_c/2$). Hence, by a sufficiently large equivalent sampling rate, the sampling theorem will also be satisfied by the natural effects of bandwidth limitation (analogue bandwidth of the T&H, limited switching speed of the M-Sequence shift register). But by omitting the anti-aliasing filter, the effective bandwidth will not really be increased since the spectral power of the stimulus beyond $f_c/2$ is quite small. Thus, the method apparently only increases the noise (supposing the same overall measurement time). However, by performing a low-pass filtering in the digital domain and a subsequent sampling rate reduction, one gains the same result as before by the basic approach (see Figure 1). But now, one has a higher degree of freedom to adapt the system performance to the actual needs by software rather than by hardware modifications.

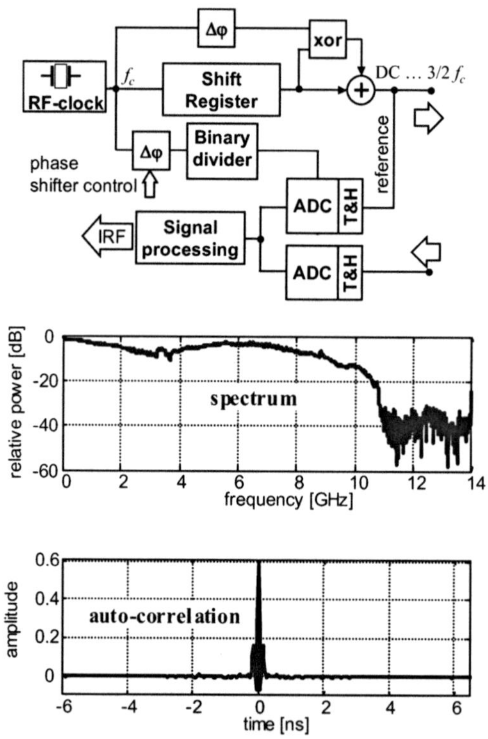

Figure 6 Bandwidth enhanced M-Sequence system using dense under-sampling. Above: block schematics. Middle: spectrum of captured reference data in the case of 7 GHz RF-clock. Below: Auto-correlation function (impulse response) of the measurement system.

The simplest way to increase the equivalent sampling rate is to repeat the measurement several times as usual but with a slightly shifted sampling instant. For this purpose, a phase shifter is introduced before the binary divider (see Figure 6). In order to gain for example an equivalent sampling rate of 4 fc, the measurements are repeated 4 times by setting the phase shifter to 0°, 90°, 180° and 270°.

As noted above, the implemented T&H circuit has an analogue bandwidth of approximately 12 GHz. Thus, the IQ-down-conversion corresponding to Figure 3 can also be replaced by the dense sampling approach connected with a digital IQ-down-conversion. Furthermore, the superposition of the baseband signal with the up-converted signal results in coverage of the spectral band from dc to about 1.5 fc which triples the available bandwidth. However, the time shape of such a test signal does no longer correspond to an M-Sequence. It is therefore recommended to capture also the transmit signal in order to be able to perform the impulse compression. Figure 6 represents a block schematic and measurement results from such a device of a tripled bandwidth. The FWHM-value of the auto-correlation function is approximately 70 ps.

Disadvantages of the presented approach are the need for a very wide receiver bandwidth and the risk of introducing jitter and drift into the system due to instabilities (drift and jitter) of the phase shifter. The results are promising so far but different technical realizations need to be compared in order to account for these issues.

6. CONCLUSION

Current radio regulation activities and new cost effective measurement principles will promote the application of ultra wideband devices for a large number of applications operating in the GHz range. The article presents three variants of an ultra wideband measurement device dealing with M-Sequence stimulation signals, which permit operation at an arbitrary centre frequency, which avoid the need for any analogue filter for the suppression of aliasing effects and which increase the bandwidth from DC to $3f_c/2$.

7. ACKNOWLEDGEMENT

The article presents a summary of results from projects such as e.g. DEMINE, DEMAND, PULSERS and WIGWAM which were funded by the EU and the German government. The Authors acknowledge IHP Frankfurt for chip fabrication and support.

8. REFERENCES

1. J. Sachs, M-sequence radar, in Ground Penetrating Radar 2nd edition, D.J. Daniels ed., *IEE Radar, Sonar, Navigation and Avionics Series 15,* pp. 225-237, 2004
2. W. Winkler, B. Heinemann, and D. Knoll, Application of SiGe:C BiCMOS to Wireless Radar, *Proceedings of the 12th GAAS Symposium,* Amsterdam, Belgium, September 2004
3. B. Heinemann, High-Performance HBT Modules in BiCMOS, *3rd IHP Workshop*, Frankfurt (Oder), Germany, September 2004
4. E. Grass, F. Herzel, M.Piz, Y. Sun, R. Kraemer, Implementation Aspects of Gbit/s Communication Systems for 60 GHz Band, *Wireless Word Research Forum,* San Diego, USA, July 2005

5. W. Ciccognani, A. Durantini, D. Cassioli, Time Domain Propagation Measurement of the UWB indoor Channel Using PN-Sequence in the FCC-Compliant Band 3.6-6 GHz, *IEEE Trans AP*, Vol. 53, No. 4, April 2005, pp. 1542-1549
6. FCC Document 02-48: Revision of Part 15 of the Commission's Rules Regarding Ultra-Wideband Transmission Systems, ET Docket 98-153
7. J. Sachs, M. Kmec, P. Peyerl, P. Rauschenbach, R. Zetik, MSCW-Radar – a novel Ultra Wideband Radar Principle, *International Radar Symposium 2005*, Berlin, Germany, September 2005
8. J. Sachs, M.Kmec, R. Zetik, Ultra Wideband Radar Assembly Kit, *International Geoscience and Remote Sensing Symposium 2005*, Soul, Korea, August 2005
9. M. Kmec
10. S. Ranvier, M. Kmec, R. Herrmann, J, Kivinen, J. Koivunen, R. S. Thomä, P. Vainikainen, MM-Wave Wideband MIMO Channel Sounding, *URSI*, New Dehli, India, October 2005
11. Y. Sun et al., A fully integrated 60 GHz LNA in SiGe:C BiCMOS technology, *BCTM*, Santa Babara, Oct. 2005, pp. 14-17.
12. S. Collonge, G. Zaharia, G. E. Zein, Influence of the human activity on wide-band characteristics of the 60 GHz indoor radio channel", *Wireless Communications, IEEE Transactions,* Vol. 3, Issue 6, p. 2396-2406, Nov. 2004
13. R. Thomä, M. Kmec, J. Sachs, R. Zetik, Architektur und Anwendung eines UWB/mm-Wellen-Channel-Sounders, *Statusseminare 2006 Mobile Kommunikation und GaN-Elektronik*, Erlangen 2006
14. J. Sachs, R. Herrmann, M. Kmec, P. Peyerl, Modified M-Sequence UWB-Radar, *German Microwave Conference - GeMiC 2006,* Karlsruhe, Germany, March 2006

ANTENNA ARRAY PROCESSING FOR RADAR APPLICATIONS WITH SUPPORT VECTOR MACHINES

Manel Martínez-Ramón[a], Nan Xu[b], Christos G. Christodoulou[b]

1. INTRODUCTION

Support Vector Machines[1] (SVM) are a good candidate for the solution of antenna array processing problems such as beamforming, estimation of angle of arrival or Ultra-Wide Band (UWB) electromagnetic design[2], because these algorithms provide superior performance in generalization ability and computational complexity. In this work we revise some applications of the SVM for antenna array processing and for target detection in UWB sea surface surveillance radar return profiles.

The first presented approach is based on the use of linear and nonlinear regressor schemes applied to antenna array processing, in particular to beamforming. The last one is related to the application of nonlinear multiclass Support Vector classification applied to radar object detection. Comparisons with conventional strategies and simulation results are provided to demonstrate the advantages of the Support Vector Machine approaches.

Since the 1990s there has been a significant activity in the theoretical development and applications of SVM. The first applications of machine learning have been related to data mining, text categorization, and other classical pattern recognition tasks. Recently, however, SVM have been applied to signal processing[3], wireless communication problems, notably spread spectrum receiver design and channel equalization.

The high-speed "learning" capability of SVMs is our main motivation for employing them in antenna problems. There are two approaches for the SVM optimization. The linear and the non-linear kernels are chosen depending on the nature of the problem. Unlike previous SVM algorithms in signal processing, in which real-valued

[a] Department of Signal Theory and Communications, Universidad Carlos III de Madrid, Avda de la Universidad, 30, 28911, Leganés, Madrid, Spain.
[b] Department of Electrical and Computer Engineering, University of New Mexico, 1 University of New Mexico, 87131-0001, NM, USA.

SVM algorithms were used to deal with complex valued data, we directly use a complex notation to solve the SVM optimization problem, which leads to functionals that are natural and straightforward extensions of the standard SVM dual optimization problems. This complex-value formulation is more suitable for handling UWB signal cases than some of the other algorithms.

2. COMPLEX SVM REGRESSOR FOR ANTENNA ARRAY PROCESSING

2.1. Temporal Reference

The temporal reference beamformer is intended to minimize the error between the output and a reference or desired signal $d[n]$. Introducing the estimation error $e[n]$, the expression for the output of the array processor can then be written as:

$$y[n] = \mathbf{w}^T \mathbf{x}[n] = d[n] + e[n] \tag{1}$$

where $\mathbf{w} = [w_1, \ldots w_M]^T$ are the weights in the adaptive array.

For a set of N observed samples of $\{\mathbf{x}[n]\}$ and when nonzero empirical errors are expected, the functional to be minimized is:

$$L_p = \frac{1}{2}\|\mathbf{w}\|^2 + C\sum_{n=1}^{N} \ell\left(\xi_n, \xi_n^{'}\right) + C\sum_{n=1}^{N} \ell\left(\zeta_n, \zeta_n^{'}\right) \tag{2}$$

with the constraints

$$\begin{aligned}
&\operatorname{Re}\left(d[n] - \mathbf{w}^T \mathbf{x}[n]\right) < \varepsilon + \xi_n \\
&\operatorname{Re}\left(-d[n] + \mathbf{w}^T \mathbf{x}[n]\right) < \varepsilon + \xi_n^* \\
&\operatorname{Im}\left(d[n] - \mathbf{w}^T \mathbf{x}[n]\right) < \varepsilon + \zeta_n \\
&\operatorname{Im}\left(-d[n] + \mathbf{w}^T \mathbf{x}[n]\right) < \varepsilon + \zeta_n^*
\end{aligned} \tag{3}$$

where ℓ is a cost function. Equation (2) is the so-called SVR primal[4,5]. The second half of Eq. (2) is intended to optimize the empirical error over the available sample set, where the first part is intended to minimize the norm of the weight vector. Provided that (as we will see later for the SVM in particular) the weight vector **w** is a linear combination of the input data[c], this provides the solution which uses the minimum amount of energy of the

[c] Any linear algorithm that can be expressed as in Eq. (1) has a solution that can be expressed as a function of the data available for training. This has a straightforward proof if one assumes that the weight vector **w** lies in the subspace spanned by the available data.

data. C is the tradeoff between both minimizations, and $\xi_n, \xi_n', \zeta_n, \zeta_n'$ are the so called slack variables. The constraints mean that for each positive (negative) real and imaginary parts of the error, the algorithm has to minimize the quantities ξ_n, ζ_n (ξ_n', ζ_n') that are the error in excess of ε. These constraints are, then, equivalent to applying a cost function that is zero if the error is less than ε, and linear otherwise.

The novelties between our approach adapted to signal processing for antenna arrays with respect to the standard SVR[4,5] are two[6,7]. First, we adopt a direct complex formulation which allows us to apply the algorithm directly to array signals, which are complex valued. Second, we give an interpretation to the numerical regularization that is usually applied to the SVM formulation to avoid numerical problems: instead of using the standard Vapnik or ε-insensitive cost function[1] that is applied to Eq. (2), we use a combination of that function and the Huber Robust cost function[8]. This provides a numerical regularization of the algorithm in a natural way and, also, provides an interpretation of this cost function that will be useful in signal processing.

The applied cost function can be expressed as

$$\ell(e) = \begin{cases} 0 & |e| < \varepsilon \\ \dfrac{1}{2\gamma C}\left(|e| - \varepsilon\right)^2 & \varepsilon < |e| < \varepsilon + \gamma C \\ \left(|e| - \varepsilon\right) - \dfrac{\gamma C}{2} & \varepsilon + \gamma C < |e| \end{cases} \tag{4}$$

The optimization of this problem using Lagrange multipliers, leads to the dual

$$L_d = -\frac{1}{2}\boldsymbol{\psi}^H\left(\mathbf{K} + \frac{\gamma}{2}\mathbf{I}\right)\boldsymbol{\psi} + \mathrm{Re}\left(\boldsymbol{\psi}^H\mathbf{d}\right) - \mathbf{1}(\alpha + \alpha' + \beta + \beta')\varepsilon \tag{5}$$

where $\boldsymbol{\psi}$ is a vector whose components are $\psi_i = \alpha_i - \alpha_i' + j\left(\beta_i - \beta_i'\right)$, and $\mathbf{K}$ is the matrix of all dot products among the training data. Equation (2) must be optimized with respect to ψ_i, and it is commonly done by quadratic programming[9]. The optimization gives also the result

$$\mathbf{w} = \sum_{i=1}^{N} \psi_i \mathbf{x}[i] \tag{6}$$

Finally, the beamformer output can be expressed as:

$$y[n] = \mathbf{w}^H \mathbf{x}[n] = \sum_{i=1}^{N} \psi_i^* \mathbf{x}^H[i]\mathbf{x}[n] \tag{7}$$

2.2. Nonlinear SVM

The nonlinear counterpart of the SVM is straightforward if one makes use of the rather old Kernel trick[10]. This trick allows us to apply a nonlinear transformation $\varphi(\mathbf{x})$ to the data which maps it into a higher dimensional (possibly infinite) Hilbert space which is provided with a dot product $K(\mathbf{x}[i],\mathbf{x}[j])=\varphi^H(\mathbf{x}_i)\varphi(\mathbf{x}_j)$ that can be expressed as a function of the input data. Then, the algorithm is functionally the same as for the linear case, but here, the components of matrix $\mathbf{K}$ in Eq. (5) are computed as

$$\mathbf{K}_{ij}=K(\mathbf{x}_i,\mathbf{x}_j) \tag{8}$$

and the expression of the array processor is expressed as

$$y[n]=\mathbf{w}^H\varphi(\mathbf{x}[n])=\sum_{i=1}^{N}\psi_i^*K(\mathbf{x}[i],\mathbf{x}[n]) \tag{9}$$

The most popular and used kernels are the polynomial and the Gaussian:

$$\begin{aligned}K(\mathbf{x}_i,\mathbf{x}_j)&=(\mathbf{x}_i^H\mathbf{x}_j+1)^p\\ K(\mathbf{x}_i,\mathbf{x}_j)&=\exp\left(\frac{\|\mathbf{x}_i-\mathbf{x}_j\|^2}{2\sigma^2}\right)\end{aligned} \tag{10}$$

2.3. Spatial reference

A nonlinear spatial reference beamformer using support vector machines has been presented[7] that combines the complex SVM strategy with the classical Minimum Variance Distortionless Response (MVDR) spatial reference beamformer[11]. The algorithm consists of minimizing the energy at the output of the beamformer while keeping constant the amplitude of the desired direction of arrival. We summarize here the nonlinear version of the algorithm. The SVM primal and constraints for this problem are

$$\begin{aligned}&L_p=\frac{1}{2}\mathbf{w}^H\mathbf{R}\mathbf{w}+C\sum_{n=1}^{N}\ell(\xi_n,\xi_n')+C\sum_{n=1}^{N}\ell(\zeta_n,\zeta_n')\\ &\mathrm{Re}(r_i-\mathbf{w}^T\varphi(r_i\mathbf{a}))<\varepsilon+\xi_i\\ &\mathrm{Re}(-r_i+\mathbf{w}^T\varphi(r_i\mathbf{a}))<\varepsilon+\xi_i^*\\ &\mathrm{Im}(r_i-\mathbf{w}^T\varphi(r_i\mathbf{a}))<\varepsilon+\zeta_i\\ &\mathrm{Im}(-r_i+\mathbf{w}^T\varphi(r_i\mathbf{a}))<\varepsilon+\zeta_i^*\end{aligned} \tag{11}$$

where $\mathbf{a}$ is the steering vector of the desired direction of arrival and r_i are a set of desired complex amplitudes, for example, the set of symbols in the signal constellation. Provided that a set of signals is collected for training issues, the solution can be expressed as

$$L_d=-\frac{1}{2}\boldsymbol{\psi}^H\left(\mathbf{K}_a^H\mathbf{K}^{-1}\mathbf{K}_a+\frac{\gamma}{2}\mathbf{I}\right)\boldsymbol{\psi}+\mathrm{Re}(\boldsymbol{\psi}^H\mathbf{r})-\mathbf{1}(\alpha+\alpha'+\beta+\beta')\varepsilon \tag{12}$$

$$y[n] = \psi^H \mathbf{K}_a^H \mathbf{K}^{-1} \mathbf{k}[n] \tag{13}$$

where $\mathbf{K}_a$ is the matrix of kernel products of the training data with vectors $r_i\mathbf{a}$ and $\mathbf{k}[n]$ is the vector of kernel products of the incoming data with the training data.

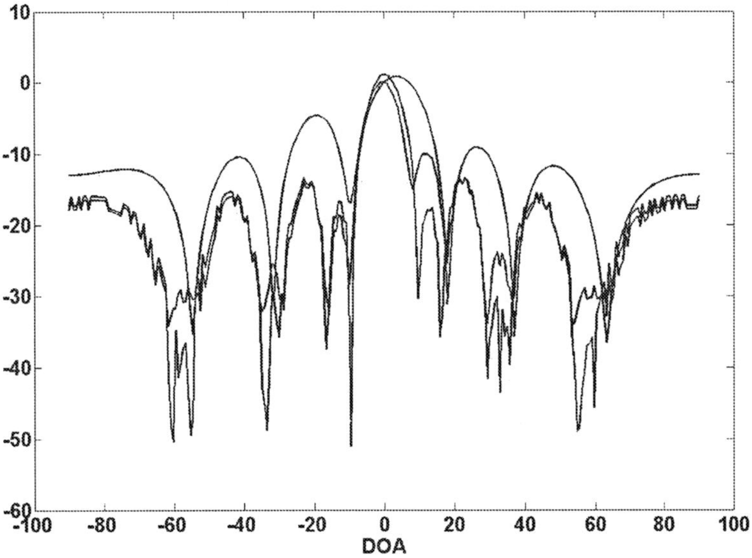

Figure 1. Approximate radiation patterns of the SVM temporal (cont. line) and spatial (dash) reference processors compared to the classical MMSE and MVDR (dots). Desired DOA at 0° and interference at -10°.

2.4. Beamforming and Tracking Example

We compare the SVM algorithms to the linear MMSE and MVDR algorithms that are the most common in antenna beamforming . The kernel that has been used in all simulations is Gaussian with its variance set to 0.1. ε = 0.1 and C=1000. The scenario consists of using signals that are emanating from a target (desired signal to track) and signals emanating from nearby clutter (intereference). The signal is corrupted by AWGN. The desired signal is a train of bursts with a preamble of 100 training and 1000 signal symbols. Figure 1 shows the approximate beam diagrams of the SVM approaches (computed by sweeping for all possible directions of arrival) compared to the classical linear algorithms. Figure 2 depicts the Bit Error Rate (BER) of the desired signal as a function of the Signal to Noise Ratio (SNR).

3. SVM CLASSIFIERS FOR RADAR TARGET DETECTION

Assume a binary classification problem for which the set $\{\mathbf{x}_i, y_i\}_{i=1}^{N}, y_i \in [-1,1]$ is available for training purposes. A classifier can be trained by finding the parameters $\mathbf{w}$ of a separating hyperplane $\mathbf{w}^T\mathbf{x}+b=0$ by optimizing a given cost function of the classification errors. The output will be $\hat{y}_i = \text{sign}\left(\mathbf{w}^T\mathbf{x}_i + b\right)$. If the input data is projected into a higher dimensional feature space, then the hyperplane becomes

$$\mathbf{w}^T\varphi(\mathbf{x})+b=0 \tag{14}$$

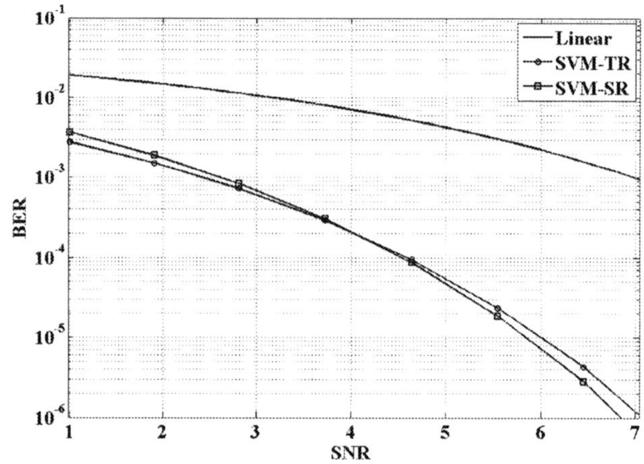

Figure 2. Bit error rate of the SVM approaches compared to those of the MMSE and MVDR (dots). Desired DOA at 0º and interference at -10º.

An SVM classifier finds a classification hyperplane that separates data classes with the largest classification margin. Similarly to the SVM regressor, SVM classifier is based on the minimization of[1]

$$L_p = \frac{1}{2}\|\mathbf{w}\|^2 + C\sum_{i=1}^{N}\xi_i \tag{15}$$

with the constraints

$$y_i\left(\mathbf{w}^T\varphi(\mathbf{x}_i)+b\right) \geq 1+\xi_i \tag{16}$$

Applying a Lagrange optimization to this problem, the following dual is obtained

$$L_d = -\frac{1}{2}\boldsymbol{\alpha}^T \mathbf{YKY}\boldsymbol{\alpha} + \mathbf{1}^T \boldsymbol{\alpha} \tag{17}$$

where $\mathbf{Y}$ is a diagonal matrix containing all labels, and $\boldsymbol{\alpha}$ contains parameters α_i. As before, this functional has to be optimized for α_i. Only a subset of parameters will be nonzero, and its associated vectors are the so called support vectors. The classifier can be written as

$$y_j = \text{sign}\left(\sum_{i=1}^{N} \alpha_i K\left(\mathbf{x}_i, \mathbf{x}_j\right) + b \right) \tag{18}$$

2.5. Example

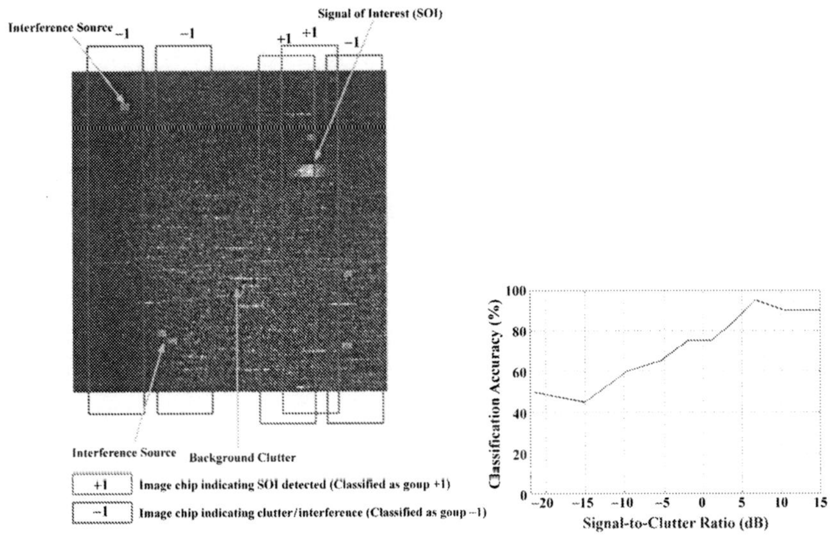

Figure 3. Example of object detection (left) and error performance of the SVM classifier applied to sea radar imaging.

A simulation experiment of SVM based automatic target detection has been implemented successfully. The experiment data are obtained based on collection of sea cluttering radar profiles. SVM based detection performance are compared under different signal-clutter ratio scenarios. Among the different used kernels, the best reported performance is the one obtained using a polynomial Kernel SVC. The patterns consisted of a set of sea clutter images with objects randomly distributed. A feature extraction is performed prior to the classification stage. The number of training images was of 100, and a leave one out test has applied to compute the error performance of the classifier. Figure 3 shows an example of radar returns with objects plus clutter noise on it. The figure also shows the accuracy rate as a function of the signal to clutter ratio.

This classifier works well with UWB scenarios since the performance of the algorithm depends on the type of data used to train the algorithm. If we use data from a an UWB environment then the algorithm will adapt to that environment.

4. CONCLUSIONS

We introduced the SVM algorithm as applied to the problem of antenna processing with application to beamforming and target detection. In order to make the regression algorithm appropriate for array processing, we applied an alternative cost function which is suitable for problems in which gaussian and other non-gaussian sources are present. This alternative cost is, as a matter of fact, just an alternative way to solve the SVM optimization that naturally includes a numerical regularization. This kind of cost functions is well suited for signal processing, in particular for UWB applications, where non Gaussian noise and interferences are present, as it is closer to the maximum likelihood cost function than the MMSE one.

Several examples have been presented, the first one being a comparison example of classical beamforming structures with the SVM solutions in temporal and spatial array processing. The second one is an example of application of nonlinear SVM classifiers to object detection in sea radar returns with clutter noise.

Acknowledgements

This work has been partly supported by grant S-0505/TIC/0223, Comunidad de Madrid, Spain.

References

1. V. Vapnik, Statistical Learning Theory, Adaptive and Learning Systems for Signal Processing, Communications, and Control, John Wiley & Sons, 1998.
2. M. Martínez-Ramón, C. Christodoulou, Support Vector Machines for Antenna Array Processing and Electromagnetics, Morgan & Claypool Publishers, CO, USA, 2006.
3. G. Camps-Valls, J. L. Rojo-Álvarez and M. Martínez-Ramón (Eds.), Kernel Methods in Bioengineering, Communications and Image Processing", Idea Group, PA, USA, 2006.
4. A. Smola and B. Schoelkopf, "A tutorial on support vector regression," NeuroCOLT, Technical Report NC-TR-98-030, Royal Holloway College, University of London, UK,1988.
5. A. Smola, B. Schoelkopf, and K. R. Mueller, "General cost functions for support vector regression," in *Proceedings of the Ninth Australian Conference on Neural Networks*, Brisbane, Australia, 1998, pp. 79–83.
6. J. L. Rojo-Álvarez, G. Camps-Valls, M. Martínez-Ramón, A. Navia-Vazquez, A. R. Figueiras-Vidal, "Support Vector Machines Framework for Linear Signal Processing", Signal Processing, Vol 85, No 12, pp. 2316-2326 , 2005.

7. M. Martínez-Ramón, C. G. Christodoulou, "Support Vector Array Processing", Antenas and Propagation Society Internacional Symposium, July 9-14, 2006, Albuquerque, NM, USA.
8. P. J. Huber, Robust Statistics: a review, vol. 43, Ann. Statistics, 1972.
9. J. Platt, Advances in Kernel Methods: Support Vector Learning (B. Schoelkopf, C. J. C. Burgues and A. J. Smola, Eds.), chapter Fast Training of Support Vector Machines Using Sequential Minimal Optimization, pp. 185–208, MIT Press, 1999.
10. M. A. Aizerman, É. M. Braverman, and L. I. Rozonoér, "Theoretical foundations of the potential function method in pattern recognition learning," Automation and Remote,Control, vol. 25, pp. 821–837, 1964.
11. J. Capon, "High Resolution Frequency-Wavenumber Spectrum Analysis", Proceedings of the IEEE, Vol. 57, No. 8, pp. 1408-1418, Aug, 1969.

RADAR SYSTEMS

PORTABLE IMAGING UWB RADAR SYSTEM WITH TWO-ELEMENT RECEIVING ARRAY

Anatoliy O. Boryssenko[1], Dmitriy L. Sostanovsky[2], and Elen S. Boryssenko[3]

1. INTRODUCTION

UWB radar (UWBR) technology enables high-resolution (i) ranging in free space[1] and in matter including ground-penetrating radar (GPR) technique[2]; (ii) sensing/imaging in medical, industry, law-enforcement, defense and other applications[3-6]. In this study, we examine the performance of two prototypes of UWBR "proof of concept" systems operating as a presence detector for through-wall sensing and imaging of individuals[7-11]. The first system[12] has one transmitter (Tx) and one receiver (Rx) assigned as 1-Tx&1-Rx. The second system[11] noted as 1-Tx&2-Rx includes an additional Rx channel creating 2-element Rx array. We implemented some advanced design solutions for antennas[13], Tx and Rx electronic circuits and radar controlling/processing software to provide versatility, low-complexity and easiness in use. In particular, all the processing/controlling functions are written and used in the Matlab environment[14]. An ordinary desktop/notebook PC is used for interfacing to UWBR via a serial port for operational control and an in-built audio card for data acquisition. The scattered data are collected by the sound card operating in asynchronous mode with respect to the internal UWBR timer that simplifies the radar-computer interface.

With this UWBR hardware and software, we examine several ideas on imaging of 2-D scenes with stationary and moving individuals by exploiting inherently high accuracy in time-of-arrival (ToA) and distance measuring, which are associated with the used impulsive signals. First, we try to track an individual behind a wall using the 1-Tx&1-Rx system. Then, we investigate the ToA difference of the received signals in two Rx channels of the 1-Tx&2-Rx system to estimate the position of scattering objects in two dimensions (2-D). For this, several Matlab-based signal/image processing algorithms[11] are tested and some related experimental results are reported.

---

[1] Antenna & Propagation Laboratory, University of Massachusetts, Amherst, MA, 01003, USA
[2] Ratio Company, Kiev, Ukraine
[3] A&E Partnership, Amherst, MA, 01002, USA

2. UWBR SYSTEM WITH ONE RX CHANNEL

The 1-Tx & 1-Rx, UWBR system prototype[12], Fig. 1a, can be applied like sketched in Fig. 1b. This prototype was developed by adopting an early designed GPR system[4,6] operating in time domain with impulsive EM signals. The radar prototype consists of two pieces of hardware, viz. the control unit and antenna block, Fig. 1. The radar controlling and signal processing are supported via an ordinary desktop or notebook PC, Fig. 1b. The SRD-based Tx unit generates the 0.8 ns rising time pulse will the equivalent 0.5-1.5 GHz bandwidth at the -10 dB reference level. The peak-to-peak voltage reaches 60V with pulse repetition rate 100 kHz. The Rx unit provides stroboscopic down-sampling of the received waveforms with about 10^6 down-converting that enables a "GHz-to-kHz" spectral transform. As a result, the VHF-UHF Rx input signals are transformed into the audio frequency band at the Rx output and can be digitized using a standard PC sound card equipped with the 16-bit analog-to-digital converter (ADC). Despite this down-conversion slows down the per-second-rate of registered waveforms, the UWBR prototype can provide up to 30 scanned waveforms per second. Each scan contains 128...1024 16-bit samples covering a certain range. The radar range can be set up to 12 m in air and less in dense media depending on their electrical features.

A set of Matlab scripts were developed to control completely the radar operation and perform digital signal/image processing. In particular, Matlab Data Acquisition Toolbox provides easy access to the sound card ADC[14]. The radar is completely controllable by a special Matlab script that communicates via a serial (RS-232) port with the radar hardware using the Matlab serial port interface[14]. A couple of samples of the radar data is shown in the oscilloscope (A-scan) mode, Fig. 2a-b. These data were captured in the test schema, Fig. 1b, with a 30-cm thick concrete wall for two radar regimes. The first regime, Fig. 2a, is a regular one where the target mark is not much visible being shadowed by other signal components. The second regime, Fig. 2b, supports real-time background compensation, which is a version of moving target indication technique[15]. Specifically, an early recorded signal scattered by a reference scene is subtracted from each new recorded signal before its displaying. As a result, a clear observable target signature appears, Fig. 2b.

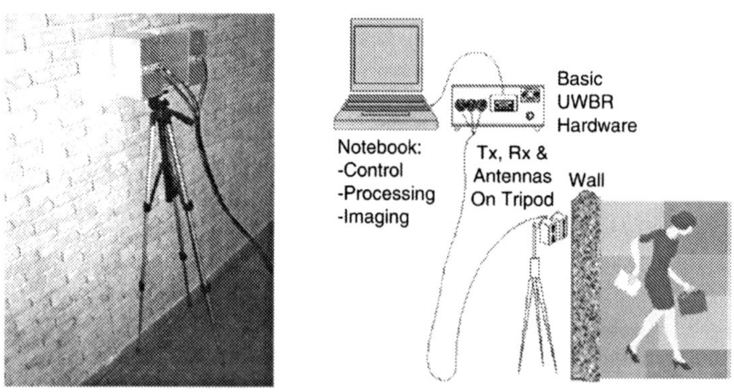

(a) (b)

Figure 1. Typical test setting for the 1-Tx&1-Rx through-wall-vision UWBR prototype: (a) Tx and Rx modules on tripod near a wall; (b) measurement schema with an individual moving behind a wall.

The B-scan data in Fig. 3 demonstrate an attempt to track a dynamic target behind a wall while using the 1-Tx&1-Rx UWBR system. In this experiment, the target is moving over a complex trajectory, Fig. 3a, that is projected at the B-scan scene image, Fig. 3b. This image allows some guessing on where the target might be positioned with respect to the radar standpoint. In this case, the target, which is moving in front of the radar, forms a synthetic aperture similarly to monostatic GPR except the radar is stationary and the target is moving for this case and vise versa for ordinary GPR survey. However, this approach is not essential for dynamic targets with arbitrary trajectories while multi-aspect observation, at least bistatic one, is required.

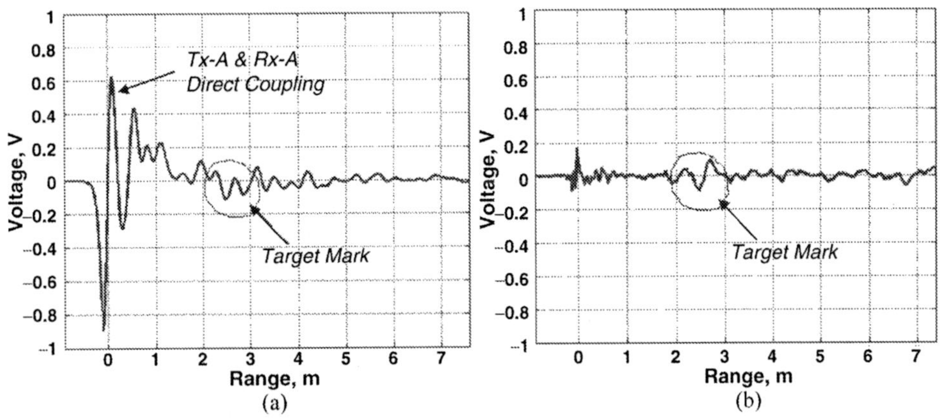

Figure 2. Rx-registered waveforms shown in the oscilloscope A-scan mode: (a) radar operates in a regular mode; (b) radar operates in change detector mode with background compensation.

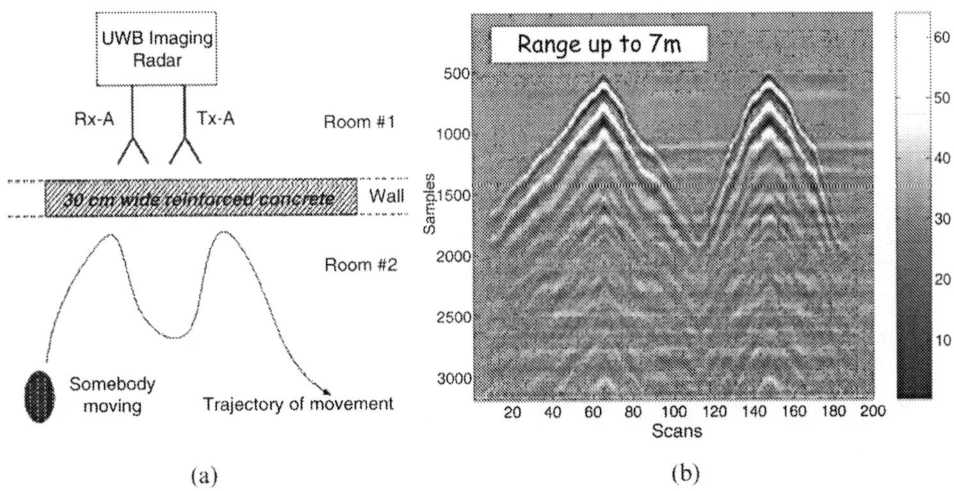

Figure 3. (a) Test scene with moving target behind a wall. (b) Resulted scene 2-D image computed by processing the Rx recorded signal vs. range ("Samples") and time ("Scans").

3. UWBR SYSTEM WITH TWO RX CHANNELS

Adding one more Rx channel would enhance the radar imaging capabilities[11] as in the 1-Tx&2-Rx UWBR whose functional diagram is drawn in Fig. 4a and its modular hardware is pictured in Fig. 4b-d. The system includes three packaged modules of the same overall size, i.e. two Rx modules, at the left and right, and one Tx module, at the centre, Fig. 4 b-d. Each module contains EM, RF and circuit parts inside. Additionally, the Tx module holds the system control electronics, interface to computer and power supply. All the Tx and Rx modules have the same antennas based on novel antenna geometry[13], Fig. 4c, that enables some radar high performed features. In particular, this vertically-polarized antenna composes two specially-shaped metallic curved flares and behaves alike TEM horn radiator but with a number of outstanding electrical features such as (i) broadband matching to 50 Ohm coaxial line without balun, (ii) high directive gain ≥5 dBi, and (iii) -20…-30 dB front-to-back ratio across > 5:1 bands, etc. These antenna electrical properties are extremely important for the demonstrated radar imaging capabilities with minimized distortions of short impulsive signals.

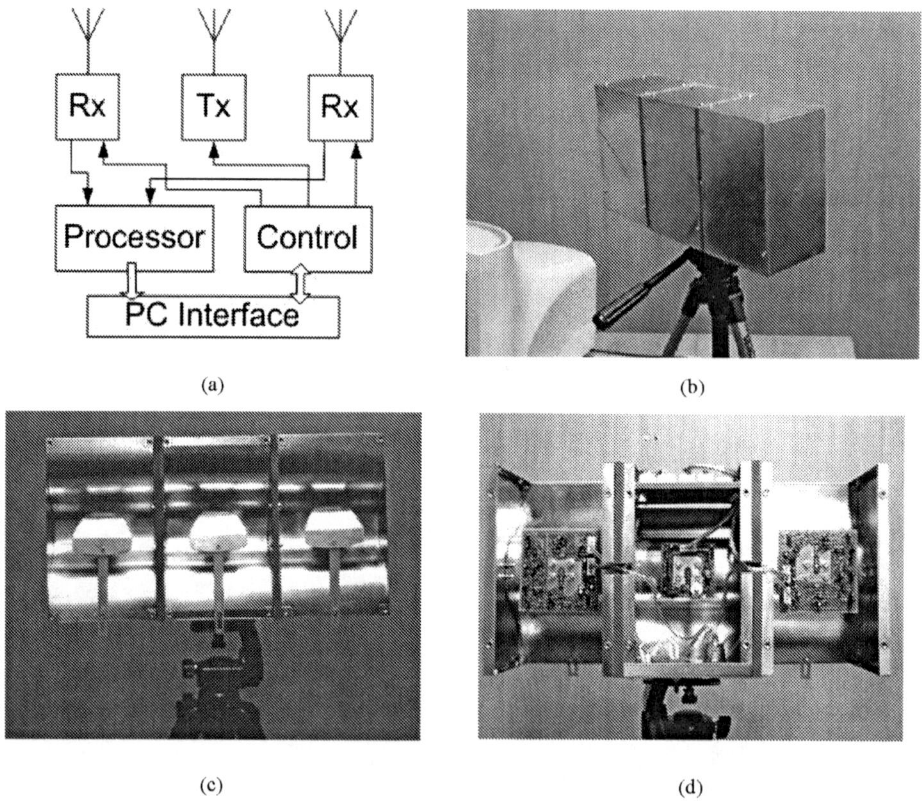

Figure 4. 1-Tx&2-Rx UWBR system: (a) block-schema; (b) assembled system installed on a tripod; (c) front - side view with open antenna radome; (d) back-side radar view with open cover and all shown PCBs.

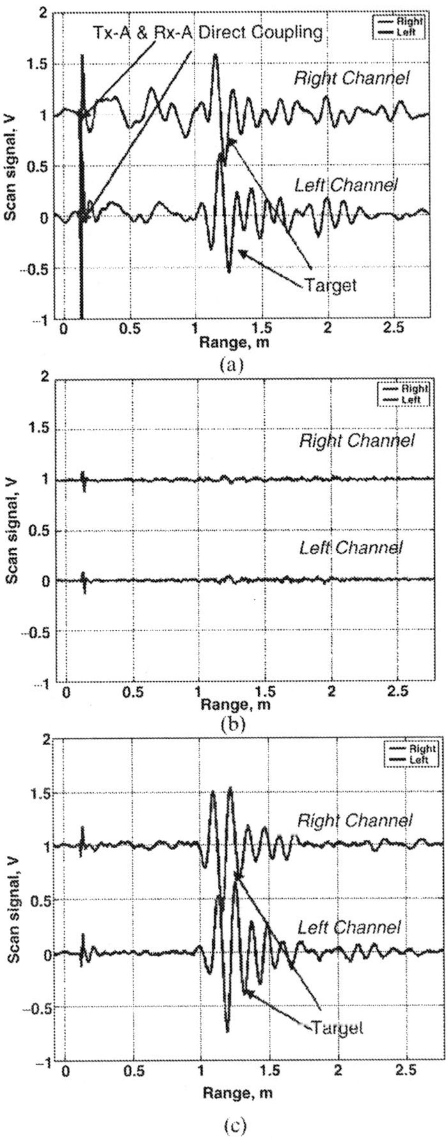

Figure 5. Rx-registered oscilloscope A-scan waveforms for two channels: (a) regular regime w/o background removing; (b) background removing over a stationary scene; (c) background removing over dynamic scene.

Similarly to the previous 1-Tx&1-Rx case, the PC Windows-compatible sound card is a key element for data collection in the 1-Tx&2-Rx radar prototype. A typical commercial sound card is capable of capturing two stereo audio channels at the linear input with low distortions and 16-bit resolution at a rate up to 100 kHz and higher. Simultaneously, very precise time synchronization is provided between the two Rx channels with the help of a card driver like that provided with Matlab Data Acquisition Toolbox[6]. This precise lockstep data capture allows a temporal comparison between the two channels, which is very important for the 2-D radar imaging algorithm discussed in Section 4.

For example, two captured radar Rx signals are plotted, Fig. 5a, in the A-scan oscilloscope mode for the regular radar regime when removing of background scattering is not applied. The target mark is caused by a person behind about the 30-cm concrete building wall. Likewise the 1-Tx&1-Rx UWBR, the 1-Tx&2-Rx radar can be run in the background compensation mode that produces nearly zero responses, Fig. 5b, at both the Rx channels for stationary scenes without dynamic targets. If something is changed over a scene under radar survey, a quite apparent target mark appears at both the Rx channels, Fig. 5c.

Note that for arbitrary located target the left and right channel responses, Fig. 5a&c, show some differences in terms of their (i) arriving time; (ii) signal peak amplitude, (iii) waveform shape and (iv) temporal duration. The first property plays a positive role for the next consideration. Other two three features are originated from aspect-depended scattering of range-distributed targets and variations in signatures of scattered signals caused by the angular-dependent antenna radiation patterns. These signal features are some kind of disturbing factors that complicate imaging with the 1-Tx&2-Rx UWBR system using the processing technique described in Section 4.

4. 2-D IMAGING OF SCENES WITH TWO-ELEMENT RX ARRAY

A simple 2-D imaging technique is studied[11], which is based on examining the ToA difference of the scattered signals at different test points in space. A number of such test points, viz. pixels, can be mapped in the radar field of view like sketched in Fig. 6a. Then, the precise difference in ToA for the Rx channels can to be converted to a DoA sort of data by using trigonometry, Fig. 6b, to determine the true location of those scattering-wave point-like sources. The trigonometry depends on several assumptions about the environment in which this system operates. It might be free-space that is an easiest case or in-matter propagation that leads to a number of complications. Also, this processing is subject to other constraints because the most of direction finding methods rely on the assumption that each wavefront impinging on the array is perfectly correlated between Rx sensors[22]. However, this condition is violated for the UWBR situations because the received signals in two channels undergo amplitude and phase variations as early discussed in connection to the experimental data in Fig. 5a&c. Thus, an angular calibration is required for the Rx channels to account for these signal angle-depended variations that is a part of our ongoing R&D efforts.

By now, we demonstrate the imaging algorithm with the linear array, Fig. 6b, to estimate the target position in 2-D plane by computing the windowed correlation of properly equalized channel signals with respect to each image pixel, Fig. 6a, to which we try to focus the image. The pixel position might correspond to the true position of a point-like target or a target scattering center. Several signal processing routines are involved to implement this algorithm including: (i) shift in time to equalize delays for each pixel in accordance to the geometrical arrangement in Fig. 6b; (ii) apply the weighted correlation to compute the pixel brightness. When the whole pixel map is plotted as in Fig. 7, the visual reception might help to decide on true target position based on the brightest-spot recognition technique that can be sufficient for simple scenes.

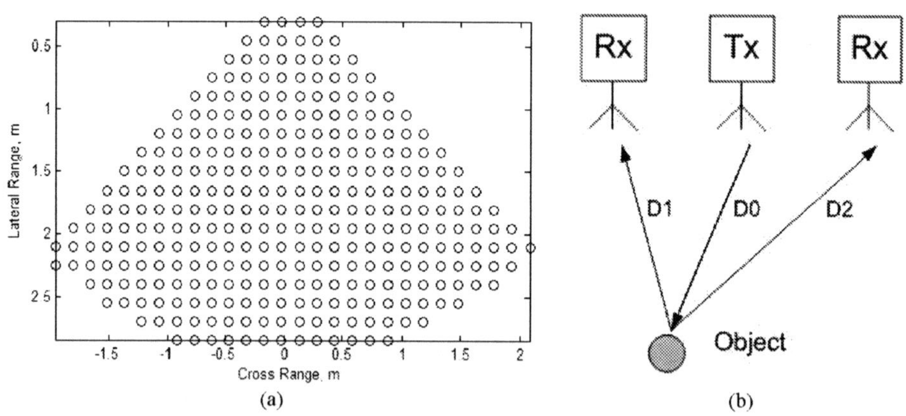

Figure 6. (a) Radar scene under survey as a numerical map of pixels. (b) Radar scene with one target to be imaged based on received waveforms at two Rx channels.

Formally, the "brightness" of any pixel *{i,j}* to which the received signals are focused can be computed as:

$$I_{i,j} = \int_{TW} W(t) \cdot s_{1,ij}(t) \cdot s_{2,ij}(t) dt \qquad (1)$$

where $W(t)$ is an appropriate weighting and time-domain windowing function[16], TW is an appropriate time gate for averaging over each pixel spot; $S_{1(2)}(t)$ are signals received at two radar channels numbered with 1 and 2:

$$s_{1(2),ij}(t) = S_{1(2)}(t - D0/c - D_{i,j}1(2)/c) \qquad (2)$$

which are equalized in time accounting for the corresponding distances in Fig. 6b, and c is the free-space speed of light. At practice, the algorithm, Eqs. (1)-(2) needs to be well-tuned by setting its proper $W(t)$ and TW parameters. For example, the TW parameter can be equal to the transmit pulse width, etc. Computing with the algorithm, Eq. (1), requires a definite amount of computer resources depending on the pixel map density and the scene size, Fig. 6a. In fact, the Matlab-based processing can support nearly-real time operation with fairly rough range/azimuth resolution cell of interest.

A sample of through-wall imaging for the same test case is shown in Fig. 7 including the synthetic data[17], Fig. 7a, and measured/computed image of the same target, Fig. 7b. This target is a stationary individual behind the 30-cm concrete wall. In fact, the target mark is widely spread in the azimuth direction because of small distance between two Rx channels that is a subject of constraint for any such a practical radar imaginary system to keep it compact in size.

5. CONCLUSIONS

We presented the UWBR hardware/software tools developed for detection and visualization of a single human subject standing or moving behind walls. We demonstrated several imaging algorithms including 2-D where resolution cell brightness is computed through windowed cross-correlation of equalized Rx channels.

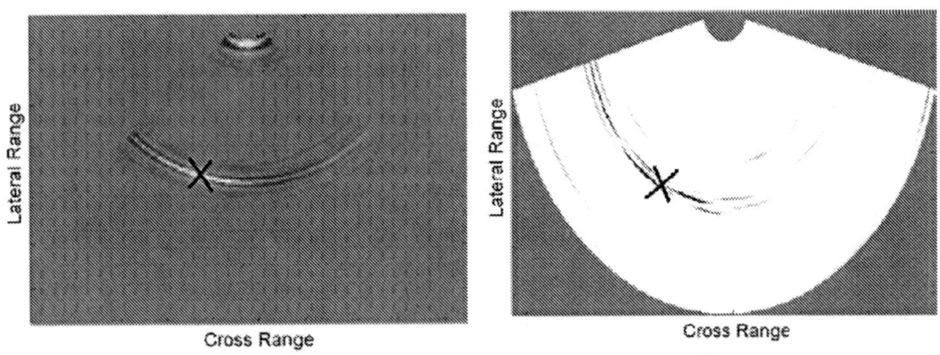

(a) (b)

Figure 7. (a) Synthetic (simulated) radar image. (b) Measured and computed radar image. Cross mark assigned the true target position at both images.

Our studies revealed that targets behind building walls made of gypsum, wood, concrete, brick, etc. can be tracked in range and azimuth direction using a set of simple hardware and software means like those used in the UWBR system with small physical aperture formed by two Rx antennas. Thus, the implemented UWBR hardware/software enables an uncomplicated solving of the inverse scattering problem. As well, apparent simplicity of the demonstrated prototypes makes them very useful as a testbed to examine new developed hardware and software radar components, and for student training. Additionally, the developed modular approach can be scaled-up to produce a multiple Tx and Rx channels.

Better radar imagery and visualization is required for effective through-the-wall radar surveillance of range/azimuth distributed targets compared to that shown with the examined UWBR prototypes. For this, more sophisticated signal/image processing algorithms and programming techniques can be used rather than those provided by Matlab. Future work will include also an in-depth study of the existing algorithms to incorporate them to process measured data for multiple-antenna and multiple-target scenarios.

6. REFERENCES

1. J. D. Taylor, Ed, *Ultra-Wideband Radar Technology*, (CRC Press, 2001).
2. J. J. Daniels, *Ground Penetrating Radar*, (IEE, 2004).
3. "RadarVision", Time Domain Corporation, www.radarvision.com.
4. A. O. Boryssenko, E. S. Boryssenko, *et al*, "Inspection of Internal Structure of Walls by Subsurface Radar", *Proceedings of SPIE*, **4084**, pp. 32-35 (2000).
5. A. O. Boryssenko, E. S. Boryssenko, *et al*, "UWB Subsurface Radar with Antenna Array for Imaging of Internal Structure of Concrete Structural Elements", *Ultra-Wideband, Short-Pulse Electromagnetics 5*, pp. 199-206 (Kluwer Academic/ Plenum Publisher, 2002).
6. A. O. Boryssenko, "Impulse Radar Emergency System to Prevent Damage due to Harmful Objects in Vegetation", *IEEE Int. Radar Conference*, pp. 116 – 121 (2000).
7. R. Benjamin, I. J. Craddock, E. McCutcheon and R. Nilavalan, "Through Wall Imaging using Real-Aperture Radar," *Proc. XXVII URSI General Assembly,* (2002).
8. S. Nag, M. Barnes, T. Payment, G.W. Holladay, "An Ultra-Wideband Through-Wall Radar for Detecting the Motion of People in Real Time", *Proceedings of SPIE*, **4744**, pp. 48-57, (2002).
9. A. R. Hunt, "A wideband imaging radar for through-the-wall surveillance," *Proceedings of SPIE,* **5403**, pp. 590-596 (2004).
10. H. E. Yacoub, T. K. Sarkar, "Monostatic Through-Wall Detection of a Metallic Sphere", *Proc. IEEE Ant. Propag. Symp.*, pp. 326-329 (2005).
11. A. O. Boryssenko, D. L. Sostanovsky, A. O. Boryssenko, "UWB Radar Imaging System with Two-Element Receiving Array Antenna", *Proc. Int. Conf. on Ant. Theory & Technique*, pp. 357-360 (2005).
12. D. L. Sostanovsky, A. O. Boryssenko, "An Experimental UWB Module-Based Sensing and Communication System", *Proc. UWB and Ultra Short Impulse Signals Workshop*, pp. 231-233 (2004).
13. A. O. Boryssenko, D. L. Sostanovsky, "A New Two-Flare-Shaped UWB Antenna Element", *Proc. Ant. Appl. Symp.*, pp. 451-471 (2005).
14. Mathworks, Inc, www.mathworks.com.
15. S. Nag, M. Barnes, "A Moving Target Detection Filter for an Ultra-Wideband Radar", *IEEE Radar Conference*, pp. 147-153 (2003).
16. E. C. Ifeachor, B.W. Jervis, *Digital Signal Processing, A Practical Approach*, (Addison-Weshley, 2001).
17. A. O. Boryssenko, "Simulation of Time-Domain Diffraction Phenomena for Subsurface Electromagnetic Sounding", *Proceedings IV Int. Seminar/Workshop on Direct and Inverse Problems of Electromagnetic and Acoustic Wave Theory*, pp. 117 – 121 (1999).

ON THE DEVELOPMENT OF A LOW-COST, COMPACT PLANAR INTEGRATED-CIRCUIT SAMPLING RECEIVER FOR UWB SYSTEMS

Jeongwoo Han, Rui Xu, and Cam Nguyen*

1. INTRODUCTION

Ultra-wideband (UWB) impulse radar and communication systems, which employ video (baseband) pulse as the transmitting signal [1]-[9], are attractive for various applications, such as pavement assessment, bridge-deck inspection, geophysical explorations, collision avoidance, fluid level sensing, detection and classification of unexploded ordnance (UXO) and land-mines, and short-range in-building communications, e.g. [10], [11].

Sampling receivers are a critical part in these UWB systems. For instance, in UWB impulse radars, such as subsurface penetrating radar, sub-sampling of the received signal by a sampling receiver is needed to extract detected information. Synchronous sampling (or equivalent-time sampling) receiver is commonly used in UWB systems due to its simple and compact structure. The synchronous sampling method has been widely used in electro-optic sampling technique to down-convert RF signal or reproduce fast transient signals on a large time scale [12]. Another application of samling receivers are for microwave instrumentation such as network analyzers, frequency counters and digitizing oscilloscopes [13]-[19]. In this application, the sampling mixer is used for sub-sampling of a fast transient signal to recover it on a large time scale and for down-conversion of a continuous-wave signal. Solid-state millimeter-wave and electro-optic samplers, which are able to sub-sample picosecond transient signals, have been developed for instruments operating in the millimeter-wave band [20]-[24]. The sampling receivers for UWB applications particularly require low conversion loss and high dynamic range because of their direct conversion operation.

In this paper, we report on the development of a new UWB synchronous receiver.

---

* Jeongwoo Han, Danam Systems, Inc., Anyang, South Korea. Rui Xu, Texas A&M University, College Station, Texas 77843. Cam Nguyen, Texas A&M University, College Station, Texas 77843

The developed receiver employs double-sided planar structure for low-cost fabrication and ease in integration with a strobe-pulse generator and a baseband circuit using hybrid MIC technique. The receiver achieves a conversion loss of 4.5-7.5 dB (without amplifier) and a conversion gain from 6.5-9.5 dB (with amplifier) over a 3-dB bandwidth of 5.5 GHz, which is, to our best knowledge, the best-reported conversion efficiency for sampling receivers at this frequency range. The receiver has a dynamic range of greater than 50 dB and a sensitivity of -47 dBm.

2. SYNCHRONOUS RECEIVER DESIGN

2.1. Receiver

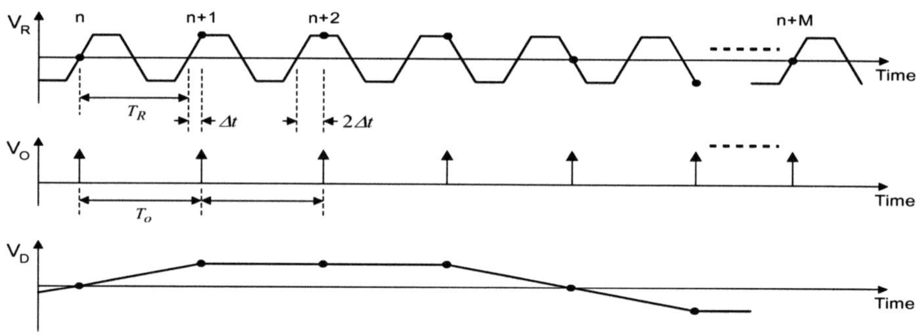

Figure 1. Illustration of the principle of synchronous sampling.

Fig. 1 illustrates the basic principle of the synchronous sampling method. In Fig. 1, V_R represents waveform of the input RF signal, V_O is the local oscillator (LO) strobe pulse signal to trigger the sampling, and V_D is the down-converted output signal through the (ideal) sample-and-hold operation of the receiver. To achieve down-conversion of the RF input V_R, the frequency of V_R and PRF of V_O should have a small difference between them. For UWB impulse radar systems, V_R is actually a pulse signal having a PRF, f_R, and an extremely large duty cycle. For down-converting or stretching of the RF pulse, the PRF of the LO strobe pulse, f_o, should have a small deviation from f_R.

Fig. 2 shows the synchronous receiver architecture together with a transmitter to demonstrate the synchronous sampling principle in an UWB system. The synchronous sampling is implemented by using two reference clock oscillators: one for the receiver and another for the transmitter, with a small frequency deviation of Δf between them. The oscillator in the transmitter determines the pulse repetition frequency (PRF) of the transmitting pulse, which is also the PRF of the RF input pulse to the receiver. The receiver's oscillator is used for the PRF of the LO strobe pulse generator of the receiver. Main components of the receiver are the sampling mixer, strobe pulse generator, reference clock oscillator, and baseband filter and amplifier. A low-noise amplifier (LNA) may be needed at the sampling mixer's input port if the received signal from the receiving antenna is too weak to be detectable or triggered in the baseband circuit.

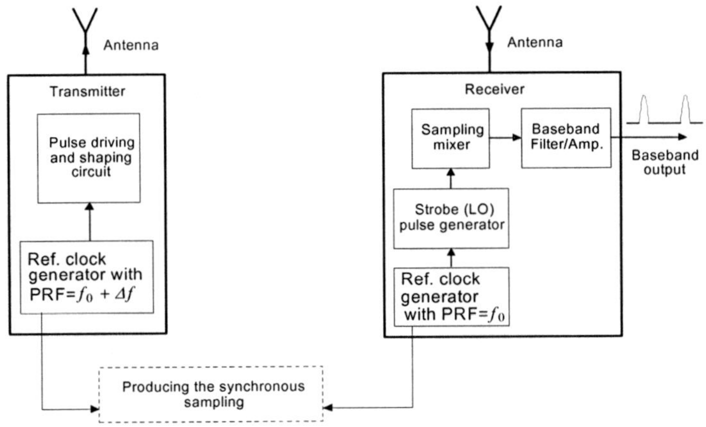

Figure 2. Block diagram of the synchronous receiver (right) together with a transmitter (left) in an UWB system.

2.2. Strobe Pulse Generator and Sampling Mixer

The strobe pulse generator and sampling mixer are the two most important components in the receiver. They are designed using step recovery diodes (SRD), Schottky diodes, microstrip line, coplanar waveguide (CPW) and coupled slotlines [25]-[26]. The reference clock generator is implemented using a commercially available low-cost VC-TCXO (Voltage Controlled Temperature Controlled Crystal Oscillator). It has about 10 ppm of frequency adjustment through voltage control, resulting in 100 Hz frequency variation needed for our design, and a very low phase noise of -135 dBc/Hz @ 1 KHz. The baseband amplifier is a commercial opamp having 14-dB gain. Table 1 summarizes the design parameters for the synchronous receiver.

Table 1. Design parameters for the synchronous receiver.

Reference clock frequency, fo	10 MHz
Equivalent sampling time interval, Δt	1 ps
Deviation frequency, Δf	100 Hz
Sample rate of ADC, fs	220 KHz
Samples/scan	2.2 K

2.2.1. Strobe Pulse Generator

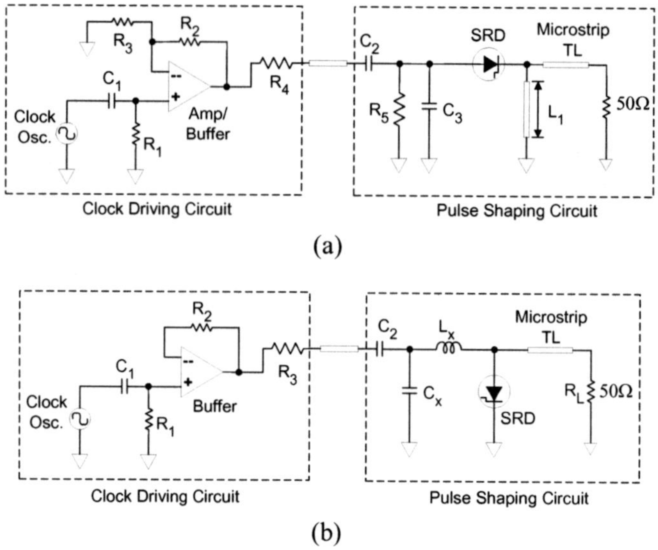

Figure 3. Circuit diagrams for (a) the delay-line and (b) shunt-mode pulse generators.

Two different types of SRD impulse generators were designed based on well-known circuit configurations [27]-[28]. One is the delay-line type and the other is the shunt-mode type. Designed circuit diagrams, including clock driving circuits, for these two different types are shown in Fig. 3. These pulse circuits were optimized to obtain high pulse power for low pulse repetition frequency (PRF) with no external DC bias required for normal operation. Low PRF is needed for various applications. For instance, in UWB radar requiring sub-sampling in the receiver, the transmitted pulse's PRF is determined by the strobe pulse's PRF in the receiver, which may be as low as 10 MHz in order to have sufficient observation time and detection range.

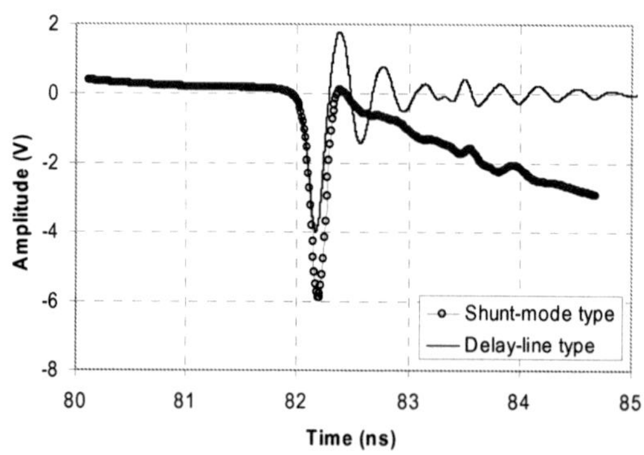

Figure 4. Measured output pulse waveforms of the delay-line and shunt-mode pulse generators.

Fig. 4 shows the measured output impulse waveforms of the delay-line and shunt-mode pulse generators fabricated using microstrip lines on RT/duroid 6010 substrate with a relative dielectric constant of 10.2, a thickness of 0.127 cm and a loss tangent of 0.0023. The FWHM (Full Width Half Maximum)'s of the generated pulses are about 150 ps. The pulse waveform of the delay-line pulse generator has large ringing even though the pulse duration is shorter than that of the shunt-mode pulse generator. Large ringing can cause some distortion on the down-converted output of the sampling mixer. Its amplitude is also smaller than that of the shunt-mode type while the slew-rates of both pulses are the same. For pulses of the same slew-rate, actual sampling aperture times, which determine the bandwidth of the sampler, are almost the same because of the clamping effect of the sampling-bridge circuit. Furthermore, for the same slew-rate pulses, a higher amplitude pulse improves the conversion loss and increases the 1-dB compression point of the sampler. Therefore, the shunt-mode type pulse generator is a better choice for the sampling mixer in all aspects.

2.2.2. Sampling Mixer

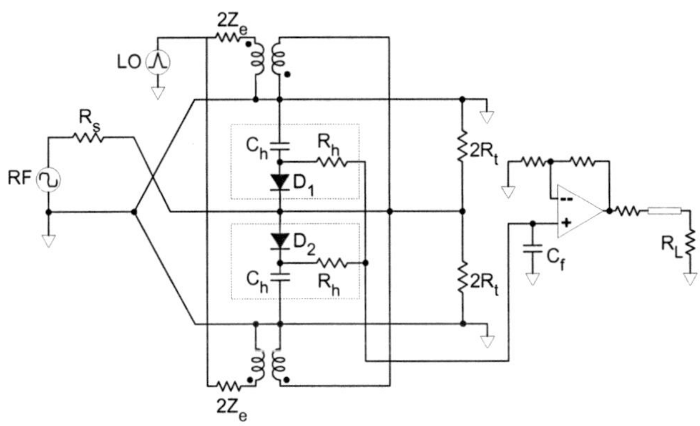

Figure 5. Equivalent circuit diagram of the sampler.

Fig. 5 shows the sampling mixer's equivalent-circuit diagram, including the baseband buffer and amplifier circuit. The equivalent circuit follows the split-ground configuration [29] for reducing the parasitic inductance of the signal transmission line. The sampling mixer is realized using the same substrate as the strobe pulse generator.

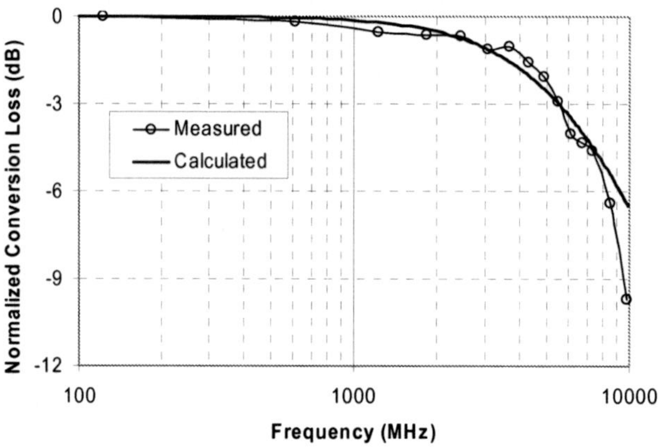

Figure 6. Measured and calculated normalized conversion losses of the sampler.

Fig. 6 shows the measured and calculated relative (or normalized) conversion loss of the sampling mixer (without the baseband amplifier). The measured conversion loss is normalized to the minimum conversion loss of the sampler, which was measured as 4.5 dB.

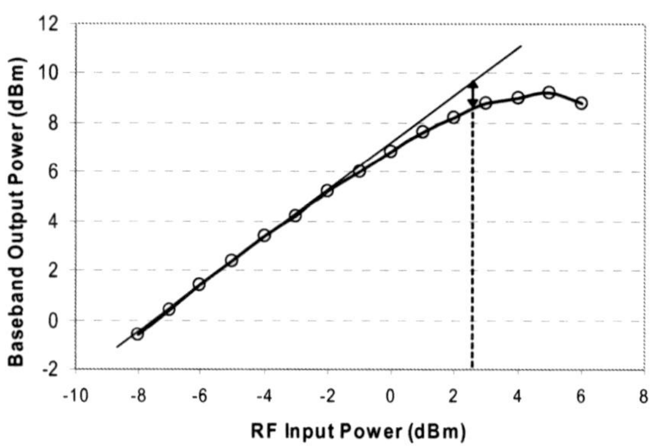

Figure 7. Measured baseband output power of the sampler.

Table 2. Summary of the performance of the designed sampler.

Conversion loss (-) and Gain (+)	-7.5 to -4.5 dB (without amp.) 6.5 to 9.5 dB(with amp.)
Baseband Amp. Gain	14 dB
RF Input Bandwidth	5.5 GHz
LO clock source PRF	12 MHz

Spurious	20 dBc typical
1-dB Compression	2.5 dBm
8-dB Sensitivity	-47 dBm
Dynamic Range	> 50 dB
RF VSWR	1.3:1

As can be seen in Fig. 6, the measured 3-dB bandwidth of the sampler is 5.5 GHz, which is in good agreement with the calculated one. With the measured 4.5-dB minimum conversion loss, the sampler exhibits a conversion loss from 4.5-7.5 dB for the RF signal from DC to 5.5 GHz and the baseband signal of 20 KHz. The sampler with the 14-dB-gain (opamp) baseband amplifier shows a measured conversion gain from 6.5 to 9.5 dB over the 5.5-GHz RF bandwidth.

Fig. 7 shows the measured baseband output power as a function of the RF input power at the RF frequency of 3 GHz, which is near the center frequency of the designed sampler. The measured 1-dB compression point is 2.5 dBm. For the RF signal of 3 GHz and 0 dBm, the level of the spurious signal, which is the second harmonic of the down-converted signal, was measured as low as 20 dBc. The sampler's sensitivity, defined as 8-dB tangential sensitivity, was determined as -47 dBm. The measured dynamic range of the sampler is more than 50 dB. A summary of the sampler's performance is shown in Table 2.

2.3. Receiver Fabrication and Measurement

The synchronous receiver was fabricated using RT/duroid 6010 substrate with a relative dielectric constant of 10.2, thickness of 0.127 cm and loss tangent of 0.0023. Fig. 8 shows a photograph of the synchronous receiver.

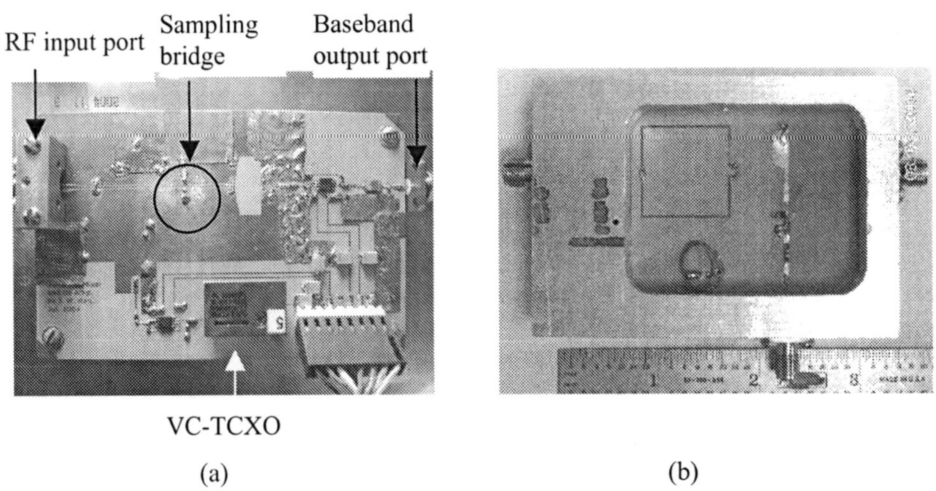

Figure 8. Photograph of the receiver top (a) and bottom (b).

Performance of the receiver is measured in terms of the quality of its reproduced baseband signal as compared with that produced by a commercial sampling oscilloscope. Fig. 9 shows the down-converted signals measured with the developed receiver and the Agilent digitized sampling oscilloscope and the error between them. As can be seen, the two down-converted signals match very well. The root mean square (RMS) error between these signals is only 0.035, signifying that the difference between the two down-converted signals is only 3.5 % of the signal's peak amplitude. These results demonstrate that the receiver can reconstruct the incoming pulse signals with negligible error.

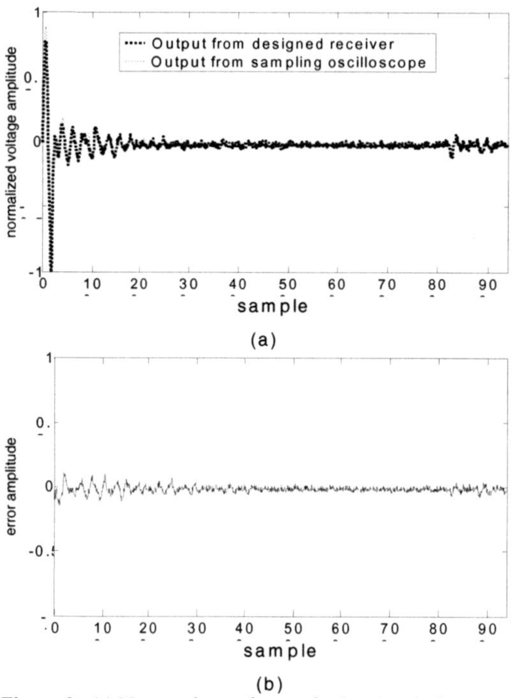

Figure 9. (a) Measured waveforms of pulse signals down-converted by the designed receiver and the digitized sampling oscilloscope. (b) Error between the two measured waveforms.

3. CONCLUSION

A new UWB synchronous receiver has been developed for UWB applications. The receiver achieves an unprecedented conversion loss of 4.5-7.5 dB (without baseband amplifier) and conversion gain from 6.5-9.5 dB (with amplifier) across a 5.5-GHz RF bandwidth, a dynamic range of more than 50 dB and low harmonic distortion in the baseband output signal. The receiver is realized in a single circuit board using microstrip line, CPW, slot line, and coupled slot lines, and is compact and low cost. Its performance in down-converting signals is comparable to a commercial digitized sampling scope with much significantly smaller size and lower cost. The low cost, high performance and small size make it very attractive for UWB applications in radar, communications, and instrumentation.

4. ACKNOWLEDGEMENT

This work was supported in part by the National Science Foundation, the Texas Transportation Institute, the Southwest Region University Transportation Center, the Texas Advanced Research Program and Dell Computer Corporation.

5. REFERENCES

1. J. S. Lee and C. Nguyen, "A low-cost uniplanar sampling down-converter with internal local oscillator, pulse generator, and IF amplifier," *IEEE Trans. Microwave Theory and Techn.*, vol. 49, no. 2, pp. 390-392, Feb. 2001.
2. J. S. Lee, C. Nguyen and T. Scullion, "Impulse ground penetrating radar for nondestructive evaluation of pavements," in *2002 IEEE MTT-S Int. Microwave Symp. Dig.*, 2002, pp. 1361-1363.
3. S. Abuasaker and G. Kompa, "A high sensitive receiver for baseband pulse microwave radar sensor using hybrid technology," in *2002 Proc. IEEE Radar Conference*, 2002, pp. 121-124.
4. J. D. Taylor and E. C. Kisenwether. "Ultra-wideband radar receivers," in *Introduction to Ultra-Wideband Radar Systems*, J. D. Taylor, Ed. Boca Raton, FL: CRC Press, 1995, Chapter 10, pp. 491-577.
5. D. J. Daniels, *Surface-Penetrating Radar*. London, U.K.: IEE, 1996, Chapter 5, pp. 104-119.
6. S. Azevedo and T. E. McEwan, "Micropower impulse radar," *Science & Tech. Review*, pp. 17-29, Jan./Feb. 1996.
7. D. M. Akos and J. B. Y. Tsui, "Design and implementation of a direct digitization GPS receiver front end," *IEEE Trans. Microwave Theory and Techn.*, vol. 44, no. 12, pp. 2334-2339, Dec. 1996.
8. A. Pärssinen, R. Magoon, S. I. Long, and V. Porra, "A 2-GHz subharmonic sampler for signal downconversion," *IEEE Trans. Microwave Theory and Techn.*, vol. 45, no. 12, pp. 2344-2351, Dec. 1997.
9. J. Han and C. Nguyen, "Integrated balanced sampling circuit for ultra-wideband communications and radar systems," *IEEE Microwave Wireless Compon. Lett.*, vol. 14, no. 10, pp. 460-462, Oct. 2004.
10. R. J. Fontana. (2000, May). "Recent applications of ultra wideband radar and communications systems," in *Ultra-Wideband, Short-Pulse Electromagnetics 5*, P. D. Smith and S. R. Cloude, Ed. New York: Kluwer Academic/Plenum Publishers, 2002, pp. 225-234.
11. Time Domain Corp., AL. PulsON Technology: Time modulated ultra-wideband for wireless applications.[Online].Available:http://www.timedomain.com/Files/downloads/techpapers/PulsONoverview.pdf.
12. M. Kamegawa, K. Giboney, J. Karin, S. Allen, M. Case, R. Yu, M. J. W. Rodwell, and J. E. Bowers, "Picosecond GaAs monolithic optoelectronic sampling circuit," *IEEE Photonics Tech. Lett.*, vol. 3, no. 6, Jun. 1991, pp. 567-569.
13. W. M. Grove, "Sampling for oscilloscopes and other RF systems: DC through X-band," *IEEE Trans. Microwave Theory and Techn.*, vol. MTT-14, no. 12, pp. 629-635, Dec. 1966.
14. J. Merkelo and R.D. Hall, "Broad-band thin-film signal sampler," *IEEE J. Solid-State Circuits*, vol. SC-7, no. 1, pp. 50-54, Feb. 1972.
15. A. Bologlu, "A 26.5-GHz automatic frequency counter with enhanced dynamic range," *Hewlett-Packard J.*, pp. 20-22, Apr. 1980.
16. B. E. Gilchrist, R. D. Fildes, and J. G. Galli, "The use of sampling techniques for miniaturized microwave synthesis applications," in *1982 IEEE MTT-S Int. Microwave Symp. Dig.*, 1982, pp. 431-433.
17. S. E. Moore, B. E. Gilchrist, and J. G. Galli, "Microwave sampling effective for ultrabroadband frequency conversion," *MSN & CT*, pp. 113-126, Feb. 1986.
18. S. R. Gibson, "Gallium arsenide lowers cost and improves performance of microwave counters," *Hewlett-Packard J.*, pp. 4-10, Feb. 1986.
19. K. Madani and C. S. Aitchison, "A 20 GHz microwave sampler," *IEEE Trans. Microwave Theory and Techn.*, vol. 40, no. 10, pp. 1960-1963, Oct. 1992.
20. R. A. Marsland, V. Valdivia, C. J. Madden, M. J. W. Rodwell, and D. M. Bloom, "130 GHz GaAs monolithic integrated circuit sampling head," *Appl. Phys. Lett.*, vol. 55, no. 6, pp. 592-594, Aug. 1989.
21. M. S. Shakouri, A. Black, B. A. Auld, and D. M. Bloom, "500 GHz GaAs MMIC sampling wafer probe," *Electron. Lett.*, vol. 29, no. 6, pp. 557-558, Mar. 1993.
22. K. J. Weingarten, M. J. W. Rodwell, and D. M. Bloom, "Picosecond optical sampling of GaAs integrated circuits," *IEEE J. Quantum Electron.*, vol. 24, no. 2, pp. 198-220, Feb. 1988.

23. M. J. W. Rodwell, M. kamegawa, R. Yu, M. Case, E. Carman, and K. S. Giboney, "GaAs nonlinear transmission lines for picosecond pulse generation and millimeter-wave sampling," *IEEE Trans. Microwave Theory and Techn.*, vol. 39, no. 7, pp. 1194-1204, Jul. 1991.
24. Y. Konishi, M. Kamegawa, M. Case, R. Yu, S. T. Allen, and M. J. W. Rodwell, "A broadband free-space millimeter-wave vector transmission measurement system," *IEEE Trans. Microwave Theory and Techn.*, vol. 42, no. 7, pp. 1131-1139, Jul. 1994.
25. J.W. Han and C. Nguyen, "Coupled-Slotline-Hybrid Sampling Mixer Integrated with Step-Recovery-Diode Pulse Generator for UWB Applications," *IEEE Trans. on Microwave Theory and Techniques*, vol. MTT-53, no. 6, June 2005, pp. 1875-1882.
26. J.W. Han, R. Xu and C. Nguyen, "Development of a Low-Cost, Compact Planar Synchronous Receiver for UWB Systems," *IEEE AP-S International Symposium & USNC/URSI National Radio Science Meeting*, Albuquerque, New Mexico, July 9-14, 2006
27. S. Hamilton and R. Hall, "Shunt-mode harmonic generation using step recovery diodes," *Microwave Journal*, pp. 69-78, Apr. 1967.
28. J. Han and C. Nguyen, "A new ultra-wideband, ultra-short monocycle pulse generator with reduced ringing," *IEEE Microwave Wireless Compon. Lett.*, vol. 12, no. 6, pp. 206-208, Jun. 2002.
29. W. M. Grove, "Sampling for oscilloscopes and other RF systems: DC through X-band," *IEEE Trans. Microwave Theory and Techn.*, vol. MTT-14, no. 12, pp. 629-635, Dec. 1966.

TIME REVERSAL WITH SINGLE ANTENNA SYSTEMS IN INDOOR MULTIPATH ENVIRONMENTS

Zhengqing Yun and Magdy F. Iskander*

1. INTRODUCTION

The time reversal (TR) technique has been well investigated for sound wave applications.[1,2] Recently, the same idea has been extended to telecommunications[3] and the electromagnetic waves.[4,5] Lerosey, et al., reported that the spatial focusing and time compression were experimentally achieved in a reverberant chamber with a one transmit and one receive antenna system (1 x 1 antenna system). TR has also been employed in other areas of research, e.g., the imaging of targets in complex environments[6,7] and microwave nulling.[8]

This research focuses on the characterization of TR with single antenna systems. To achieve energy focusing using a single antenna system, wideband signals and multipath environments are required. We examine the TR property in indoor multipath environments which have typical propagation features such as the waveguide effect of hallways, multiple reflection and transmission of slab walls, diffraction from edges (especially metal edges), and scattering from various small structures. The signal waveform is a wideband monocycle. The two-dimensional finite-difference time-domain (FDTD) method is employed for the simulation of the wave propagation in an indoor environment. The effect of furniture and other scattering objects in the region of interest is investigated. To quantitatively characterize the spatial energy focusing, a spatial energy focusing factor (SEF) is defined. It is found that spatial energy focusing can be achieved in the indoor environment and that the scatterers can significantly improve the spatial energy focusing.

We also employed the ray-tracing method to simulate the spatial focusing of TR signals. Since our ray-tracing algorithm is in frequency domain, the simulation scheme is different from that of the FDTD which is a time domain method. First, frequency

* University of Hawaii, Hawaii Center for Advanced Communications, 2640 Dole St. Holmes 240, Honolulu, HI, 96822

sweeping is needed to get wideband signal responses. Second, the time-reversal process is also carried out in frequency domain by conjugating the phases of the received signal at a receive location. Time domain impulse response can be obtained via the inverse Fourier transform. One of the advantages of the ray-tracing method is that it can simulate problems with large areas such as urban and other outdoor/indoor regions where the FDTD method becomes computationally prohibitive. We have developed new ray-tracing algorithms based on space division which have been shown accurate and most importantly efficient.[9] It can be easily modified to accommodate the frequency sweeping needed for TR simulation.

2. THE MULTIPATH ENVIRONMENTS

We first consider a two-dimensional indoor environment (38m x 28m) as shown in Fig. 1. The walls are assumed homogeneous slabs with the same material (ε_r = 3) and thickness (6 cm). Two cases are considered. In the first case, only the walls are present (called the walls-only case). In the second case, in addition to the walls, metal furniture including bookcases, flat panel computer monitors, metal door frames, and cubical support metal bars are considered (called the full-scattering case). The dimensions of these metal structures are shown in Fig. 2.

The transmit antenna Tx is located in a hallway as shown in Fig. 1 (the square). Thirteen receiver (Rx) locations are selected to represent hallways, small rooms, big rooms, and different distances to the transmitter.

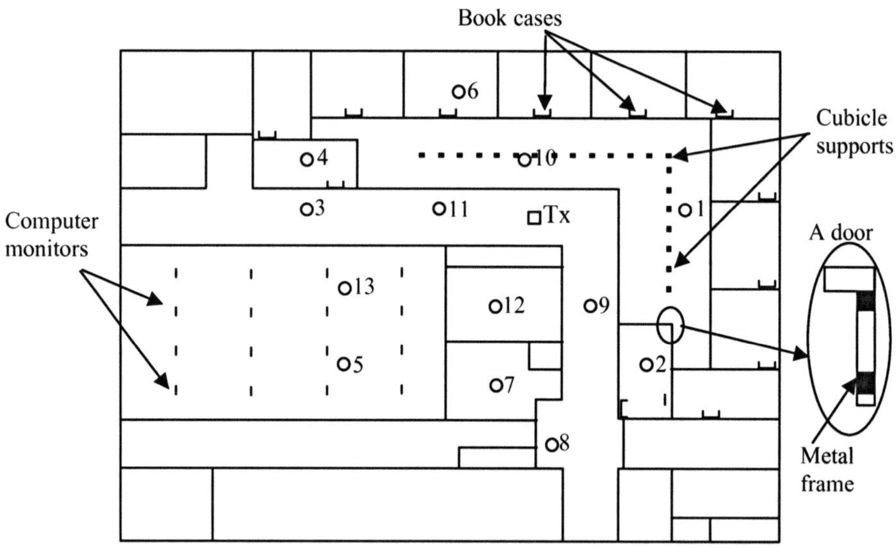

Figure 1. Floor plan of a realistic office building with one transmitter Tx (square) and 13 receivers (circles). The receivers are placed at different locations to help examine the feasibility of using single-antenna TR systems in different multipath environments. Besides walls, other metal scattering objects such as computer monitors (flat panels), door frames, book cases, and cubicle supports are also shown.

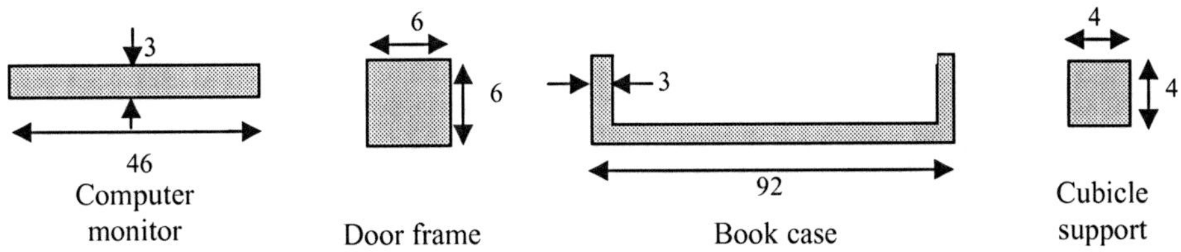

Figure 2. Geometries and dimensions of the metal scatterers. All dimensions are in centimeters.

3. SIGNALS SENT FROM TX AND RECEIVED AT RX

A Gaussian monocycle is employed as the source pulse to be sent from Tx, which can be represented by $(1-t^2/\sigma^2)\exp(-t^2/2\sigma^2)$ with $\sigma = 0.4$ ns. The -3 dB bandwidth of the pulse is found to be 463 MHz and the center frequency is 562.7 MHz. We selected this frequency band to make sure that a larger region can be simulated using the FDTD method.

The entire region is discretized with square cells. The cell dimension is 1 x 1 cm^2. Since the wavelength of the center frequency is 53.3 cm, the cell size can guarantee accurate results.[10] The received signal at each of these 13 Rx locations is obtained using the FDTD method. In Fig. 3, typical received signals are shown for Rx location # 1 (Rx1) for the walls-only and the full-scattering cases. It can be seen from Fig. 3 that the multipaths for the full-scattering case is different from that of the walls-only case. Furniture and other objects in the full-scattering case usually increase the number of multipaths and thus provide a richer multipath environment, which will help the energy focusing of the time-reversed signal.

To quantitatively characterize the multipath, we calculated the r.m.s. delay spread for the received signals at 13 receiver locations. It is found that in all of the 13 locations, the furniture objects increase the delay spreads. The average delay spread for the walls-only cases is 0.12 μs; while for the full-scattering cases, it is 0.14 μs.

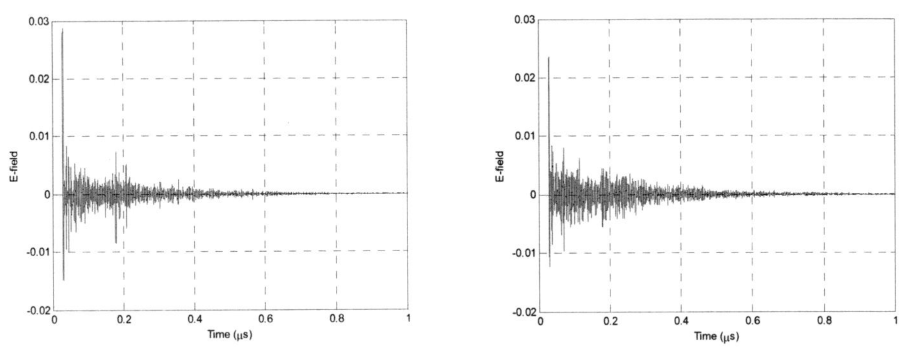

(a) Walls only (b) Full scattering

Figure 3. Typical signals received at Rx1.

4. FDTD SIMULATION RESULTS

The received signal at a receiver is time reversed and sent into the same multipath environment from the receiver. The time-reversed signals are recorded at the transmit antenna Tx. In the following, we will first show the impact of receiver locations on the shape of the time-reversed signals received at the transmitter location. Then we will show how the scatterers help focus the energy of the time-reversed signal at Tx.

There are 13 receiver locations (Rx1 to Rx13 as shown in Fig. 1) each with different surrounding environments which provide different multipath scenarios. Some typical time-reversed signals received at Tx are shown in Fig. 4 where signals in walls-only and full-scattering cases are compared with the source waveform. Note that we set the time equal to zero when maximum value of the received signal at Tx is attained and that the magnitudes of all the waveforms are normalized.

It can be seen from Fig. 4 that the TR waveforms recorded at Tx are symmetric about the time center ($t = 0$). It can also be seen that when more scattering objects are present (the full-scattering case) the shapes of the TR signals received at Tx are more similar to the shape of the source waveform (monocycle). Note that this improvement of the signal shape is dependent on the multipath environments surrounding the receiver locations. The improvement for the case when the receiver is located at Rx7 is less than the case when the receiver is located at Rx1 as shown in Fig. 4. This can be explained by identifying where the sctterers are introduced in the walls-only environment. It can be seen from Fig. 1 that scatterers are directly added between Rx1 and Tx which create a significant impact on the multipath environment. It can also be seen that the scatterers only have small impact on the multipath environment between Rx7 and Tx, leading to negligible improvement of the shape of the time-reversed signal.

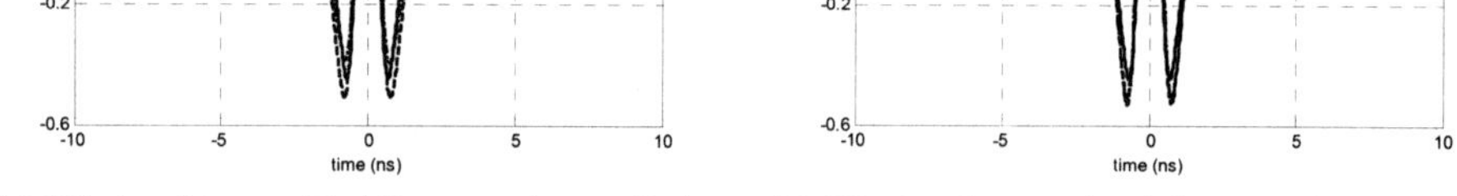

(a) TR signal transmitted from receiver at Rx1 (b) TR signal transmitted from receiver at Rx7

Figure 4. Received signal at Tx for the time interval [-10, 10] ns.

Fig. 5 (a) and (b) show the spatial field distribution in the neighborhood of Tx which is a square of the size 2m x 2m when the receiver at Rx1 is transmitting. It can be seen that scatterers significantly improve the field (or energy) focusing at Tx. It should be noted that the high field in the walls-only case formed a ridge in the Tx neighborhood which represents the wave front emanated from Rx1. For the full scattering case, this wave front is not as obvious as in the walls-only case, indicating the rays arriving at Tx are more uniformly spreading out in the domain of angle of arrival.

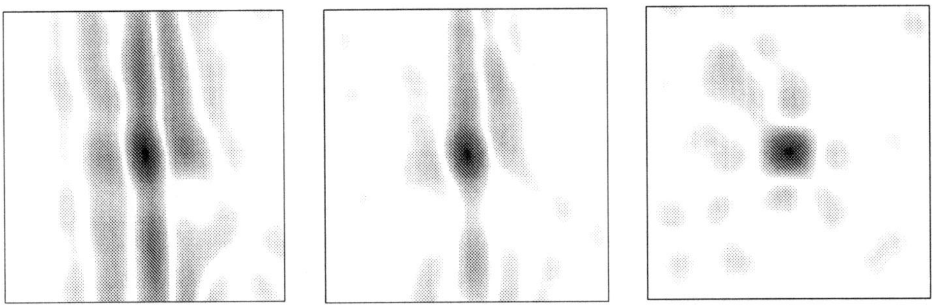

(a) Walls only (transmitting from Rx1) (b) Full scattering (transmitting from Rx1) (c) Full scattering (transmitting from Rx5)

Figure 5. Spatial field distribution around Tx when Rx1 is transmitting (darker areas represent higher electric fields and lighter areas represent lower electric fields).

Fig. 5(c) shows the field distribution when the TR signal is sent from Rx5 for the full-scattering case. It can be seen that the spatial focusing is better than the case where the TR signal is sent from Rx1 location. We could not see any ridge-like wave front. This can be explained by noticing that more scattering objects are present between Tx and Rx5 than that between Tx and Rx1. A more uniform distribution of angle-of-arrival of rays arriving at Tx when receiver at Rx5 is sending the TR signal, thus the better spatial focusing of energy.

To quantitatively characterize the spatial focusing, we define the spatial energy focusing factor (SEF), S_F. First, we can calculate the energy in a circular region centered at Tx and with radius r as $E_S(r) = \iint_{S(r)} E^2(x, y)dxdy$ where $S(r)$ is the circular region of radius r. The spatial focusing factor is then defined by $S_F(r) = E_S(r) / E_S(R)$ where R = const. is the radius of a reference circular region (and $0 \le r \le R$). We also define the relative half energy radius (HER), h_{er}, as follows. First, we find a radius $r_{50\%}$ that satisfies $S_F(r_{50\%}) = 0.5$; then the relative half energy radius is calculated as $h_{er} = r_{50\%}/R$. Obviously, the smaller the HER, the more focused the energy at Tx.

Selecting $R = 1$ m, Fig. 6 (a) compares the spatial focusing factors for walls-only and full-scattering cases when Rx1 is transmitting. It can be seen that when scatterers are present, 50 % of the energy is stored in the range of $r = 0$~20 cm, which gives $h_{er} = 0.2$; while for the walls-only case, 50 % of the energy is stored in the region with $r = 0$~45 cm, which gives $h_{er} = 0.45$. This result indicates that scatterers significantly help the spatial energy focusing for the time-reversed signals.

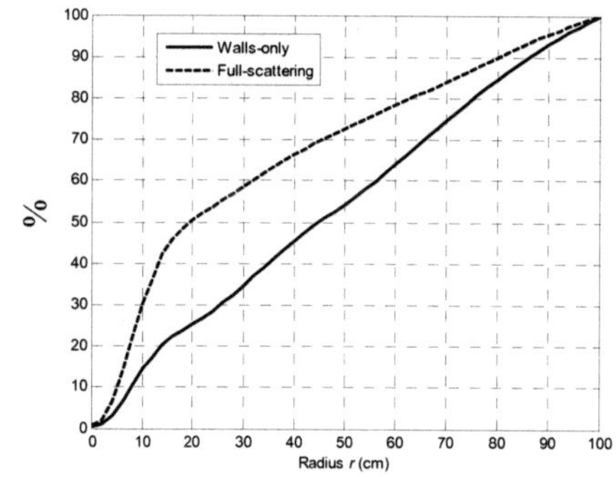

(a) When receiver at Rx1 is transmitting

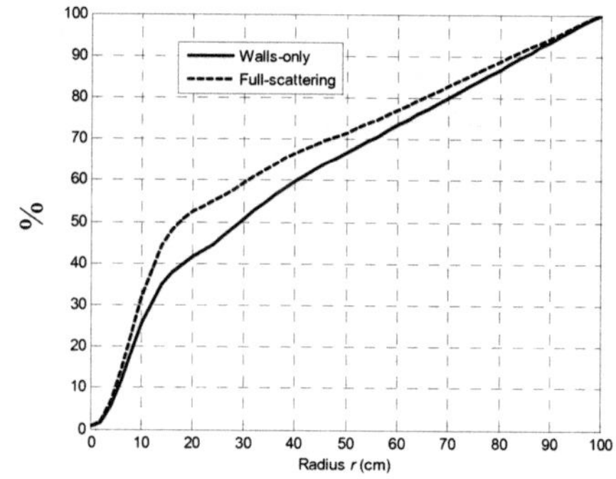

(b) Average value over all 13 Rx locations

Figure 6. The spatial energy focusing factors

Fig. 6 (b) shows the average values of the spatial energy focusing factor (SEF) for all of the 13 Rx locations. It can be observed that scatterers always improve the spatial focusing of the time-reversed signals. The HER for the full-scattering case is 0.18 while for the walls-only case it is 0.3 which is 66.7 % larger.

5. SIMULATION OF TIME REVERSAL USING RAY-TRACING METHOD

The FDTD method can efficiently simulate environments with small areas. For urban and other situations with large areas, ray-tracing method provides a more computationally efficient approach. Recent advances in ray-tracing methods have further improved its computational efficiency using techniques such as the space-division method.[9] In this section we show the application of ray-tracing method in the characterization of time reversal signals. Note that in our simulations, we use an indoor environment to show the capability of the ray-tracing method. The ray-tracing method can be used for urban and larger areas without fundamental difference.

The procedure using ray-tracing method to calculate the TR signal at a transmitter Tx is as follows. First, determine all the rays arriving at a receiver location Rx. These ray paths are independent of frequencies. But the field associated with each ray is frequency dependent. Second, sweep the frequency and calculate the fields at Rx as a function of frequency, f. Suppose for each frequency component, the source strength is E_0 and there are N rays arriving at Rx, the field at Rx can be expressed as

$$E(f) = \sum_{i=1}^{N} A_i(f)e^{-j\varphi_i(f)}E_0 = A(f)e^{-j\varphi(f)} \tag{1}$$

where A_i is the field gain and φ_i the phase shift of the ith ray; and A is the magnitude and φ the phase of the resultant field received at Rx. Third, time reverse the received signal at Rx. In frequency domain, this process is equivalent to the inverse of the complex field

$E(f)$. The time-reversed signal to be sent from receiver Rx can thus be represented by $E_{tr}(f) = 1/E(f) = [1/A(f)]e^{j\varphi(f)}$. Note that the magnitude is inversed and the phase is conjugated. When E_{tr} is sent from Rx, it will travel through the N rays back to the transmitter. Since the gain and the phase shift for each ray will be the same as before, the TR signal received at Tx can be written as $E_{tr}^{Tx}(f) = \sum_{i=1}^{N} A_i(f)e^{-j\phi_i(f)}E_{tr}(f)$ which is, noting that $E_{tr}(f)$ can be factored out from the summation, $E_{tr}^{Tx}(f) = 1/A(f)e^{j\varphi(f)}\sum_{i=1}^{N} A_i(f)e^{-j\phi_i(f)}$. From (1) we have $\sum_{i=1}^{N} A_i(f)e^{-j\varphi_i(f)} = A(f)e^{-j\varphi(f)}/E_0$ and thus $E_{tr}^{Tx}(f) = E_0$, which is exactly the original source magnitude.

Fourth, to characterize the spatial focusing, it is necessary to calculate the time-reversed signal in the neighborhood of Tx. In this case, the ray trajectories will be different and a full ray-tracing process should be carried out to calculate the field at a nearby location from Tx. Finally, time-domain impulse response can be calculated using the inverse Fourier transform.

An example: Fig. 7(a) shows an indoor environment with the locations of the transmitter (Tx) and receiver (Rx). A nearby location, p, is 14 cm to the Tx. Using our ray-tracing algorithm,[9] received signals at Rx are calculated in the frequency band of 3 to 10 GHz with frequency step of 0.2 MHz. The total number of frequency points is 35,001. Then the received signal is time reversed (in frequency domain) and sent from Rx. The frequency-domain TR signals at Tx and Rx are then calculated.

Applying the inverse Fourier transform to the frequency-domain signals, the impulse responses of the time reversed signal at Tx and p can be obtained as shown in Fig. 7(b). It can be seen that the magnitude of the impulse response at Tx is much more significant than that at p, showing the spatial energy focusing of the TR signal at Tx.

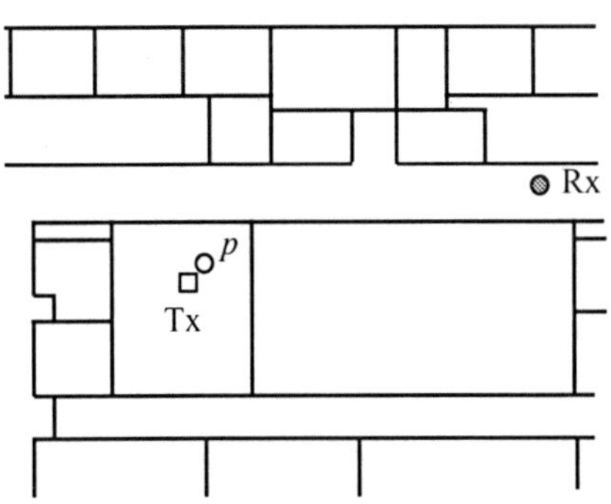

(a) An indoor environment

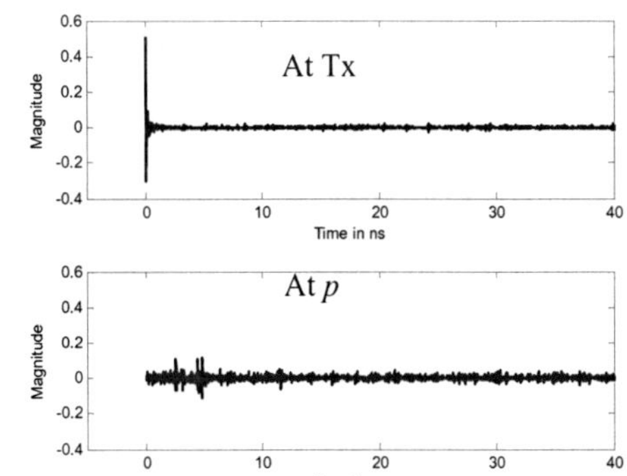

(b) Impulse responses at Tx and p.

Figure 7. Ray-tracing results.

6. CONCLUSIONS

FDTD simulation results show that spatial energy focusing can be achieved in indoor environments using 1 x 1 antenna systems. When the multipaths are richer, spatial focusing will be better achieved. Scatterers such as metal door frames, computer monitors, bookcases, desks, and cubicle frames can all improve the spatial focusing by introducing more multipaths.

The spatial focusing effect can be measured by the SEF factor which describes how concentrated the energy is around Tx. The relative half energy radius which indicates the radius of a circle centered at Tx on which half of the energy in a larger given circle is distributed. It is found that at all 13 Rx locations, the furniture pieces help concentrate the energy around Tx and, on average, the relative half energy radius for the walls-only cases is 66.7 % larger than that for the full-scattering cases.

The ray-tracing method is successfully used in the calculation of time-reversed signals in frequency domain. The spatial energy focusing is also observed for the single antenna systems. Since the FDTD method can only deal with problems with small areas, ray-tracing method will be of great value in characterizing the TR signal for urban and other large propagation environments.

REFERENCES

1. Derode, A. Tourin, and M. Fink, Random multiple scattering of ultrasound. I. Coherent and ballistic waves, *Physical Review E*, **64**, 036605-1~7 (2001).
2. Derode, A. Tourin, and M. Fink, Random multiple scattering of ultrasound. II. Is time reversal a self-averaging process? *Physical Review E*, **64**, 036606-1~13 (2001).
3. G. Montaldo, G. Lerosey, A. Derode, A. Tourin, J. de Rosny, and M. Fink, Telecommunication in a disordered environment with iterative time reversal, *Waves in Random Media*, **14**, 287-302 (2004).
4. G. Lerosey, J. de Rosny, A. Tourin, A. Derode, G. Montaldo, and M. Fink, Time reversal of electromagnetic waves, *Physical Review Letters*, **92** (19), 193904-1~3 (2004).
5. L. Borcea, G. Papanicolaou, C. Tsogka, and J. Berryman, Imaging and time reversal in random media, *Inverse Problems*, **18**, 1247-1279 (2002).
6. C. Oestges, A. D. Kim, G. Papanicolaou, and A. J. Paulraj, Characterization of space-time focusing in time-reversed random fields, *IEEE Trans. Antennas Propag.*, **53** (1), 283-293 (2005).
7. D. Liu, G. Kang, L. Li, Y. Chen, S. Vasudevan, W. Joines, Q. Liu, J. Krolik, and L. Carin, Electromagnetic time-reversal imaging of a target in a cluttered environment, *IEEE Trans. Antennas Propag.*, **53** (9), 3058-3066 (2005).
8. A. G. Cepni, and D. D. Stancil, Single antenna microwave nulling using time-reversal techniques, *Proceedings of IEEE IMS'05*, 1723-1726 (2005).
9. Z. Yun, Z. Zhang, and M. F. Iskander, A Ray-Tracing Method Based on Triangular Grid Approach and Application to Propagation Prediction in Urban Environments, *IEEE Transactions on Antennas and Propagation*, **50** (5), 750-758 (2002).
10. A. Taflove, and S. C. Hagness, *Computational Electrodynamics, the Finite-Difference Time-Domain Method*, (Artech House, Boston, 2000).

FURTHER DEVELOPMENTS IN HIGH-VOLTAGE UWB DIRECTIONAL COUPLERS

Lanney M. Atchley, Leland H. Bowen, Everett G. Farr, and William D. Prather*

1. INTRODUCTION

Directional couplers in Ultra-Wideband (UWB) High-Voltage (HV) radar systems allow the possibility of using a single antenna for both transmission and reception. In this paper, we continue the development of HV UWB coupled-line directional couplers reported earlier in [1], and first suggested by Baum in [2]. We investigate smoothing the impedance profile, increasing the isolation, and reducing the transmission loss, while maintaining high voltage standoff and high mechanical reliability. Note that a more detailed version of this paper appears in [3].

In [1], we built two oil-filled coupled-line directional couplers, in which we achieved a isolation of 20 dB, and a transmission loss of 1.5 to 1.9 dB above the expected values. Furthermore, the impedance profiles ranged from 30 Ω to 75 Ω for a 50 Ω input impedance. Here, we investigate two new prototypes in an effort to smooth the impedance profile, with the hope that this will increase the isolation. We describe here our best air-filled and best oil-filled prototypes, after much experimentation.

The couplers described here incorporate a number of advances over those in [1]. First, the coupling factor was smaller than that previously used, –9.5 dB instead of –4.4 dB. This was changed because we found it challenging to maintain mechanical tolerances on two long transmission lines that are very closely spaced. Furthermore, we hoped the larger separation would reduce transmission losses. Next, we eliminated the angled feed sections, and instead positioned the connectors directly in contact with the coupled lines. In doing so, we simplified the design, eliminated impedance discontinuities, and made it possible to machine to tighter tolerances. Finally, we used spring-loaded pressure

* Lanney M. Atchley, Leland H. Bowen, and Everett G. Farr, Farr Research, Inc. Albuquerque, NM, 87123; William D. Prather, AFRL/DE, Kirtland AFB, NM 87117.

contacts at the joints between the end of the transmission line and the connector. We did so in order to increase the mechanical reliability of the connection.

We begin now by reviewing the operation of directional couplers. Following that, we describe the fabrication and testing of the two best new prototypes, and we compare their results to those in [1].

2. BACKGROUND

We review here the operation of UWB HV directional couplers. To understand the need for a directional coupler, we provide a diagram of a UWB radar system with a single antenna, shown in Figure 2.1. The diagram shows a UWB source (Port 1) connected to an antenna (Port 2) through a directional coupler of length ℓ. When a signal is received back from a target in Port 2, an exact replica of the received signal is coupled into Port 4, where it is detected by an oscilloscope. The signal at Port 4 is an exact replica of the received signal for a time equal to the round-trip transit time of the coupler [1,2], or $2\ell\varepsilon_r^{1/2}/c$. A signal conditioner and/or a limiter may be used to protect the oscilloscope from spurious high-voltage signal that can leak through the coupler due to its finite isolation.

In Figure 2.2 we show a generic outline of a coupled-line directional coupler. Any port can be the driven port, but for this discussion we will define the driven port to be Port 1. The signal at Port 1 passes through the coupler to Port 2 with little attenuation. The signal from Port 1 is coupled to Port 3 but is attenuated by approximately 9.5 dB in the design used here. No signal should be coupled to the isolated port (Port 4) but in actual measurements the leakage signal at this port is down by approximately 20-30 dB from the source signal. This figure is referred to as the isolation of the directional coupler. When comparing Figures 2.1 and 2.2, one should note that during transmission Port 1 in Figure 2.1 is the driven port but during reception Port 2 in Figure 2.1 becomes the driven port with respect to the return signal.

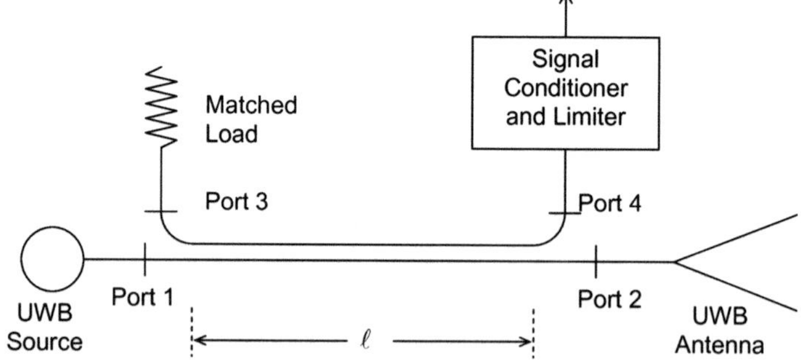

Figure 2.1. UWB Radar system with single antenna.

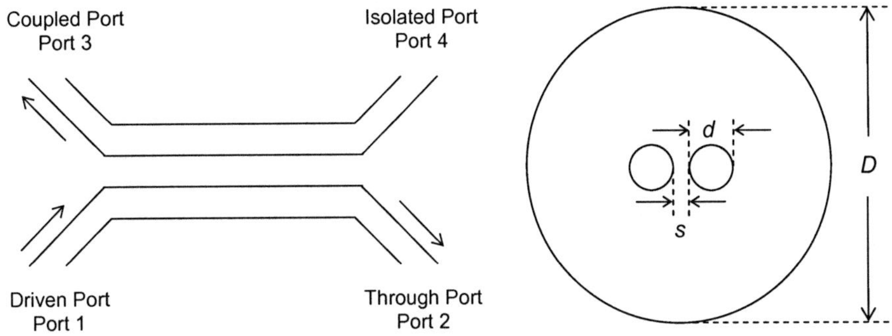

Figure 2.2. The Coupled-Line Directional Coupler, left, and a typical cross section right.

The theory and calculations for the directional coupler design are provided in [1], along with predictions of the performance in the time domain, and voltages at all four ports as a function of the even-mode and odd-mode impedances. Also included in [1] is the theory to optimize the impedances to maximize the coupling.

Let us consider now the importance of the various figures of merit for directional couplers. First, a uniform 50 Ω impedance profile is important in UWB radar systems in order to minimize reflections that later appear at the digitizer or oscilloscope. These coupler-created reflections are usually larger than the desired signals, and they significantly complicate data processing.

Next, a high isolation is necessary to ensure that only a very small signal couples directly from the source port into to the isolated port. This is critical when a high voltage source is used, because the leaked voltage will couple directly into the oscilloscope, which may cause damage. Finally, the transmission loss in a directional coupler reduces the voltage available to the antenna, and therefore should be minimized.

3. AIR-FILLED COUPLER: PROTOTYPE 8

To improve the isolation and reliability of our directional couplers, we built two designs with three major new features. Prototype 8, which is air-filled, is described in this section. Prototype 9, which is oil-filled, is described in the next section. The two coupled lines in Prototype 8 are 17 mm (21/32 in) in diameter, approximately 30.5 cm (12 in) in length, and are separated by 1.8 mm (0.072 in). The outer conductor is 63.3 mm (2.49 in) in diameter. Thus, Prototype 8 has even-mode and odd-mode impedances of 50 Ω, resulting in a –9.5 dB coupler. Each port of Prototype 8 has an N-Type connector.

Prototypes 8 and 9 incorporated three major improvements over earlier versions. First, we removed the angled feed sections between the connector and coupled lines. Instead, we attached the connector directly to the coupled line. Second, we made the end of the center conductors square, to improve the isolation between the driven port and the isolated port. Third, we shortened the transition from the connectors to the coupled lines as much as possible, while maintaining voltage standoff. Finally, we replaced the solder joints between the connectors and coupled lines with spring-loaded pressure contacts. This to addressed the problem of loose solder joints observed in earlier designs.

In Figure 3.1 we show the square ends of Prototype 8. There are small springs behind the pins that connect the tubes to the N-type connectors. The springs eliminate the problem of fragile solder joints, but they make it necessary to use plastic spacers to hold the tubes in place. One of the spacers can be seen in the figure to the right of the end plugs.

Figure 3.1. Prototype 8 showing square ends.

We measured a simple TDR of Prototype 8, while the other three ports were terminated in 50 ohms, and the results are shown in Figure 3.2. This impedance profile is considerably more smooth than those of earlier air-filled couplers. We also measured the port voltage outputs when Port 1 is driven by the Kentech model ASG1 pulser, and the results are shown in Figure 3.3. Here, we observe our lowest voltage to date at the Isolated Port, and we have achieved 33 dB of isolation between the driving signal and the output of the Isolated Port.

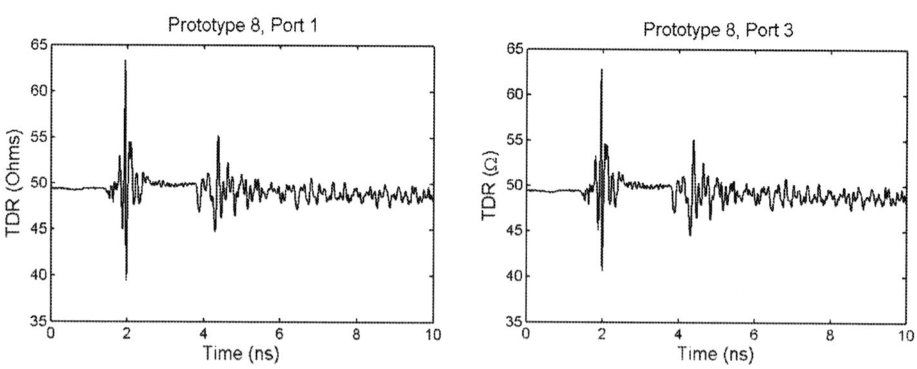

Figure 3.2. Simple TDRs of Prototype 8.

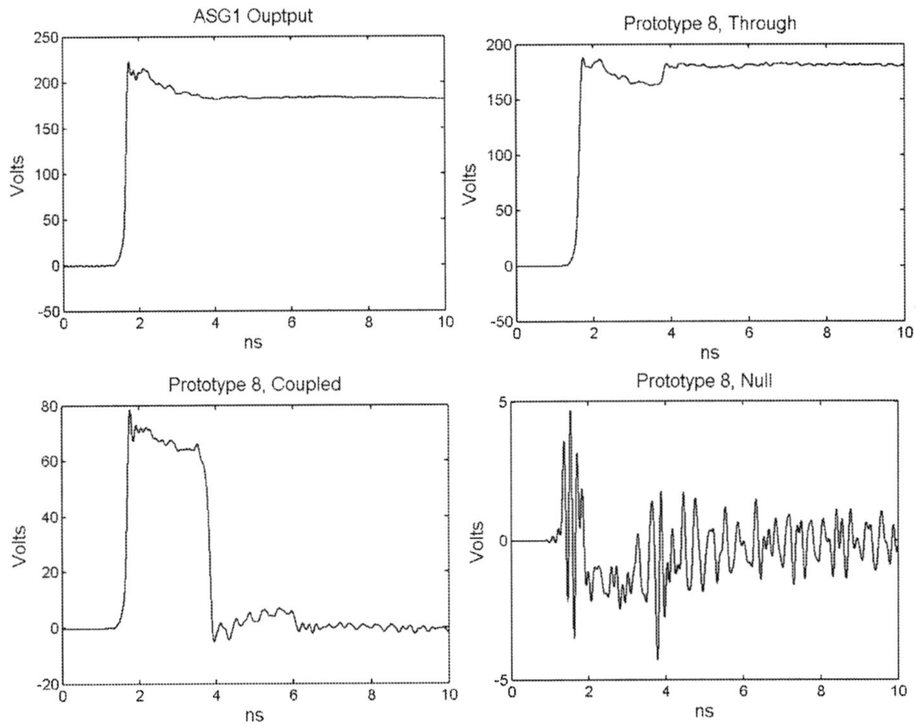

Figure 3.3. Port Voltages of Prototype 8 when driven with a Kentech model ASG1 pulser.

Figure 4.1. Prototype 9 directional coupler.

4. OIL-FILLED COUPLER: PROTOTYPE 9

Next, we built an oil-filled coupler, Prototype 9. As with Prototype 8, the ends of the center conductors are square, the transition from the connectors to the coupled lines is shortened as much as possible, and the connections are spring-loaded pressure contacts instead of solder joints. Prototype 9 is approximately 30.5 cm (12 in) long. The two coupled-lines are 1 cm (13/32 in) in diameter and separated by 2.3 mm (0.089 in). The outer conductor is 6.19 cm (2.44 in) in diameter. Thus, the even- and odd-mode impedances are 50 Ω, resulting in a –9.5 dB coupler.

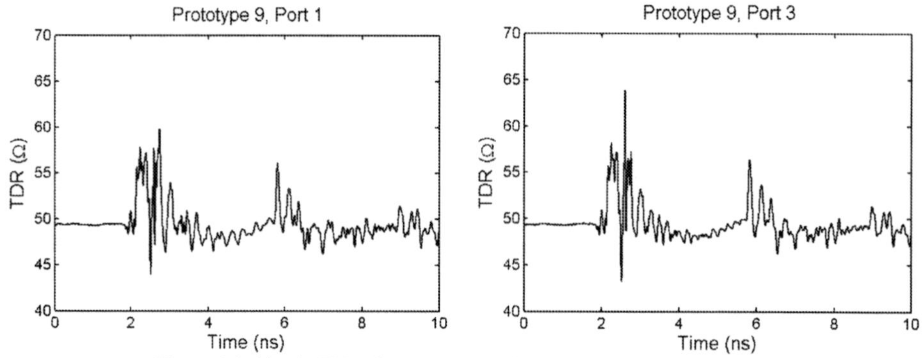

Figure 4.2. Simple TDR of Prototype 9 with three ports terminated in 50 Ω.

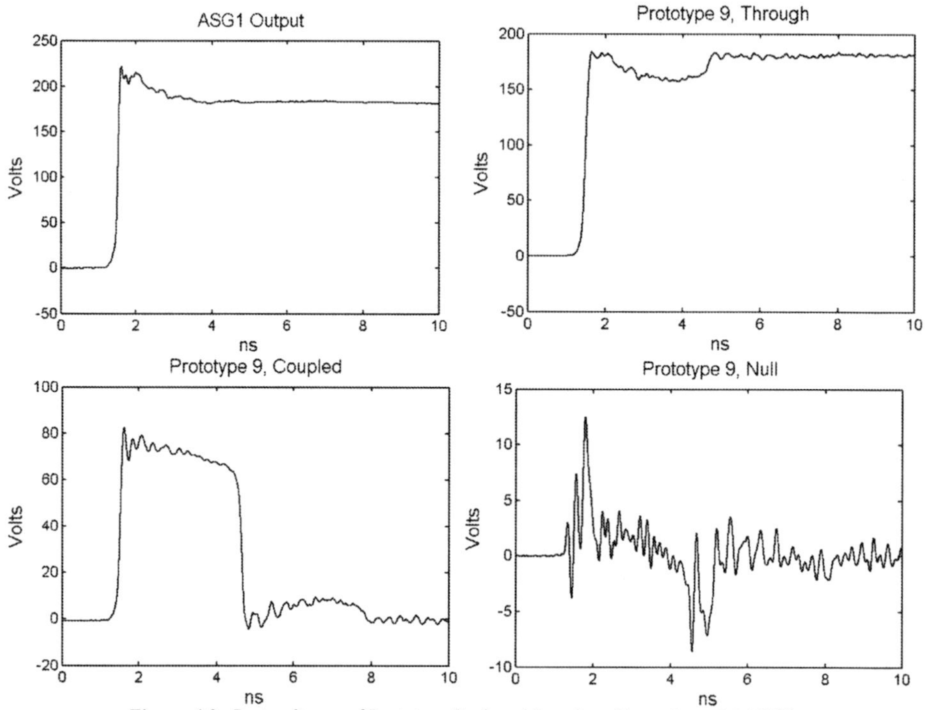

Figure 4.3. Port voltages of Prototype 9 when driven by a Kentech model ASG1.

In Figure 4.1 we show the ends of the center conductors for Prototype 9. As with Prototype 8, we used springs to hold the tubes in place. However, in this case, large springs are used inside the tubes to push the end plugs outward. This is done because the pins connecting the ends to the HN-type connectors are located right at the edge of the ends, so no spring can be located behind the pins. As with the Prototype 8, we used spacers to hold the tubes in place. Since the materials used for the spacers have about the same dielectric constant as the oil, the spacers have little effect on the impedance match.

We measured the impedance profile of Prototype 9 with a simple TDR, as shown in Figure 4.2. The impedance profile is considerably more smooth than earlier oil-filled designs. We also measured the port voltages when Port 1 is driven by the Kentech model ASG1 pulser, and the results are shown in Figure 4.3. In this case, the voltage at the Isolated Port is down by 25 dB from the source.

5. DISCUSSION

We began this paper by noting the relevant specifications of previous directional couplers. We now consider whether we have made any progress on those numbers.

First, we consider the impedance profiles of the new couplers, as quantified by the range of the high and low TDR values. In Prototypes 1 and 2 we achieved impedance values ranging between 30 and 75 ohms, while in Prototypes 8 and 9 the impedance values ranged between 44 and 63 ohms. Thus we were successful in achieving somewhat smoother impedance profiles in the new couplers, although there remains some room for improvement. We expected this would lead to higher isolation, and we were pleased to observe that we achieved isolations of 33 dB and 25 dB in Prototypes 8 and 9, respectively, as opposed to 20 dB in Prototypes 1 and 2

We have also achieved a lower transmission loss in Prototypes 8 and 9 than in Prototypes 1 and 2. We define transmission loss as the difference between the measured and expected voltages at the Through Port. We found transmission losses of 0.4 and 0.6 dB in Prototypes 8 and 9, as opposed to 1.5 and 1.9 dB in Prototypes 1 and 2.

Let us consider now the consequences of having only a modest isolation in a directional coupler. The prompt response at the Isolated Port is 25 to 33 dB lower than the source waveform. Most oscilloscopes incur damage with inputs larger than 5 volts or so, so the source voltage is limited to 89 – 223 V peak without additional protection. (Our design goal was 30 kV.) To use a larger source voltage, one must use either a limiter, an attenuator, or a switch to protect the oscilloscope. However, there are significant challenges associated with commercially available limiters, as described in [4]. Furthermore, it is unclear that there is any advantage in using attenuators over simply using a lower source voltage. Finally, there has been very little work done on switches, which might be implemented with either a PIN diode or a GaAs FET. Thus, the value of driving the system at 30 kV, which was the goal of our oil-filled design, remains unclear.

Note that our TDR measurements were carried out with a fast sampling oscilloscope, the Tektronix model TDS8000B with 80E04 sampling head, with a source risetime of 28 ps. So our TDR data have considerably more detail than what is needed for our intended high-end bandwidth of a few gigahertz. Thus, it might be appropriate to apply a risetime filter to smooth out the TDR data, which would tend to smooth out the data, but we have not done so here because the smoothed data could conceivable obscure important trends.

Next, we consider the voltage capacity of these couplers. The connectors are the limiting component in both designs. We used HN-Type connectors in all the oil-filled couplers (Prototypes 1, 2, and 9), and N-Type connectors in the air-filled coupler (Prototype 8). We do not yet have good data on how high a voltage can be tolerated by these connectors with impulse-like sources. But we were able to test the HN-Type connectors as high as 25 kV for short impulses without observing flashover, using an FID model FPG 30-1KM pulser. We suspect that we could go higher if we had the right source available for testing.

6. CONCLUSIONS

To improve the properties of the UWB HV directional couplers described in [1], we have implemented a number of refinements to smooth their impedance profile, increase their isolation, and reduce their transmission loss. We have built and tested new couplers that have implemented a number of changes. These changes include using a larger separation between the coupled lines to reduce transmission loss, smoothing the rough spots in the impedance profile, and simplifying the design to increase the isolation. We have experimented with the shape of the ends of the center conductors, and we have improved the reliability by replacing solder joints with spring-loaded pressure contacts.

As a result of our modifications, we have reduced the transmission loss by 1 to 1.5 dB, we have increased the isolation to 25 to 33 dB (from 20 dB), and we have smoothed the impedance profile significantly. We have also demonstrated that our directional couplers can tolerate 25 kV for very short (3 ns) pulses. Despite these gains, one would really like to have a higher isolation than 33 dB in such systems.

To make use of a high-voltage UWB directional coupler with modest isolation, it will be necessary to use it with either a limiter or switch to protect the oscilloscope. A limiter would likely require a custom design, since commercially available designs have limitations. The transient response of PIN diode and GaAs FET switches is an area that remains unexplored, so it might prove to be a fruitful area for future research.

We have used a number of the directional couplers described here in UWB radar experiments, and the results of those measurements were reported in a separate paper [5].

REFERENCES

1. L. M. Atchley, E. G. Farr, D. E. Ellibee, and D. I. Lawry, A High-Voltage UWB Coupled-Line Directional Coupler, Sensor and Simulation Note 489, April 2004. Also appeared as E. G. Farr *et al*, "A High-Voltage UWB Coupled-Line Directional Coupler for Radar," *Ultra-Wideband, Short-Pulse Electromagnetics 7*, Proceedings of the Conference in Magdeburg, July 2004.
2. C. E. Baum, Coupled Transmission Lines as a Time-Domain Directional Coupler, Sensor and Simulation Note 451, October 2000.
3. L. M. Atchley, L. H. Bowen, E. G. Farr, and D. E. Ellibee, Further Developments in High-Voltage UWB Directional Couplers, Sensor and Simulation Note 513, March 2006.
4. L. M. Atchley, E. G. Farr, and W. D. Prather, The Response of Commercial Limiters to Transient Signals, Measurement Note 59, April 2005.
5. L. M. Atchley, E. G. Farr, and L. H. Bowen, Experiments with a High-Voltage Ultra-Wideband Radar Using a Directional Coupler, to appear in early 2006 as a Sensor and Simulation Note.

SCATTERING

RESONANCE BEHAVIOR OF A DIELECTRIC TARGET IN A HALF-SPACE USING THE CNR (COMPLEX NATURAL RESONANCE) METHOD

S.K. Padhi[1] and N.V. Shuley

1. INTRODUCTION

Due to the advent of short-pulse technology (UWB), there has been a considerable amount of interest shown in the application of the transient or impulse response of a radar target. A typical application for such a system is the use of an UWB illuminating pulse in ground penetrating radar (GPR). However, the detection and identification of buried and subsurface objects still remains a difficult problem, as there is no single sensor or diagnostic tool which can provide sufficient information about the object. The singularity expansion method (SEM) has been used for a long time in the detection and recognition of targets in free space. Whilst this method is usually used to represent the scattered signal from the object in terms of a sum of complex exponentials, it also contains the impulse response of the system which can be obtained via a deconvolution process if the incident signal is known. A similar methodology is very popular in GPR systems in the tracking and identification of buried landmines. Although the resonance behavior of the target occurs in the late time portion of the transient signal, both early and late time portions of the signature are useful for extracting the features of the target[1-4]. We primarily focus on the late time response of the target as it contains target specific information.

[1] The authors are with EMI Group, School of ITEE, University of Queensland, St Lucia, QLD-4072, Australia, Email: padhi@itee.uq.edu.au.

The primary objective of this research work is to detect and identify a dielectric target buried below a homogeneous half-space from the backscattered time domain signature. It is well known that the transient scattering of electromagnetic waves from a conducting or dielectric target contains information about the resonance behavior of the target itself as quantified by the complex poles or complex natural resonances (CNR) of the target. The CNRs are theoretically aspect independent, depending only on the shape and electrical properties of the target such as permittivity and conductivity. For a target embedded below an interface, the transient response of target itself can be recovered from the system response using a deconvolution technique. Many deconvolution techniques are described and applied in radar signal processing and image processing fields both in the time domain and frequency domain. However, the time domain approach is found to be superior compared to current frequency domain approaches[2-5]. Similarly, there are many techniques for extracting the CNRs and, in this context; we use the matrix pencil method to extract the poles from the late time impulse response of target[6].

2. THEORY

In this study the finite difference time domain (FDTD) method is used for obtaining the transient signatures. The simulation setup is shown in Fig. 1(a). A dielectric cylinder with radius *d*, length *h*, permittivity ε_{r2} and conductivity σ_2 is embedded inside a homogeneous interface media of permittivity of ε_{r1} and conductivity σ_1. A plane wave excitation using a temporal variation that is the first derivative of a Gaussian pulse illuminates the target and the scattered waves are collected in the forward direction. The functional form of the input pulse and its frequency content are shown in Fig 1(b). This is a half-space problem, where the target is buried inside an interface of known electrical properties. The system response and target response can be defined as the scattered waveform from the interface without and then with target present.

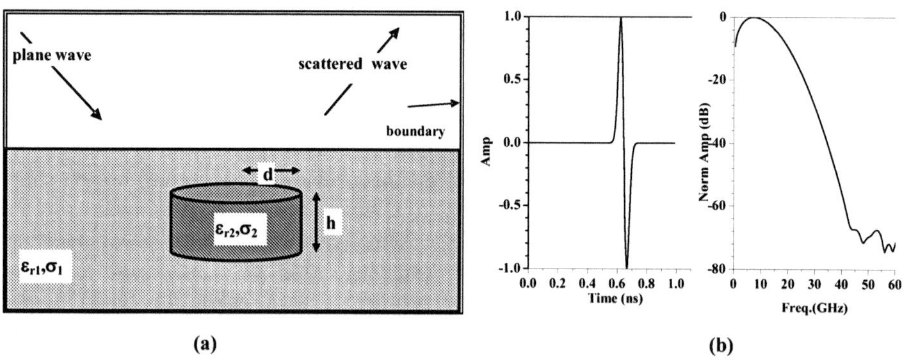

Fig. 1. (a) The geometry of problem. The other parameters are: d=1 cm, h=3.0cm, ε_{r1}=4.0, ε_{r2}=8.0, σ_1=0.01, σ_2=0.3; (b): The pulse and its frequency spectrum.

Further, the target response can be expressed as convolution of system response and impulse response of target (h(t)). The impulse response can be extracted by deconvolving the target and system responses. The impulse response of different targets is extracted by deconvolution process for two different cases; when the interface is homogeneous dielectric media (Case A) and when the interface is free space (Case B).

2.1. Deconvolution:

The singular value decomposition (SVD) method is used in the deconvolution process to calculate the impulse response $h(t)$ of target. Let the system response with and without the target present be denoted $y2(t)$ *and* $y1(t)$ respectively. The convolution process can be written as;

$$y1(t) * h(t) = y2(t) \qquad (1)$$

The aim of deconvolution process is to extract the unknown h from given or measured $y1$ and $y2$ data sets. The detailed procedures of estimating $h(t)$ are outlined in[7]. The estimated integrity of the impulse response can be characterized by calculating the maximum difference norm *(Err)* of $y2_{est}(t)$ and $y2(t)$ as:

$$Err = \| y2 - h_{est} * y1 \| \qquad (2)$$

2.2. Pole extraction:

The singularity expansion method (SEM) is a technique of approximating the pulsed EM response of a scatterer with a series of damped exponentials. The technique was initially developed by Baum and has been used by others in extracting the CNRs from the transient signature[2-5]. It is shown that the late time response can be approximated by

$$f(t) = \sum_m R_m \exp(s_m t) \qquad (3)$$

Where, R_m and s_m ($s_m = -\alpha_m \pm j\omega_m$) are the residues and CNRs of the scatterer respectively. The early time response contains local high frequency reflections and diffractions from the scatterer and does not concern us here. However, the late time response contains the CNR information and this is extracted from the signature using the matrix pencil of functions algorithm (MPOF) [6].

3 NUMERICAL RESULTS

3.1. Finite difference time domain method:

In order to test the deconvolution algorithm, a finite difference time domain code is used to calculate the time domain response of the target embedded in a half space as shown in Fig. 1(a). A plane wave (TM polarization) using a temporal shape that is the first derivative of a Gaussian pulse is incident on the target plus the interface. The specularly scattered waves at a far-field point are recorded. The cell size Δx is chosen as 1 mm, and the time step Δt as 1.926 *ps*. The pulse and its frequency spectrum are shown in Fig 1(b). To avoid dispersion, the criteria $\lambda > \Delta x$, is satisfied throughout the spectrum and the lattice has 100x100x100 cells. The extraction process for obtaining the impulse response is as follows; first the scattered waves are collected at a far-field distance first with and then without the target present. Let these signals be *y2(t)* and *y1(t)* respectively sampled at 2000 points.

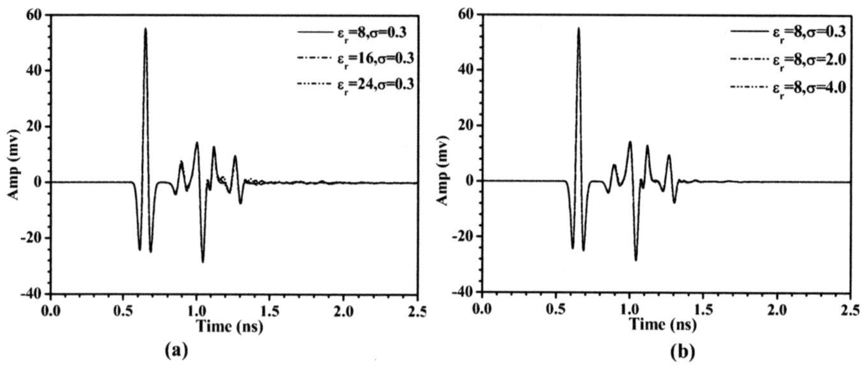

Fig. 2. The far-field response of target as a function of (a): permittivity, (b): conductivity. The incident pulse is at θ=45° and ϕ=0°.

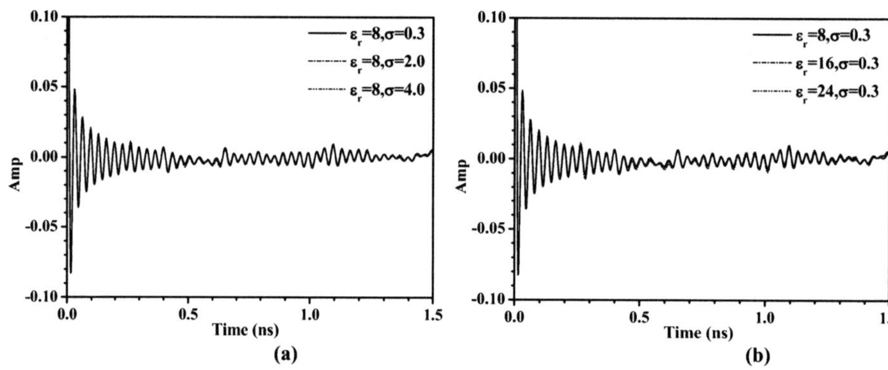

Fig. 3. The impulse response of target as a parameter of (a): permittivity and (b): conductivity.

The far field pattern of the system response with dielectric permittivity and conductivity of the target as a parameter are shown in Fig 2. As can be seen, the specular reflection (early time response) from the scatterer is independent of conductivity. Minor variations with regard to conductivity are seen only in late time. The singular value decomposition method is used to extract the impulse response. In the process, the first 147 consecutive singular values are included in the estimation of the impulse response of target. These values are chosen by correlating their frequency content. The extracted impulse response of the target parameterized by its permittivity and conductivity are shown in Fig 3. The maximum difference norm (Err) for the estimated impulse response is also shown in Table-1. The late-time response of the extracted impulse response is again sampled and input into matrix pencil algorithm to extract the CNRs (Case A).

Next, we would like to verify the extracted poles, by repeating the same procedure for the same target but now in free space, where the constitutive parameters of the interface media have been replaced by air (Case B). In this case the first 115 singular values were used in the extraction process and again the sampled late time responses were used to calculate the CNRs. The extracted CNRs for both cases (Case A and Case B) are tabulated in Table 1. As can be seen from tabulated data, the mean value of the maximum difference for the estimated impulse response is $3.2\text{x}10^{-4}$. The extracted poles in Case B can be compared with those of Case A. In both cases, the variation of the CNR's with respect to the constitutive parameters is very similar. As the conductivity increases, the damping coefficient (α_m), as expected, moves towards the left in the S-plane.

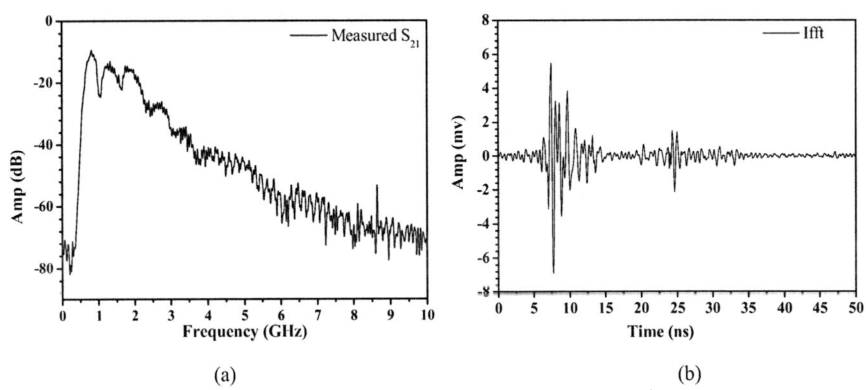

Fig. 4. Measured performances of cylindrical target (Cylinder '1'); (a): measured S_{21} and (b): time domain response using IFFT routine.

3.2. Measurements:

In order to partially validate our calculated data, bistatic measurements were performed. Two ultra-wideband TEM horn antennas were used in a bistatic transmitter-receiver configuration. Metallic targets were used in the experiment. An Agilent 8510 VNA was used to measure the scattered signals. The transmitting antenna was excited through the VNA and a linear 30dB LNA was connected to the receiving antenna terminals to amplify the received signals. The operating frequency range was 45 MHz to 10 GHz in 512 steps. The target was placed on top of a Styrofoam podium and the scattered signals were collected both with and without the target placed on top of the stand. The frequency response data was transformed to the time domain data by using an IFFT routine and deconvolution and pole extraction processes were carried out as described in Section 2. The measured frequency response of metallic cylinder 'A' and its corresponding time domain responses are shown in Fig 4.

The extracted CNRs for various canonical targets are tabulated in Table 2. The theoretical CNR value of various targets were calculated from tabular data from[8] and compared with measured data. The agreement is very close. The measured CNRs of Sphere '1' compares very well with measured published data[9-10].

4 CONCLUSION

This paper describes an approach to model the impulse response of a dielectric target in a half-space using a time domain deconvolution technique. A finite difference time domain code is used to calculate the time domain responses. A singular value decomposition algorithm is used to estimate the impulse response of target and a robust algorithm such as MPOF is used to calculate the CNR's of the scatterer. The extracted poles are compared with the poles scattered by the target in free space. Bistatic measurements were performed to calculate the CNRs of a target in a free-pace environment. The extracted CNRs agree closely with the published data. This experiment technique demonstrates that it has wide scale application in finding the signatures of dielectric target buried in half-space. The introduced experimental technique can be further modified to enhance the ability to detect the signature from dielectric target in a dispersive and lossy medium.

5 REFERENCES

1. E.J. Rothwell, and W. Sun, "Time Domain Deconvolution of Transient Radar Data", *IEEE Trans. on Antennas and Propag,* **38** (4), 470-475 (1990).
2. C.E. Baum, "On the Convergence and Numerical Sensitivity of the SEM Pole Series in Early Time Scattering Response", *Electromagnetics,* (1), 209-228 (1981).
3. L.W. Pearson & R. Roberson, "The Extraction of the Singularity Expansion Description of a Scatterer from Sampled Transient Surface Current Response", *IEEE Trans. Antennas and Propagation*, **28** (2), 182-190 (1980).
4. K.S. Cho and J.T. Cordaro, "Calculation of the SEM Parameters from the Transient Response of a Thin Wire", *IEEE Trans. on Antennas and Propagation*, **28** (6), 921-924 (1980)
5. C.C. Chen and L.J. Peters, "Buried Unexploded Ordnance Identification via Complex Natural Resonances', *IEEE Trans. on Antennas and Propagation*, **45** (11), 1645-1654 (1997).
6. T.K. Sarkar and O. Pereira, "Using the Matrix Pencil Method To Estimate the Parameters of a sum of Complex Exponentials", *IEEE Trans. Antennas and Propagation Magazine*, **37** (1), 48-55 (1995).
7. S.K. Padhi, N.V. Shuley, H.S. Lui and S. Crozier, "Time domain characterization of dielectric target in a half-space using complex natural resonance method", Proceeding of EMC Zurich-2006, 109-112, (2006).
8. L. Marin, "Natural-mode representation of transient scattering from rotationally symmetric bodies", *IEEE Trans. on Antennas and Propag,* **22** (2), 266-274 (1974).
9. F.I. Tseng and T.K. Sarkar, "Experimental Determination of Resonant Frequencies by Transient Scattering from Conducting Spheres & Cylinders", *IEEE Trans. Antennas and Propagation*, **32** (9), 914-918 (1984).
10. F.I Tseng and T.K. Sarkar, "Deconvolution of the Impulse Response of a conducting Sphere by the Conjugate Gradient Method", *IEEE Trans. on Antennas and Propag,* **35** (1), 105-110 (1987).

TABLE 1: EXTRACTED POLES FROM DIFFERENT CONSTITUTIVE PARAMETERS FOR CASE A AND CASE B.

Target #		Case A		Case B		Err:
ε_r	σ	α_m (10^9Np/s)	f_r(GHz)	α_m (10^9Np/s)	f_r(GHz)	($\times 10^{-4}$)
8	0.3	-0.26	7.94	-2.94	8.00	2.85
	2.0	-0.43	7.31	-13.8	7.06	3.33
	4.0	-0.30	7.34	-15.9	6.80	3.48
16	0.3	-0.067	7.26	-5.14	6.52	3.03
	2.0	-0.18	7.36	-7.14	6.29	3.22
	4.0	-0.48	7.35	-13.6	6.32	3.33
24	0.3	-0.96	7.19	-0.12	8.21	3.26
	2.0	-1.22	7.25	-2.01	8.44	3.13
	4.0	-6.87	8.09	-5.02	8.55	3.19

TABLE 2: EXTRACTED RESONANT FREQUENCIES FOR DIFFERENT TARGETS (SPHERE '1': 1.5" DIA; SPHERE '2': 2" DIA; CYLINDER '1': 0.75" DIA AND ASPECT RATIO 0.5)

Target #	Theory/Experiment	Resonant Frequencies (GHz)		
Sphere 1	Theory	6.79	10.39	*
	Experiment	7.06	9.12	*
Sphere 2	Theory	3.4	5.18	6.98
	Experiment	3.29	5.29	7.02
Cylinder 1	Theory	7.28	8.37	9.6
	Experiment	7.40	8.54	9.58

ANALYSIS OF THE LATE-TIME TRANSIENT FIELD SCATTERED BY A LINE SOURCE ABOVE A GROUNDED DIELECTRIC SLAB

Gary D. Dester and Edward J. Rothwell *

1 INTRODUCTION

Layered materials are often interrogated using transient electromagnetic signals for the purpose of materials characterization. It is thus useful to understand the temporal behavior of the reflected field, particularly the late-time component.

Baum[1] first proposed a model for the late-time radar target response using the aspect-independent resonance frequencies of the target. The late-time signal is described by

$$R(t) = \sum_{n=1}^{N} a_n e^{\sigma_n t} \cos(\omega_n t + \phi_n), \quad t > T_L \tag{1}$$

where T_L denotes the beginning of the late time signal, and N modes are required to describe the time signal. The natural modes consist of amplitude (a_n) and phase (ϕ_n) terms, and the complex resonant frequencies ($\sigma_n + j\omega_n$). This representation has formed the basis for several different target identification schemes [2]–[4].

Building from this model, several studies have been undertaken to investigate the transient response of layered materials due to an incident plane wave in both the frequency and time domains. It has been shown[5] that for a transient plane wave incident upon a planar slab, the late-time portion of the reflected field can be characterized as a natural mode series. Another study[6] has obtained a natural mode series approximation for a transient plane wave incident upon a curved, coated surface. These results have been used as a basis for diagnosing changes to material layers using the E-pulse technique[7]. In practical situations, the interrogating field produced by an antenna will not have a planar wavefront. Thus, a localized source (such as a line source) might be a better model to study the late-time responses of these materials. In order to investigate the late-time response of a layered medium to a non-planar wavefront, this paper examines the fields scattered by a cylindrical wave incident upon a simple planar layered medium consisting of a grounded

* Department of Electrical and Computer Engineering, Michigan State University, East Lansing, MI 48824, USA desterga@msu.edu, rothwell@msu.edu

dielectric slab. The principal goal is to show that the late-time component of these scattered fields can be characterized by a natural mode series and to investigate whether this representation bears any relationship to the natural mode response of the material to a plane wave by moving the source far from the planar slab.

2 PROBLEM DESCRIPTION

Figure 1 shows the geometry of the problem under consideration. An x-directed magnetic line current is located in free space a height h above a planar slab of thickness d. The conductivity, permittivity and permeability of the dielectric slab are given by σ, ϵ_1, and μ_o. Although both the TE and TM cases have been examined, only the results for the TM case are presented.

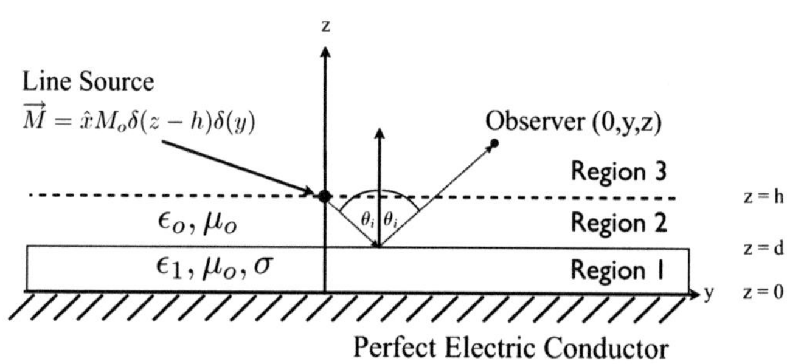

Fig. 1. Geometry for cylindrical-wave reflection from a conductor-backed lossy dielectric slab

Assume e^{jwt} time dependence and let $\overrightarrow{F}$ be the electric vector potential, which describes the electric and magnetic fields through the operations $\overrightarrow{E} = -\frac{1}{\epsilon}\nabla \times \overrightarrow{F}$ and $\overrightarrow{H} = -j\omega\overrightarrow{F}$. By defining the spatial Fourier transform pairs as

$$F_x(y,z) = \frac{1}{2\pi}\int_{-\infty}^{\infty} \tilde{F_x}(k_y,z)e^{jk_y y}dk_y, \tag{2}$$

$$\tilde{F_x}(k_y,z) = \int_{-\infty}^{\infty} F_x(y,z)e^{-jk_y y}dy \tag{3}$$

it is possible to solve the homogeneous wave equation in each of the three regions shown in Fig. 1. Applying the boundary conditions in each of the source-free regions, and solving for the constants, gives

$$\tilde{F}_x(k_y, z) = \frac{\epsilon_o M_o}{2jp} e^{-jp|h-z|} - \frac{\epsilon_o M_o R_M}{2jp} e^{j2pd} e^{-jp(h+z)}, \tag{4}$$

where

$$R_M = \frac{\epsilon_o q \sin qd + jp\epsilon \cos qd}{\epsilon_o q \sin qd - jp\epsilon \cos qd} \quad , \quad r = \sqrt{(z-h)^2 + y^2}. \tag{5}$$

Here the first term is the direct potential of the line source and the second term is the potential due to the field reflected from the layer. The quantities p and q are the wave numbers in the z direction in free space, and the dielectric slab, respectively:

$$p = \pm\sqrt{k_o^2 - k_y^2} \quad , \quad q = \pm\sqrt{k^2 - k_y^2}. \tag{6}$$

From (2) and (4) the scattered magnetic field is given by

$$\begin{aligned} H_x^s(y, z) = {} & \frac{\epsilon_o M_o \omega}{4\pi} \int_{-\infty}^{\infty} \frac{e^{-jp(h+z-2d)}}{p} e^{jk_y y} dk_y \\ & + \frac{\epsilon_o M_o \omega}{4\pi} \int_{-\infty}^{\infty} \frac{j2\epsilon \cos qd}{\epsilon_o q \sin qd - jp\epsilon \cos qd} e^{-jp(h+z-2d)} e^{jk_y y} dk_y. \end{aligned} \tag{7}$$

Since the frequency and time-domain responses of the electric and magnetic fields are similar, only the magnetic field is considered in the following discussion.

The representation of the scattered magnetic field consists of two integrals from which the fields can be numerically computed as a function of source and observer location as well as the properties of the dielectric slab. The first integral of (7) is the expression for a Hankel function which can be easily evaluated numerically. The second integral can be evaluated by performing a real-line integration along the k_y axis, since the dielectric slab is lossy and therefore all the singularities of the integrand lie off the real axis.

3 TIME DOMAIN ANALYSIS

To obtain the time-domain response of the material slab, (7) was evaluated as a function of frequency from 10 MHz to 30 GHz at 3000 points for a fixed source/observer location and with fixed parameters for the dielectric, and then the inverse fast Fourier transform (FFT) was applied. Because the amplitude of the field grows with frequency, the field must be appropriately windowed before applying the FFT.

Figure 2 shows the time-domain scattered field found using a Gaussian window function with three different source and observer locations chosen

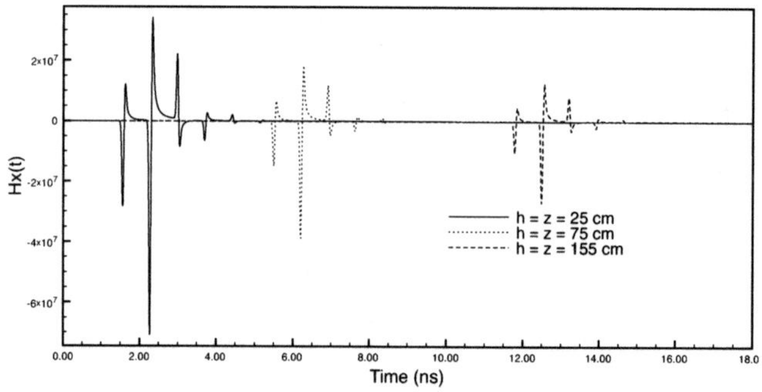

Fig. 2. Scattered magnetic field in the time domain at differing source and observer heights. ($\theta_i = 30.0°$, $\sigma = .01\, S/m$, $\epsilon_r = 4.8$, $d = 5\, cm$)

such that the "angle of incidence" (the angle θ_i in Fig. 1) is constant for each location. The responses consist of a number of temporal events, each representing a reflection from the air-material or air-conductor interface. Since these events are similar to those that occur when a conducting body is illuminated by a transient wave, this suggests that the response of the material layer can possibly be written as a natural mode series. Note that the amplitude of the response decreases and the timing of the events is delayed as the source and observation point move farther from the slab.

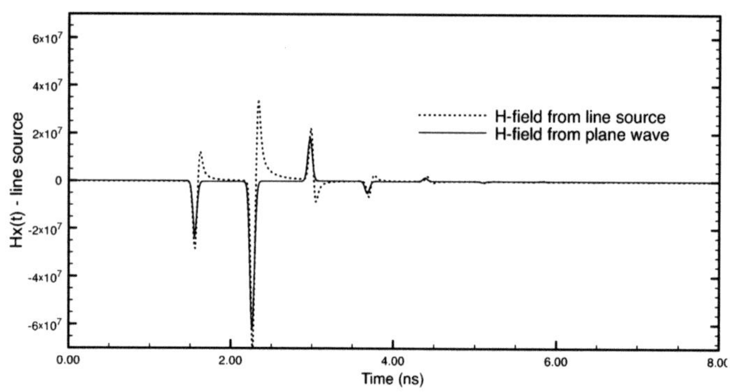

Fig. 3. Comparison of time-domain reflected waveforms for a plane wave and line source excitation. ($\theta_i = 30.0°$, $h = z = y = 25\, cm$, $\sigma = .01\, S/m$, $\epsilon_r = 4.8$, $d = 5\, cm$)

Further justification for a natural-mode expansion can be obtained by comparing the results to those produced by a planar wavefront excitation, which are known to have an explicit natural-mode representation. Figure 3 shows the scattered magnetic fields for both plane and cylindrical-wave excitation. The angle of incidence of the plane wave is chosen to match the source/receiver geometry of the cylindrical excitation. Note that the temporal events have nearly identical time positions in each case, although their shapes are different. The difference in shapes is due to the "tail" produced by the transient radiation of a line source, which is not present in the plane-wave excitation.

4 NATURAL MODE EXTRACTION

Using the E-pulse method [8],[9], the natural resonance frequencies were extracted from the late-time portion of the time-domain response, which begins with the first reflection from the material/conductor interface. Using a 3.6

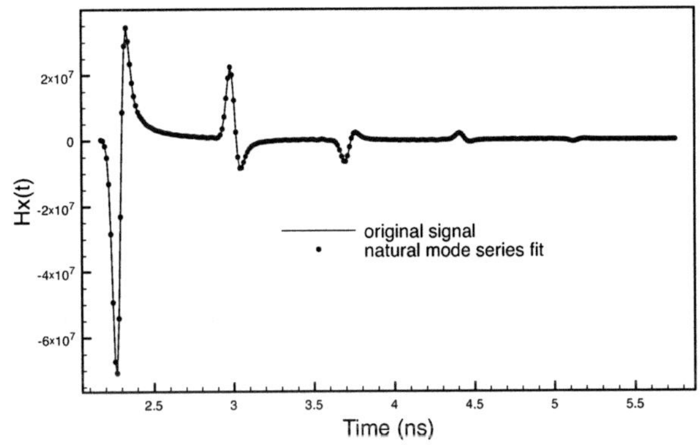

Fig. 4. Comparison of the computed late-time magnetic field signal and the signal reconstructed from the extracted natural mode series. ($\theta_i = 30.0°$, $h = z = y = 25\,cm$, $\sigma = .01\,S/m$, $\epsilon_r = 4.8$, $d = 5\,cm$)

nanosecond window (which captures 5 reflection events within the planar slab), the first 13 natural modes were extracted from the late-time signal. The 3.6 ns extraction window starts at the beginning of the late time reflection. An example of the original and reconstructed signal is shown in Fig. 4. As can be seen, the natural mode reconstruction produces an excellent representation of the late-time portion of the scattered field.

5 COMPARISON TO PLANE WAVE EXCITATION

Intuitively, one would expect that as the source and observer are moved further from the slab while keeping the angle of incidence constant, the natural modes extracted with a cylindrical-wave excitation would approach those extracted with a plane-wave excitation. As the distance increases, the cylindrical wavefronts become increasingly planar as they intersect with the surface of the slab. To investigate this, the source and observer locations were varied (while keeping the angle θ_i in Fig. 1 constant) and the previous simulations were carried out to obtain the natural modes for each source/observer location.

The natural modes from the plane wave case were rather straightforward to calculate. From [7],

$$R(\omega) = \frac{\Gamma(\omega) - P^2(\omega)}{1 - \Gamma(\omega)P^2(\omega)} \tag{8}$$

describes the reflection coefficient for a plane wave reflecting from a conductor-backed dielectric slab, where

$$P(\omega) = e^{-jk_z d}, \quad k_z = \sqrt{k^2 - k_o^2 \sin^2\theta_i}, \quad k_o = \omega/c, \quad k^2 = \omega^2\mu\epsilon\,(1 - j\sigma/\omega\epsilon)$$

and Γ is the interfacial reflection coefficient for a plane wave incident on an interface between two media. Using (8), an expression was obtained for the reflected wave as a function of frequency. Following the same procedure described above, the reflected plane wave was calculated from 10 MHz to 30 GHz, windowed, and an inverse FFT was performed on the signal. Then the natural modes were extracted from the time-domain signal. Thus it is possible to compare the natural frequencies derived from the plane-wave excitation with the natural frequencies due to cylindrical-wave excitation.

Figure 5 shows the first 3 natural frequencies for cylindrical-wave excitation and the natural frequencies derived for the plane-wave excitation. Note that for each of the modes, there is a general migration of the natural frequencies as the source height is increased, and especially for modes 2 and 3 it seems that the migration is in the direction of the plane wave natural frequency. For the first mode, the general direction of the natural frequency movement appears to be away from the plane wave case, however it is possible that as h increases, the modes begin to move toward the first plane wave natural frequency.

A few comments are in order. At first glance, it appears that there is quite a large spread among the natural frequencies as the height is increased. However, it should be noted that the imaginary (ω) axis is expanded. Figure 6 shows the first 5 natural modes on one plot to give a better perspective. The spread in Fig. 5 is due to numerical inaccuracies introduced by the varying position of the extraction window used in determining the natural modes.

A final comment relates to the limits of this simulation. Numerical noise becomes a factor when the source height is increased beyond approximately 150 cm. Noise is introduced during the numerical integration to obtain the magnetic field, during the windowing and Fourier transform to the time domain, and especially during the natural mode extraction. Therefore, while

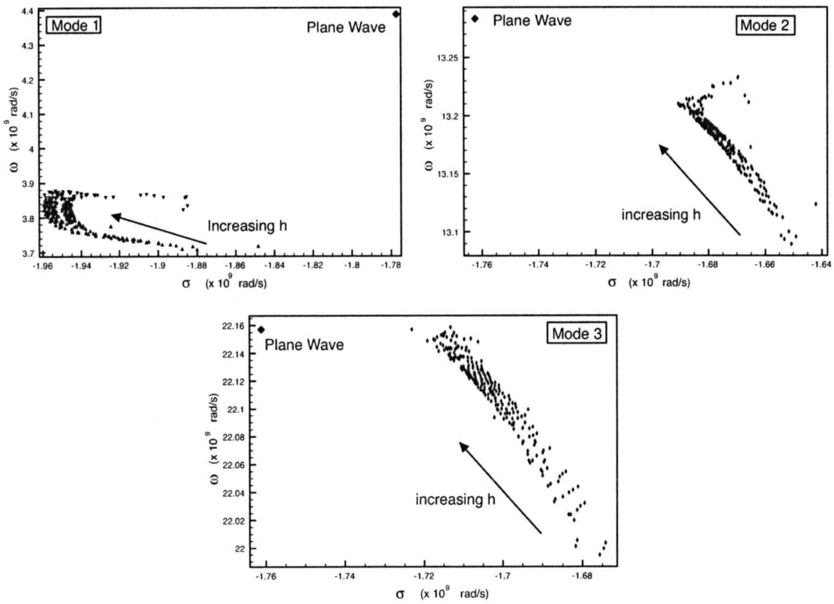

Fig. 5. Display of natural frequencies for modes 1-3. Source heights $h = 25 - 140\,cm$. $4\,cm$ step size.($\sigma = .01\,S/m$, $\epsilon_r = 4.8$, $d = 5\,cm$)

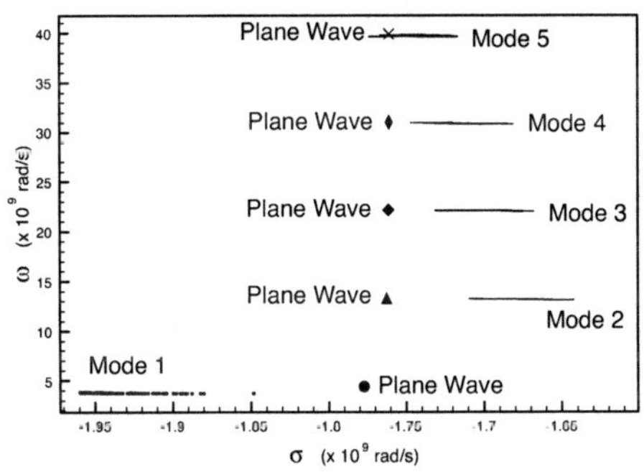

Fig. 6. Comparison of natural frequencies for modes 1-5 for plane-wave and cylindrical-wave excitation. Cylindrical-wave source heights $h = 25 - 140\,cm$.

it appears that the natural modes of cylindrical excitation approach those for plane-wave excitation as the height of the source is increased, further exploration may be required to verify this observation.

6 CONCLUSION

The time-domain waveform for the fields scattered from a magnetic line source above a grounded dielectric slab were obtained by using Fourier integration on space coordinates and then using an FFT in time. Through a natural mode extraction algorithm, it is possible to reconstruct the late-time reflected signal and express the signal as a natural mode series.

As the source and receiver are moved further from the slab, the extracted natural frequencies appear to move toward those for plane-wave excitation. The extraction process is limited by numerical noise, preventing a definitive statement regarding the relationship between the natural modes due to a cylindrical wave derived through this method and those determined for the plane-wave excitation. Further investigation is underway to examine the poles of the magnetic field in the Laplace domain for comparison with these results.

References

1. C. E. Baum, Emerging technology for transient and broad-band analysis and synthesis of antennas and scatterers, *Proc. IEEE.* **64**, 1598-1616 (1976).
2. C. E. Baum, E.J. Rothwell, K. M. Chen and D. P. Nyquist, The singularity expansion method and its application to target identification, *Proc. IEEE* **79**, 1481-1492 (1991).
3. Joon-Ho Lee, In-Sik Choiand Hyo-Tae Kim, Natural frequency-based neural network approach to radar target recognition, *IEEE Trans. Signal Processing* **51**, 3191-3197 (2003).
4. Chi-Chih Chen and L. Peters, Buried unexploded ordnance identification via complex natural resonances, *IEEE Trans. Antennas and Propagat.* **45**, 1645-1654 (1997).
5. J. C. Oh, E. J. Rothwell, and D. P. Nyquist, Natural resonance representation of the transient field reflected by a conductor-backed lossy layer, *J. of Electromagn. Waves and Appl.* **17**, 673-694 (2003).
6. H. Vollmer and E. J. Rothwell, Resonance series representation of the early-time field scattered by a coated cylinder, *IEEE Trans. Antennas and Propagat.* **52**, 2186-2189 (2004).
7. G. Stenholm, E.J. Rothwell, D.P. Nyquist, L.C. Kempel, and L.L. Frasch, E-pulse diagnostics of simple layered materials, *IEEE Trans. Antennas and Propagat.* **51**, 3221-3227 (2003).
8. E. J. Rothwell and K. M. Chen, A hybrid E-pulse/least squares technique for natural resonance extraction, *Proc. IEEE* **76**, 296-298 (1988).
9. E. Rothwell, K. M. Chen, and D. P. Nyquist, Extraction of the natural frequencies of a radar target from a measured response using E-pulse techniques, *IEEE Trans. Antennas and Propagat.* **35**, 715-720 (1987).

AN ANALYSIS OF TIME-DOMAIN DORT METHOD FOR ULTRAWIDEBAND PROBING OF EMBEDDED OBJECTS IN DISPERSIVE AND RANDOM MEDIA

Mehmet E. Yavuz and Fernando L. Teixeira*

1. INTRODUCTION

Time-reversal (TR) techniques were first developed in acoustics[1] and they involve the re-transmission of signals acquired by a set of transceivers in a time-reversed fashion. In lossless and stationary media, time-reversed signals yield optimal focusing around the original scatter locations due to the TR invariance of the wave equation. An important application of TR is the *selective* focusing on desired scatterer(s) in media with multiple scatterers. This involves the eigenvalue decomposition (EVD) of the TR operator (TRO) of a TR antenna array (TRA) and uses these eigenvectors as the TRA excitations. This TR-based technique is known as the DORT[2,3] method (French acronym for *decomposition of the time reversal operator*), which was first applied to electromagnetic (EM) waves by using time-harmonic signals and a full aspect sensor configuration[4]. Then, extensions to limited aspect configuration and ultrawideband (UWB)[5,6] signals have been considered. Most of the previous works on TR have assumed lossless media. However, when the intervening medium is dispersive and/or lossy, TR invariance is broken and compensation techniques become necessary for satisfactory TR operation. A Short Time Fourier Transform (STFT) based method has recently been introduced to compensate for the cumulative and frequency-dependent attenuation in homogeneous dispersive background media[7]. In this work, this compensation technique is extended for the application of the time-domain (TD)-DORT method in both dispersive and random inhomogeneous media. For this end, the effects of dispersion and conductivity on the eigenvalue and eigenvector structure of the TRO are studied first. Then, time-dependent wideband inverse filters are designed and applied for compensation. Additionally, a numerical compensation method proposed for conductive medium only[8] is extended to both dispersive and conductive media. The simulation scenarios considered are based on typical subsurface sensing scenarios, where UWB signals under limited aspect sensor

* ElectroScience Laboratory, Department of Electrical and Computer Engineering, The Ohio State University, 1320 Kinnear Road, Columbus, OH 43212 USA. E-mail: {yavuz.5,teixeira.5}@osu.edu

configurations are employed. The random media models are based on inhomogeneous soil models with fluctuating dielectric permittivity having prescribed correlation functions and variances.

2. TIME-DOMAIN DORT IN RANDOM DISPERSIVE MEDIA

DORT is developed by considering the N×N multistatic data matrix (MDM) $\mathbf{K}(\omega)$ from a TRA of N transceivers, where ω is the frequency of operation. In the frequency domain, TR of $\mathbf{K}(\omega)$ is represented by its Hermitian conjugate $\mathbf{K}^H(\omega)$ and the TRO is defined by $\mathbf{T}(\omega) = \mathbf{K}^H(\omega)\mathbf{K}(\omega)$. The singular value decomposition (SVD) of the MDM is given by $\mathbf{K}(\omega) = \mathbf{U}(\omega)\mathbf{\Lambda}(\omega)\mathbf{V}^H(\omega)$, where $\mathbf{U}(\omega)$ and $\mathbf{V}(\omega)$ are unitary matrices and $\mathbf{\Lambda}(\omega)$ is real diagonal matrix of singular values. The EVD of the TRO can be written as $\mathbf{T}(\omega) = \mathbf{V}(\omega)\mathbf{S}(\omega)\mathbf{V}^H(\omega)$, where $\mathbf{S}(\omega)= \mathbf{\Lambda}^H(\omega)\mathbf{\Lambda}(\omega)$ is the diagonal matrix of eigenvalues. The columns of the unitary matrix $\mathbf{V}(\omega)$ are normalized eigenvectors of the TRO ($\mathbf{v}_m(\omega)$, m = 1, ..,N). For isotropic scattering from well-resolved point-like scatterers, each significant eigenvalue of the TRO is associated with a particular scatterer. Subsequent back-propagation of the corresponding eigenvector yields a wavefront focusing on that scatterer[3]. Therefore, selective focusing on the mth scatterer is achieved by exciting the TRA with N×1 column vector $\mathbf{r}_m(\omega)$ generated from $\lambda_m(\omega)$ and eigenvector $\mathbf{v}_m(\omega)$ via $\mathbf{r}_m(\omega) = \lambda_m^{-1}(\omega)\mathbf{K}^H(\omega)\mathbf{v}_m(\omega)$. For ultrawideband signals, eigenvalue decomposition can be applied at all the available frequencies and a time-domain signal can be generated via $\mathbf{r}_m(t) = IFT(\lambda_m^{-1}(\omega)\mathbf{K}^H(\omega)\mathbf{v}_m(\omega))$, where *IFT* denotes the inverse Fourier transformation. Backpropagation of these time-domain signals characterizes the TD-DORT method[9].

3. SIMULATION SETUP AND DISPERSIVE, RANDOM MEDIA MODELS

Two-dimensional (2D) finite-difference time-domain (FDTD) simulations with grid size $N_x \times N_y$ = 360×360 are used to synthesize the TRO in dispersive and conductive random media (with uniform spatial cell size of Δs=1.0 cm). TM_z polarization is considered. Frequency dispersion in the intervening medium is modeled by an $M_s(=2)$ species inhomogeneous Lorentz model for the complex permittivity:

$$\begin{aligned}\varepsilon(\vec{r},\omega) &= \varepsilon_0\left(\varepsilon_\infty(\vec{r}) + \chi(\vec{r},\omega)\right) - \sigma/(j\omega) \\ &= \varepsilon_0\left(\varepsilon_\infty(\vec{r}) + \sum_{p=1}^{Ms}\left(\frac{\left(\varepsilon_s(\vec{r}) - \varepsilon_\infty(\vec{r})\right)G_p\omega_p^2}{\omega_p^2 - i2\omega\alpha_p - \omega^2}\right)\right) - \frac{\sigma}{j\omega}\end{aligned} \tag{1}$$

where $\vec{r}=x\hat{x}+y\hat{y}$ denotes spatial position, ω_p($=2\pi f_p$ with f_1=130.39 MHz, f_2=330.58 MHz) is the resonant frequency for the pth species, α_p ($=1.414\omega_p$, p=1,2) is the corresponding damping factor, G_p's (G_1 = 0.75, G_2 = 0.25) are constants with unity summation and $\varepsilon_s(\vec{r})$, $\varepsilon_\infty(\vec{r})$ are the static and infinite frequency permittivities, respectively. $\varepsilon(\vec{r}) = \varepsilon_m + \varepsilon_f(\vec{r})$ where ε_m(=3.54= $\langle \varepsilon_\infty(\vec{r}) \rangle$) is the average relative infinite frequency permittivity and $\varepsilon_f(\vec{r})$ is a zero mean Gaussian random variable. $\langle \varepsilon_s(\vec{r}) \rangle$ = 4.25 and the soil has also a static conductivity of σ=0.397 mS/m. The mean parameters considered here correspond to those of a realistic soil types (Puerto Rico type clay loam of moisture of 2.5%)[10]. The Gaussian distribution is characterized by uniform variance (δ) and uniform correlation length (l_s). The two-point correlation function is given by:

$$C(\vec{r}_1 - \vec{r}_2) = \delta \exp\left(-\frac{|\vec{r}_1 - \vec{r}_2|^2}{l_s^2} \right) \quad (2)$$

At each spatial location, the difference between $\varepsilon_s(\overline{r})$ and $\varepsilon_\infty(\overline{r})$ is chosen constant and positive, i.e. $\varepsilon_s(\overline{r}) = \varepsilon_\infty(\overline{r}) + \Delta\varepsilon$ and $\Delta\varepsilon > 0$. A uniform linear TRA lying parallel to the x-direction (with aperture length of a = 160Δs=1.6 m) at y=0 m is used with N=11 z-directed dipole transceivers starting from (x_{t1}, y_{t1}) = (-0.8,0) m. The UWB excitation pulse is the first derivative of the Blackmann-Harris pulse centered at f_c = 400 MHz. Same setup is also used for the nondispersive and lossless (σ = 0) case to provide a reference solution where the permittivity is given by $\varepsilon_r(\overline{r}) = \varepsilon_\infty(\overline{r})$.

4. RESULTS AND DISCUSSION

We first compare the TRO obtained for a single cylindrical perfectly electrical conducting (PEC) scatterer with radius r = 4.0 cm buried at y=1.35 m below the central antenna in both dispersive-conductive (DPC) and nondispersive-lossless (NDL) homogeneous media (δ = 0). The first two singular value distributions of the MDM and the phase and magnitude distribution of first eigenvector are shown in Fig. 1 and Fig. 2, respectively. Dispersion effects and losses yield smaller singular values, particularly at higher frequencies. On the other hand, ratios of the first and second singular values along the bandwidth remain almost the same. Despite the decrease on the singular value magnitudes, relative phase distribution among the array elements is preserved over the frequency band. Therefore, cross-range scatterer locations can still be estimated from these distributions[3] Once these singular values and eigenvectors are employed in TD-DORT method to generate the time-domain signals to be backpropagated from the TRA, resulting focusing quality is degraded as compared to the ideal NDL case.

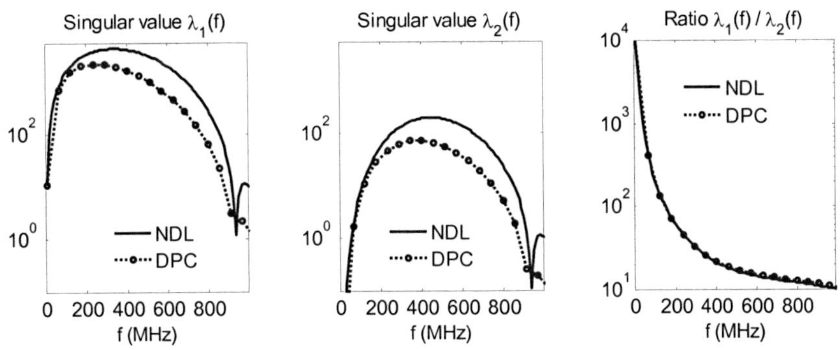

Figure 1. First and second dominant singular values and their ratios obtained in NDL and DPC media

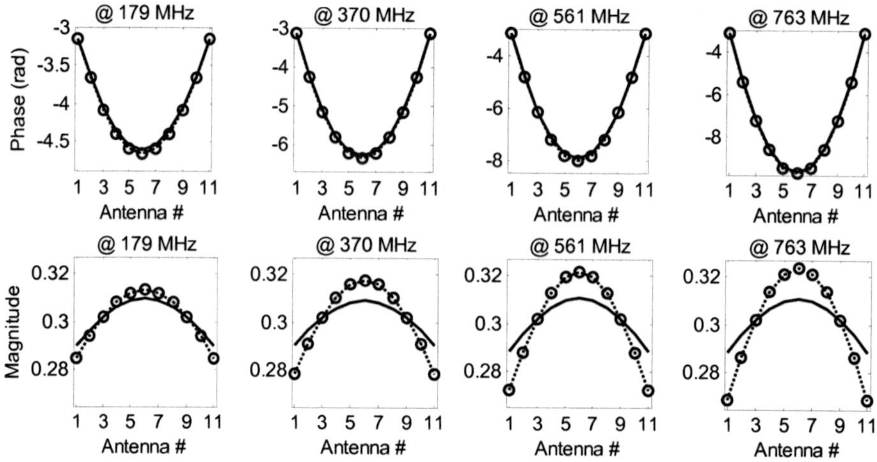

Figure 2. Phase and magnitude distribution of the first eigenvector along the bandwidth (solid line: NDL case, dotted line: DPC case)

In subsurface scenarios involving deeper targets and/or stronger attenuation, focusing of the time-reversed signals or time-domain eigenvectors will be further degraded unless dispersion and/or loss compensation is applied on the signals received at the TRA before backpropagation. A compensation technique was previously proposed for homogeneous background media and utilizes time-dependent inverse filters based in Short Time Fourier Transforms[7] Such filters can easily be constructed for homogeneous media; however, for random media difficulties arise since attenuation depends on position. Unless the characteristics of the medium are known pointwise, it is not possible to find exact filters to compensate for losses. Therefore, approximate filters should be employed instead. To construct these filters, we assume that only first and second order statistics of the dispersive and/or conductive random medium are known (i.e. variance and the correlation length). To simulate the attenuation undergone in the original random medium due to dispersion and/or losses, multiple realizations of a test random medium are used. The following procedure is repeated for both nondispersive-lossless and dispersive-conductive random test media. An ultrawideband signal is transmitted from a point source at a certain location in the test media and propagating signals are recorded at increasing distances with fine sampling. We denote the resulting distance versus time curves as *DvT plots*. For each realization, the total electric field at distance $\bar{r}$ can be written as

$$\vec{E}(\bar{r}') = \vec{E}_{inc}(\bar{r}') + \int_V d\bar{r}' \bar{\bar{G}}(\bar{r}, \bar{r}').o(\bar{r}')\vec{E}(\bar{r}') \tag{3}$$

where $\vec{E}_{inc}(\bar{r})$ is the incident field (invariant for all realizations), $\bar{\bar{G}}$ is the Green's function for an homogeneous medium (with static conductivity present) of relative dielectric permittivity ε_m and $o(\bar{r}) = \omega^2\mu_0(\varepsilon(\bar{r})-\varepsilon_m)$ is the contrast function, which depends on the random fluctuation of each particular realization. Then, an average DvT plot can be obtained via ensemble averaging of DvT plots obtained for different realizations of test random media. The ensemble averaged field at receivers is written as:

$$\left\langle \vec{E}(\vec{r}) \right\rangle_{Ne} \approx \vec{E}_{inc}(\vec{r}) + \int_V d\vec{r}' \bar{\bar{G}}(\vec{r},\vec{r}').\left\langle o(\vec{r}')\vec{E}(\vec{r}') \right\rangle_{Ne} \tag{4}$$

where $\langle . \rangle_{Ne}$ denotes ensemble statistical average over N_e realizations. In the Born approximation regime (weak fluctuations), the ensemble averaged electric field becomes:

$$\left\langle \vec{E}(\vec{r}) \right\rangle_{Ne} \approx \vec{E}_{inc}(\vec{r}) + \int_V d\vec{r}' \bar{\bar{G}}(\vec{r},\vec{r}').\left\langle o(\vec{r}') \right\rangle_{Ne} \vec{E}_{inc}(\vec{r}') \tag{5}$$

This approximation linearizes the dependency of $\langle \vec{E}(\vec{r}) \rangle_{Ne}$ on $\langle o(\vec{r}) \rangle$. In this regime, the ensemble average of the random media can be used to obtain the average electric field response. For large number of realizations, the average response in this approximation approaches that of a homogeneous medium with $\varepsilon_{sens} = \langle \varepsilon_s(\vec{r}) \rangle$ and $\varepsilon_{\infty ens} = \langle \varepsilon_\infty(\vec{r}) \rangle$. Therefore, DvT plot of a homogeneous medium having such parameters can be used to obtain approximate time-dependent filters for any realization of a given dispersive and/or lossy random medium, as long as background fluctuations are weak. A similar procedure is also used to obtain the DvT plot for nondispersive and lossless case. Once DvT plots are obtained, compensation filters can be constructed in a usual fashion[7].

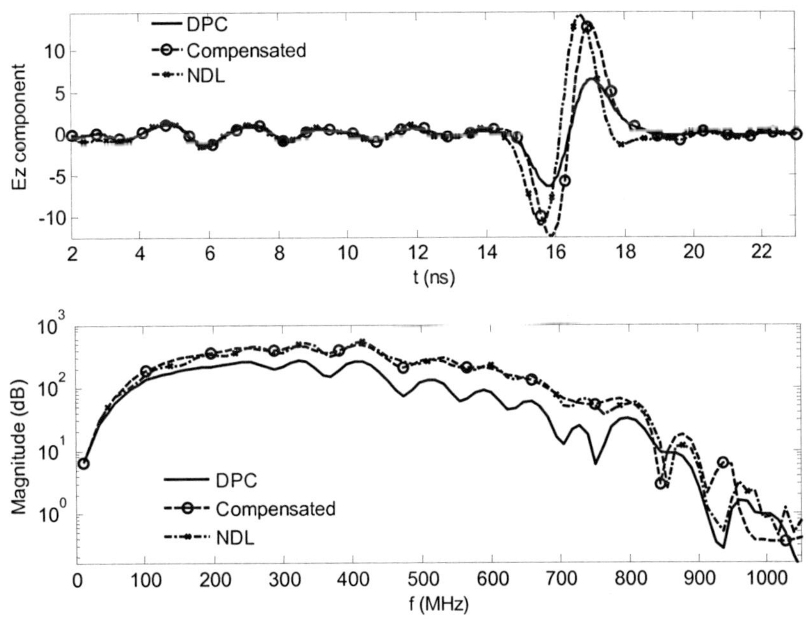

Figure 3. Time (top) and frequency (bottom) domain comparisons of the received signal at one of the TRA antennas in dispersive random medium (with variance $\delta = 0.005$ and correlation length $l_s = 3\Delta_s$) without compensation, its compensated version, and a reference signal that would have been received if the medium were nondispersive. Hamming windows of length $256\Delta t$ and overlapping factor of 0.5 are used[7].

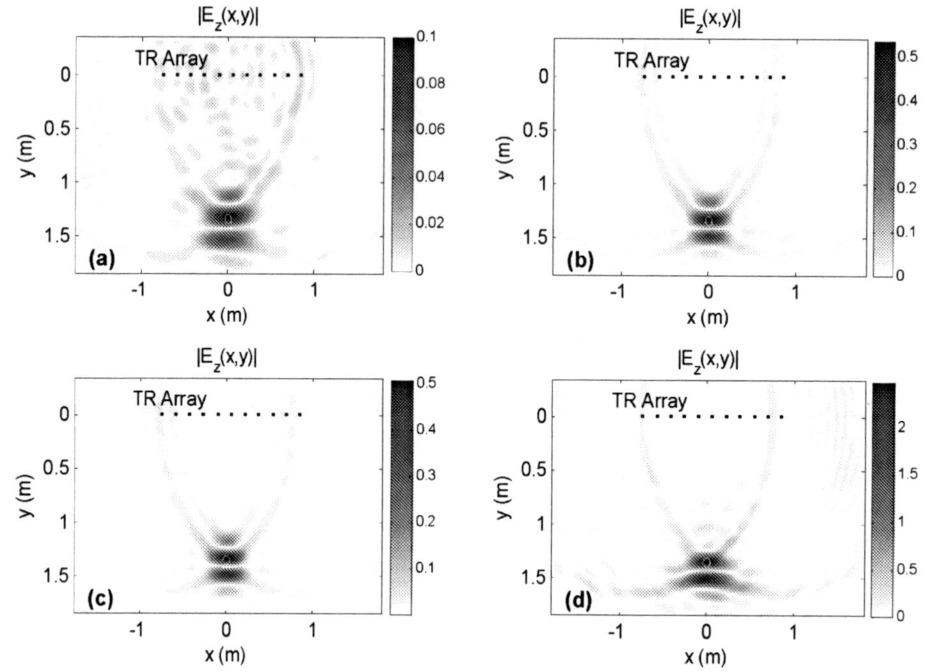

Figure 4. Snapshots of the spatial distribution of $|E_z|$ on the xy-plane in NDL and DPC (with and without compensation) random media. (a) DPC without compensation, (b) NDL, (c) DPC with compensation using inverse filters, (d) in modified medium using numerical compensation algorithm.

Applying these filters to the received signals partially compensates for losses as shown in Fig. 3. The compensation filters are zero-phase filters since phase delay is automatically compensated by TR. Since any additive noise present in the system would also be amplified, compensation filters are not of use when signal-to-noise ratio is below acceptable levels (which may depend on the application considered). We note that only impenetrable scatterers are considered in the analysis here. The method should be modified for penetrable scatterers where reflectivity is frequency dependent. In Fig. 4, magnitude of the focused E_z field components in dispersive-conductive and nondispersive-lossless random media are shown with and without compensation. The cross-range resolutions for each case are also provided in Fig. 5. Note that performance varies for each realization, but compensation yields better performance than the case without compensation. The overall performance also depends on the moisture content of the medium, distance of the scatterer to the TRA and the amount of fluctuations[9].

5. NUMERICAL COMPENSATION

Recently, another alternative algorithm that compensates for medium losses has been discussed[8]. In this approach, the compensation is performed using synthetic data from FDTD simulations where the sign of the conductivity term is reversed (active-like medium). This approach is unstable and is limited for small conductivity values because of the ill-posedness of the resulting initial-value problem, and requires point-wise knowledge of the medium conductivity at all spatial points. Despite the disadvantages, this compensation technique can still be extended to Lorentz or Debye dispersive media by modifying the medium susceptibility function $\chi(\omega)$ (Eq. 1) accordingly. In the Lorentz case, this corresponds to reversing the sign of the damping factors α_p. As long as the original attenuation is small and required backpropagation distance is not large, this method can compensate for the dispersive and conductive effects. However, instability and high sensitivity to numerical noise (ill-posedness) again exist since the entire media amplifies the signal. We apply this technique for the same dispersive random medium of the previous subsection. The time-domain signals (with compensation) are transmitted into the modified medium and the resulting spatial distribution of the focused E_z field component is given in Fig. 4(d). Also corresponding cross-range is compared with others in Fig. 5(b). It is observed that modified medium amplifies the signals much more than the reference case which is not realistic. Also, long time instability exists for this simulation. Note finally that, this compensation can only be applied numerically, i.e. it cannot be implemented in practice since it is impossible to change the medium characteristics. On the other hand, the compensation via filters is suitable for practical implementation since only the received signals are modified.

6. SUMMARY AND CONCLUSIONS

TD-DORT method has been studied for detection and localization of impenetrable scatterers in dispersive and conductive random media. Effects of dispersion and losses on the TRO have been analyzed. It has been observed that although dispersion and/or losses result in decrease on the TRO eigenvalues, localization information can still be extracted from the phase information. Using phase information along with attenuation compensation results in a focusing performance closer to that achieved in a nondispersive and lossless case.

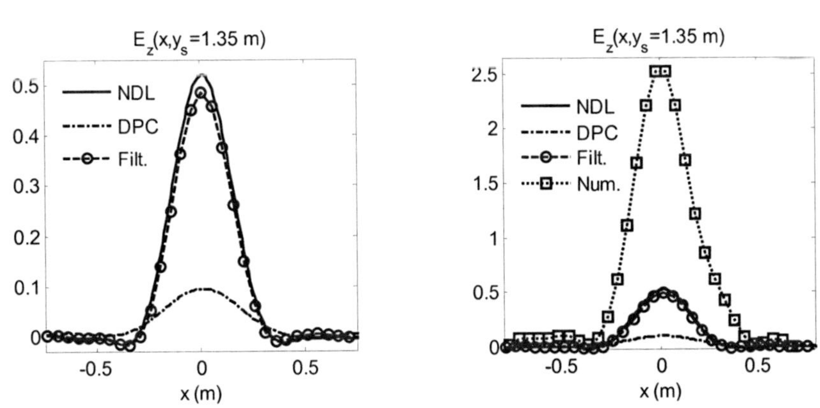

Figure 5. Cross-range field distributions at y_s=1.35 m for comparing (left) the compensation using inverse filters in DPC media (Filt.) with NDL and DPC case, (right) numerical compensation (Num.) with others

For cases where background medium is inhomogeneous and dispersive (and/or lossy) and only statistical information available it is not possible to exactly compensate for the dispersive and/or loss effects. However, a satisfactory approximate compensation can be achieved in the regime of low fluctuations and losses (under Born approximation) by using ensemble averages. Additionally, an alternative numerical compensation method has been extended to both dispersive and/or lossy case.

7. ACKNOWLEDGMENTS

This work has been supported in part by National Science Foundation under grant ECS-0347502, and Ohio Supercomputer Center under grants PAS-0061 and PAS-0110.

8. REFERENCES

1. M. Fink, D. Cassereau, A. Derode, C. Prada, P. Roux, M. Tanter, J. Thomas, and F. Wu, Time-reversed acoustics, *Rep. Prog. Phys.* **63**, 1933–1995 (2000)
2. C. Prada and M. Fink, Eigenmodes of the time reversal operator: a solution to selective focusing in multiple-target media, *Wave Motion* **20**, 151–163 (1994)
3. C. Prada, S. Mannevile, D. Spoliansky, and M. Fink, Decomposition of the time reversal operator: detection and selective focusing on two scatterers, *J. Acoust. Soc. Am.* **99**(4), 2067–2076 (1996)
4. H. Tortel, G. Micolau, and M. Saillard, Decomposition of the time reversal operator for electromagnetic scattering, *J. Electromagn. Waves Appl.* **13**, 687–719 (1999)
5. G. Micolau, M. Saillard, and P. Borderies, DORT method as applied to ultrawideband signals for detection of buried objects, *IEEE Trans. Geosci. Remote Sensing* **41**(8), 1813–1820 (2003)
6. M. E. Yavuz and F. L. Teixeira, A numerical study of time reversed UWB electromagnetic waves in continuous random media, *IEEE Antennas and Wireless Prop. Lett.*, **4**, 43–46 (2005)
7. M. E. Yavuz and F .L. Teixeira, Frequency dispersion compensation in time reversal techniques for UWB electromagnetic waves, *IEEE Geosci. Remote Sensing Lett.* **2**(2), 233–237 (2005)
8. P. Kosmas and C. M. Rappaport, Time reversal with the FDTD method for microwave breast cancer detection, *IEEE Trans. Microwave Theory Tech.*, **53**(7), 2317–2323 (2005)
9. M. E. Yavuz and F. L. Teixeira, Full time-domain DORT for ultrawideband electromagnetic fields in dispersive, random inhomogeneous media, *IEEE Trans. Antennas and Propag.* (accepted)
10. F. L. Teixeira, W.C. Chew, M. Straka, M. L. Oristaglio, and T. Wang, Finite-difference time-domain simulation of ground penetrating radar on dispersive, inhomogeneous and conductive soils, *IEEE Trans. Geosci. Remote Sensing*, **36**(6), 1928–1937 (1998)

APPLICATION OF THE METHOD OF SUBREGIONS TO MEASUREMENT OF LAYERED MATERIALS

Bradley Perry, Steven Cossmann, Leo Kempel, and Edward Rothwell *

1 INTRODUCTION

Layered material coatings are often applied to surfaces in order to reduce the scattered field strength within specific frequency bands. A full understanding of the transient behavior of the reflected field from layered materials can be used to diagnose the composition of these coatings. It has been shown previously that the transient field reflected from a two-layer conductor-backed geometry can be broken into responses from subregions of the material stack [1]. Experimental validation of the form of the reflected field from an N-layered material structure is of interest here. The method of subregions is used to show that the transient plane-wave field reflected from an N-layer planar stack of materials consists of a series of late-time responses of subregions of the stack, with the i-th subregion consisting of the first i-material layers backed by a semi-infinite region with the material parameters of region (i+1). Each of these late-time responses exists during time periods corresponding to the excitation of the different layers of the N-layered medium.

Measurements are made using the MSU Reflectivity Arch Range which is a circular structure 6.096 meters in diameter, standing 1.219 meters high with horn antennas mounted around the periphery. These antennas can be placed in various configurations around the arch, but for these measurements are placed side by side in order to approximate a mono-static arrangement. Scattering from the material stack, which is mounted in the center of the circular structure and at the same height as the horn antennas, is measured using a true time domain system consisting of a digital sampling oscilloscope with time-domain reflectometry plug-in and a pulse generating network. A PEC plate calibration is used to characterize the system (cabling, horn antennas, etc.) in order to attain the desired response. Measured results are compared in the time domain with responses of various subregions of the material stack in order to show that each subregion can be characterized individually, disregarding the composition of the rest of the material stack, whose response will turn on once the incident wave has propagated further into the layered

* Department of Electrical and Computer Engineering, Michigan State University, East Lansing, MI 48824, USA perrybr3@msu.edu, rothwell@msu.edu

medium. This result is useful for evaluation of layered coatings on bodies whose underlying composition may be unknown or complex.

2 METHOD OF SUBREGIONS

In the method of subregions, a representation of the reflected field from an N-layered material stack is found in terms of the temporal responses from subregions of the material stack during certain time periods. These time periods correspond to the time that it takes for a wavefront to propagate through each material region and return to the observation plane, which is located at the interface between free space and the material stack. Note that the reflection from an N-Layered material stack can be written in terms of the various subregion responses as

$$\begin{aligned} \Gamma_N(t) &= [R_1(t)u(t)]u(\tau_1 - t) + [\Gamma_1(t)u(t-\tau_1)]u(\tau_2 - t) \\ &+ \ldots + [\Gamma_{N-1}(t)u(t-\tau_{N-1})]u(\tau_N - t) + \Gamma_N(t)u(t-\tau_N) \end{aligned} \quad (1)$$

where τ_α is the two-way transit time of the arbitrarily incident planar wavefront from the first interface to the α-th material interface, and is given by

$$\tau_\alpha = \sum_{i=1}^{\alpha} \frac{2\Delta_i}{c} \sqrt{\epsilon_{ir}\mu_{ir} - sin^2\theta}. \quad (2)$$

Here, each subregion reflection coefficient, $\Gamma_i(t)$, is the temporal response from an i-layered subregion backed by a half-space with the material parameters of region (i+1), and θ is the incidence angle at the first interface. To illustrate this, examine Figure 1. In Figure 1(a), an N-layered material stack is shown. Each subregion of the stack is shown in Figures 1(b)-(e). During the earliest time of the response, the interfacial reflection between free space and the first material region is representative of the response of the material stack; this geometry is shown in Figure 1(b). Once the incident wavefront has propagated to the interface between the first and second layers and the reflected field from this interface has returned to the observation plane, the response from the first subregion turns completely off and a new response turns on. This is the response from the geometry of Figure 1(c). This progression of responses continues with the next subregion shown in Figure 1(d) and on up to the response of the (N-1) layer geometry shown in Figure 1(e). Finally, the response in the late-time period is given by the full, N-layer geometry response.

3 EXPERIMENTAL PROCEDURES

In this section, experimental parameters and procedures are discussed, along with calibration of the time domain measurement system. Included in the

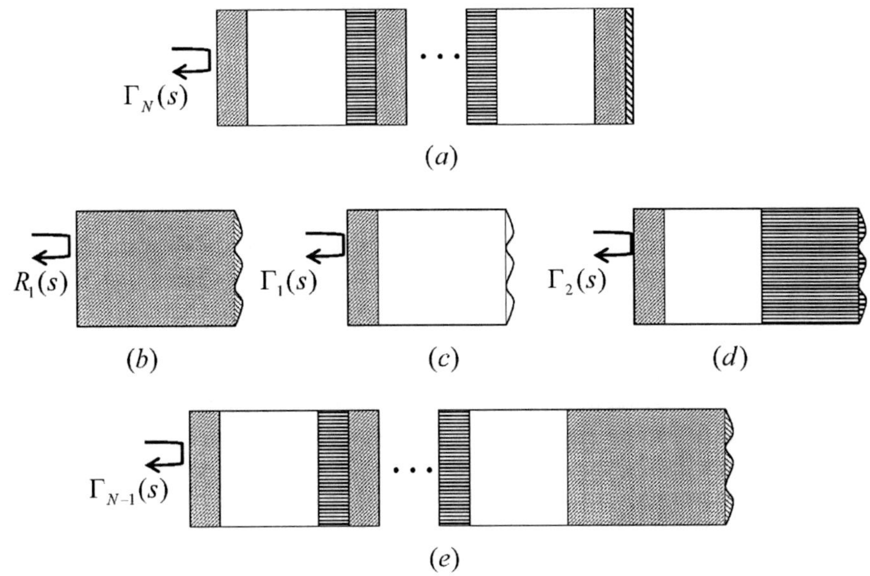

Fig. 1. Subregions of an N-layered material stack:(a) complete geometry (b) interfacial reflection (c) single layer subregion (d) two-layer subregion and (e) (N-1)-layer subregion

calibration procedures are necessary measurements and weighting techniques needed to obtain the desired response from raw measured data.

3.1 Measurement System Description

As stated previously, measurements are made using the MSU reflectivity arch range and a true-time domain measurement system. This system consists of an HP 84750A digital sampling oscilloscope with HP 54753A TDR plug-in, a Picosecond Pulse Labs 4015B step generator, a PSPL 4015RHP remote pulse head, and a 5208 pulse generating network. The transmitting and receiving horn antennas have a 2-18 GHz bandwidth, and are manufactured by American Electronics Laboratories, model H-1498. The target, in this case a stack of material samples, each measuring 2 feet on a side and approximately 1 inch thick, is placed in the center of the circular arch range.

3.2 Determination of Experiment Parameters

Experiment parameters were selected to either maximize the signal-to-noise ratio (SNR) or minimize the effect of mutual interactions on the measurements. Increasing SNR provides for more accurate results in arch range measurements, while minimizing the mutual interactions in the measurements allows the system response and the desired target response to be obtained using the calibration procedures outlined below.

By maximizing signal-to-noise ratio, more accurate results can be obtained from the arch range measurements. In general, increasing the number of averages will increase the SNR [2]; however, when the background noise is high, the subtraction of the background noise measurement can worsen the results due to small time drifts in longer sampling processes. It has been found experimentally, using the measurement system described here, that 1024 averages provide a good SNR without an excessively long sampling process. [3]

In order to minimize the affect of mutual interactions on the measurement, absorbing foam and time-gating are used. Based on the duration of the target response, 10 ns windows were chosen for measurements taken with the time-domain measurement system. This allows the response of the material to the incident field to be measured, while the effect of mutual interactions between the material and surrounding objects is removed when they are sufficiently separated in time. It also allows the strong direct interaction between the horns, which is difficult to completely remove through subtraction, to be windowed out, since it occurs much earlier in time than the target response. Although the metallic pedestal needed to support the material samples is close in proximity to the target, measurements have shown that the interactions with the pedestal are negligible due to the focusing of the field by the dielectric lenses [4].

3.3 Arch Range Measurement and Calibration

In order to obtain the desired reflected-field response, the system response and the effects of the surrounding environment need to be removed through an appropriate calibration. The techniques described in [3]-[7] are employed. The system response is a function of the cables, antennas, antenna coupling, dielectric lenses, mutual interactions, and arch-range clutter. For the arch range system, the measured waveform of an unknown target can be formulated as

$$\begin{aligned} R_T(f) &= E(f)H_T(f)H_R(f)\{H_A(f) \\ &+ H_{TL}(f)H_{RL}(f)[H_S^T(f) + H_{SC}^T(f) + H_C(f)]\} + N(f). \end{aligned} \tag{3}$$

Here $H_S^T(f)$ is the unknown response of the target and $N(f)$ is the background noise present in the environment. Examining Equation 3, the measured waveform is formulated beginning with the source pulse, $E(f)$. This pulse arrives at the transmitting horn antenna, which has a transfer function of $H_T(f)$. Next, the field is radiated by the horn antenna and is either passed through a collimating lens, with a transfer function of $H_{TL}(f)$, or it directly couples to the receiving antenna, with this path represented by the transfer function $H_A(f)$. The portion of the field passing through the collimating lens is then either reflected by the target directly, which is the scattered response $H_S^T(f)$, or it is involved in interactions with objects in the laboratory environment. Interactions with the environment are either direct, which are referred to as arch-range clutter with transfer function $H_C(f)$, or they are mutual interactions between the target and the environment, represented by the transfer function $H_{SC}^T(f)$. With all three of these paths, the signal returns through another collimating lens, which has a transfer function of $H_{RL}(f)$. Finally, the signal passes through the receiving horn, with transfer function $H_R(f)$, and arrives at the receiver. Noise that will also be present at the receiver is included through the term $N(f)$.

Several measurements are required to eliminate undesired components of the reflected field response. The first of these measurements is used to eliminate the effects of the laboratory environment, and is carried out by taking a measurement with the material holder left empty. This background measurement includes the same responses as the measurement target, with the exception of the scattering and mutual interaction terms. Thus the background measurement is given as

$$R_b(f) = E(f)H_T(f)H_R(f)\{H_A(f) + H_{TL}(f)H_{RL}(f)H_C(f)\} + N(f). \quad (4)$$

By taking this background measurement immediately following other measurements, the error introduced by the background noise, $N(f)$, is minimized. By subtracting the background response from the target response, the following result is obtained:

$$\begin{aligned} R_{T-b}(f) &= E(f)H_T(f)H_R(f)H_{TL}(f)H_{RL}(f)[H_S^T(f) + H_{SC}(f)] \\ &= S(f)[H_S^T(f) + H_{SC}^T(f)]. \end{aligned} \quad (5)$$

This result is in terms of a transfer function of the measurement system, $S(f)$, the desired target response $H_S^T(f)$, and a mutual interaction term, $H_{SC}^T(f)$, which represents interactions between the target and objects in the environment, such as the metal stand used to hold the material sheets. Thus the desired target response can be obtained if the system transfer function is known and the mutual interaction terms are assumed negligible. (In fact, these terms are often quite important and must be removed using time gating). To acquire the system transfer function, the response of a calibration target must be measured. The response of the calibration measurement with the background measurement removed is found to be

$$R_{C-b}(f) = S(f)[H_S^C(f) + H_{SC}^C(f)]. \quad (6)$$

Here $H_S^C(f)$ is the theoretically known response of the calibration target. By assuming the mutual interaction between the calibration target and the surrounding environment, $H_{SC}^C(f)$, can be neglected or removed, the system response is obtained as

$$S(f) = \frac{R_{C-b}(f)}{H_S^C(f)}. \quad (7)$$

This system transfer function is used to obtain the unknown target response as

$$H_S^T(f) = \frac{R_{T-b}(f)}{S(f)}. \tag{8}$$

Using the developments of (3) through (8), the calibrated response from the time-domain measurement system is obtained using three measurements: measurement of the material, a background measurement, and a calibrator measurement. The calibrator used for this work is an aluminum conducting plate of the same size as the material samples to be measured.

4 MEASURED RESULTS

Measurements of various stacks of materials were taken in order to show the decomposition of the time domain responses in terms of various subregions. The measurements in Figures 2-4 were obtained using the calibration method described in Section 3.3, utilizing a cosine taper weighting function. This weighting function was chosen because it preserves more energy in the signal than a gaussian centered at the middle of the frequency band. This choice does, however, introduce more oscillations in the temporal response, which can be seen in the measured results of Figures 2-4.

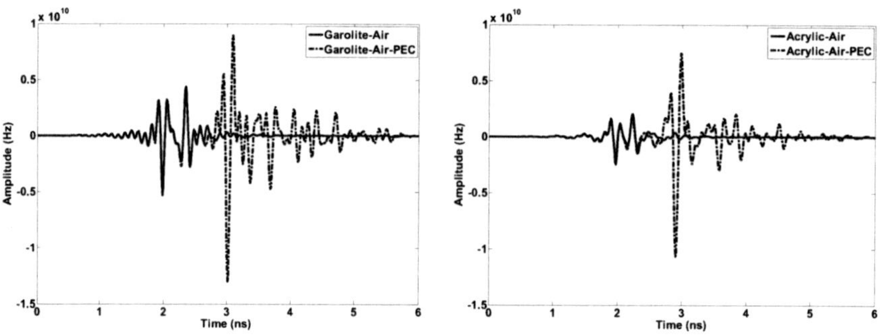

Fig. 2. Comparisons of measured time domain responses from single, air-backed material layers, and two-layered, PEC- backed geometries

Figure 2 shows two different comparisons of single, air-backed material layers with two-layered, PEC-backed geometries. The figure on the left shows measurements of an air-backed garolite ($\epsilon_r = 4.75$, $\mu_r = 1$) layer and a conductor backed two-layer stack of garolite and styrofoam. It can be seen that the response from the air-backed layer is identical to the response from the two-layer geometry until the point in time at which the response from the conductor is observed at the input to the material stack; this occurs at around 3 ns. The figure on the right shows an identical result for the comparison of an air-backed cast acrylic ($\epsilon_r = 2.5$, $\mu_r = 1$) layer with a two-layer, conductor-backed acrylic-styrofoam geometry.

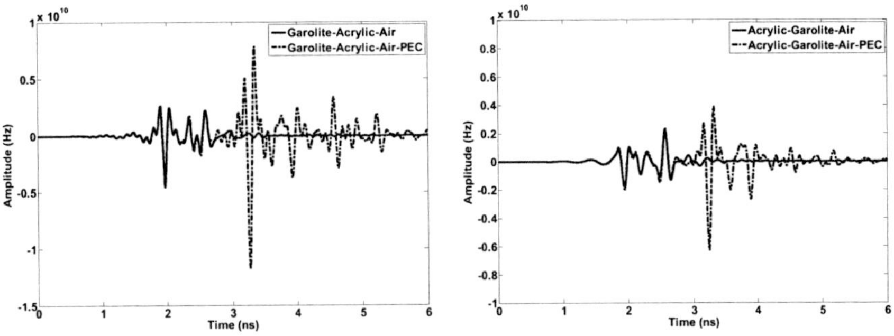

Fig. 3. Comparisons of measured time domain responses from two-layered, air-backed material stacks, and three-layered, PEC- backed geometries

The responses from various two-layer, air-backed geometries compared with three-layer geometries consisting of the same material layers backed by a thick layer of low density styrofoam, and finally a conducting plate are shown in Figure 3. The figure on the left is a garolite-acrylic-styrofoam stack with both air- and conductor-backing, while the figure on the right is an acrylic-garolite-styrofoam stack. Here it can be seen that for both cases, the subregion responses match well with the complete temporal response of the PEC backed geometries until the response from the conductor is observed, as was the case for the two-layered, conductor-backed geometries.

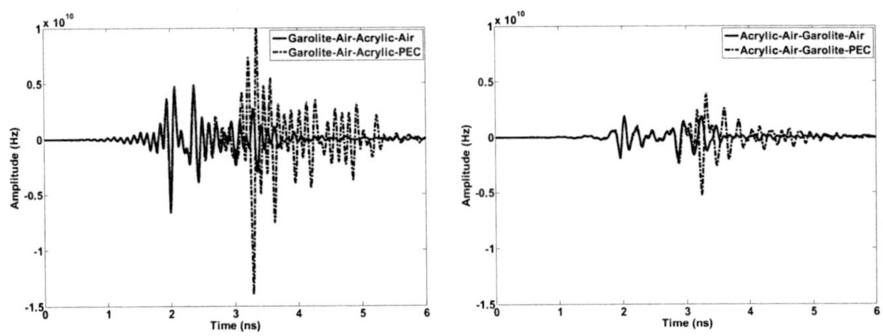

Fig. 4. Comparison of measured time domain responses from three-layer, air- and PEC- backed material stacks

The method of subregions is also useful for geometries which do not have an air layer backed by a conductor. To see this, examine Figure 4, where several three-layer stacks are compared, with and without a conductor backing. The measured response from the three-layer stacks, which are shown in the left and right portion of this figure, are a garolite-styrofoam-acrylic and an acrylic-styrofoam-garolite stack, respectively. Here it can be seen that the temporal responses agree until the response from the last interface between either air, or the conducting plate, is seen at the observation plane, occurring at around 3.5 ns.

5 CONCLUSIONS

Measured results were presented to introduce the method of subregions for measurement of layered materials. Experimental results have shown that the response from a stack of layered materials can be represented by the temporal responses of subregions of that stack for certain time periods related to the propagation of a wavefront through the material stack. These results prove useful in applications where measurement of layered materials become difficult, such as when backing materials are unknown, or complex, and interest is only in the response from the layered region.

References

1. B.T. Perry, "Natural Resonance Representation of the Transient Field Reflected From a Multilayered Material," 2005 IEEE Ant and Prop Society Symp and URSI/ANSI National Radio Science Mtg, Monterrey, CA.
2. Michael A. Morgan, "Ultra-Wideband Impulse Scattering Measurements," IEEE Transactions on Antennas and Propagation, vol. 42, no.6, pp. 840-852, June 1994.
3. A. Kizilay, "A perturbation method for transient multipath scattering from targets above periodic surfaces," Ph.D. Dissertation, Michigan State University, East Lansing, MI, 2000.
4. B. Perry, "Natural Resonance Representation of the Transient Field Reflected From a Multilayered Material," Ph.D. Dissertation, Michigan State University, East Lansing, MI, 2005.
5. J. Ross, III. "Application of transient electromagnetic fields to radar target discrimination," Ph.D. Dissertation, Michigan State University, East Lansing, MI, 1992.
6. J. C. Oh, E. J. Rothwell, and D. P. Nyquist, Natural resonance representation of the transient field reflected by a conductor-backed lossy layer, *J. of Electromagn. Waves and Appl.* **17**, 673-694 (2003).
7. M.A. Morgan and B.W. McDaniel, "Transient Electromagnetic Scattering: Data Acquistion and Signal Processing," IEEE Transactions on Intstrumentation and Measurement, vol. 36, no.2, pp. 263-267, June 1994.

A NUMERICAL STUDY ON THE SENSITIVITY OF TIME-REVERSAL IMAGING METHODS AGAINST CLUTTER, NOISE AND MODEL PERTURBATIONS

Mehmet E. Yavuz and Fernando L. Teixeira*

1. INTRODUCTION

Time-reversal (TR) techniques using acoustical waves have shown to provide promising results for the detection and localization of scatterers in multiple scattering environments by providing super-resolution and statistical stability[1]. These techniques exploit the TR invariance of the wave equation in lossless and stationary media, and involve the retransmission of scattering signals acquired by a set of receivers in a time-reversed fashion. Back-propagated TR signals tend to focus around the original scatter location(s). Many TR applications require the analysis of the TR operator (TRO) obtained from the multistatic data matrix (MDM) of a TR array (TRA). In particular, the eigenvalue decomposition (EVD) of the TRO forms the basis for both the so-called DORT (decomposition of the TRO)[2,3] and TR-MUSIC (TR multiple signal classification)[4] methods. These two approaches are examples of *TR imaging* algorithms that rely on the use of complementary subspaces of the TRO. Specifically, DORT exploits the signal subspace (SS), whereas TR-MUSIC employs the null subspace (NS). For well-resolved point-like scatterers, location and strength information of the scatterers are associated with eigenvalues and respective eigenvectors of the SS of the TRO in a one-to-one fashion[2]. Therefore, backpropagation of these eigenvectors yields images of the primary targets. This constitutes the basis for TR imaging via DORT. On the other hand, regardless of well-resolvedness criteria, the NS space of the TRO is orthogonal to SS, i.e. the projection of any vector formed by the linear combination of SS eigenvectors onto the NS should (ideally) be zero. This is the basis of TR-MUSIC imaging. In this work, we investigate the performance of both DORT and TR-MUSIC against perturbations produced by clutter, model mismatch, and TR invariance breaking (due to losses in the background media), in both narrowband (NB) and ultrawideband (UWB) cases. We focus on frequency bands and sensor deployment scenarios typical of

* ElectroScience Laboratory, Department of Electrical and Computer Engineering, The Ohio State University, 1320 Kinnear Road, Columbus, OH 43212 USA. E-mail: {yavuz.5,teixeira.5}@osu.edu

subsurface sensing problems. Clutter is produced from subsurface inhomogeneities modeled by continuous random media models based on soil parameters having spatially fluctuating Gaussian random permittivities and prescribed (also Gaussian) correlation functions. Perfectly conducting objects (PEC) embedded in the random media are considered as primary scatterers.

2. TIME-REVERSAL IMAGING VIA DORT AND TR-MUSIC

Both DORT and TR-MUSIC start by obtaining an N×N symmetric MDM $\mathbf{K}(\omega)$ from a TRA of N transceivers, where ω is the frequency of operation. In the frequency domain, the TR of $\mathbf{K}(\omega)$ is represented by its Hermitian conjugate $\mathbf{K}^H(\omega)$ and the TRO is defined by $\mathbf{T}(\omega)=\mathbf{K}^H(\omega)\mathbf{K}(\omega)$. The singular value decomposition (SVD) of the MDM is given by $\mathbf{K}(\omega)=\mathbf{U}(\omega)\mathbf{\Lambda}(\omega)\mathbf{V}^H(\omega)$, where $\mathbf{U}(\omega)$ and $\mathbf{V}(\omega)$ are unitary matrices and $\mathbf{\Lambda}(\omega)$ is real diagonal matrix of singular values. The EVD of the TRO can be written as $\mathbf{T}(\omega) = \mathbf{V}(\omega)\mathbf{S}(\omega)\mathbf{V}^H(\omega)$, where $\mathbf{S}(\omega)= \mathbf{\Lambda}^H(\omega)\mathbf{\Lambda}(\omega)$ is the diagonal matrix of eigenvalues. The columns of the unitary matrix $\mathbf{V}(\omega)$ are normalized eigenvectors of the TRO ($\mathbf{v}_m(\omega)$, m = 1, ..,N). Ideally, SS of the TRO is formed by eigenvectors with nonzero eigenvalues whereas NS is formed by those with almost zero eigenvalues, i.e. $SS(\omega)=\{\mathbf{v}_1(\omega),..., \mathbf{v}_{Ms}(\omega)\}$ with $\lambda_1>..>\lambda_{Ms}>0$ and $NS(\omega)=\{\mathbf{v}_{Ms+1}(\omega),..., \mathbf{v}_N(\omega)\}$ with $\lambda_{Ms+1}\approx..\approx\lambda_N\approx0$, where M_s is the number of significant eigenvalues. For UWB signals in cluttered environments, non-isotropic scattering or extended targets, M_s can be different at each frequency. For imaging purposes, background Green's function vector (or steering vector) $\mathbf{g}(\mathbf{X}_s,\omega)$ at each search point in the probed domain is required, and defined as

$$\mathbf{g}(\mathbf{X}_s,\omega) = [\ G(\mathbf{X}_s, \mathbf{R}_1,\omega), .. , G(\mathbf{X}_s, \mathbf{R}_N,\omega)\ ]^T \qquad (1)$$

where $G(\mathbf{X}_s,\mathbf{R}_i,\omega)$ is the Green's function of the medium between vector positions of the search point ($\mathbf{X}_s$) and the ith antenna location of the TRA ($\mathbf{R}_i$, i=1,..,N) and T is the transpose operation. Imaging is achieved via the following inner product for p=1,.., M_s:

$$\mathbf{D}_p(\mathbf{X}_s,\omega) = |\langle\mathbf{g}(\mathbf{X}_s,\omega), \mathbf{v}_p^*(\omega)\rangle| = |\mathbf{v}_p^T(\omega)\,\mathbf{g}(\mathbf{X}_s,\omega)| \qquad (2)$$

Note that if the central frequency (CF) of operation is employed only, this is denoted as CF-DORT which provides narrowband imaging. For UWB signals, time-domain pulses can be produced via $\mathbf{r}_m(t) = IFT(\ \lambda_m^{-1}(\omega)\mathbf{K}^H(\omega)\mathbf{v}_m(\omega)\ )$, where *IFT* denotes the inverse Fourier transformation. This characterizes the time domain (TD)-DORT[5].

The functional given in Eq. (1) is the point spread function (PSF) of the TRA and peaks around the pth scatterer location $\mathbf{X}_p$ to form the image for that scatterer. In homogeneous media (HM), the lateral size (cross-range) of the main lobe of the PSF depends on the classical diffraction limit, which is proportional to wavelength and propagation distance and inversely proportional to aperture length of the TRA[6] (co-range resolution, on the other hand, is inversely proportional to the bandwidth of the pulse). In inhomogeneous media, classical TR can beat this limit (to some extent) due to the increased effective aperture length caused by multipath[6]. However, for strongly interacting scatterers where well-resolvedness is lost, the eigenvectors of the SS become linear combinations of the Green's function vectors connecting the scatterers to each TRA element. Imaging using these eigenvectors may create wavefield interference, which degrades scatterer location estimates. However, even for closely-spaced scatterers, NS is still orthogonal to the SS. Therefore, inner products of the steering vectors with the NS eigenvectors would vanish only at the scatterer locations. This is the basis for TR-MUSIC (or simply MUSIC), where the following functional (pseudospectrum), with sharp peaks at the scatterer locations, is employed:

$$\mathbf{M}(\mathbf{X}_s,\omega) = \left(\sum_{i=M_s+1}^{N} \left| \left\langle \mathbf{g}(\mathbf{X}_s,\omega), \mathbf{v}_p^*(\omega) \right\rangle \right| \right)^{-1} \quad (3)$$

Similarly, when the CF is employed only, algorithm is known as the CF-MUSIC. As shown in prior works, CF-MUSIC resolution is superior to DORT. However, CF-MUSIC requires a non-null NS, i.e. number of significant eigenvalues should be less than the number of array elements ($M_s<N$). Furthermore, it is not always possible to obtain the exact background steering vectors for an inhomogeneous media since pointwise information is often not available in practice. Therefore, approximate vectors should be employed instead. In this work, we use the homogeneous background medium Green's function with the mean properties of random media to obtain (approximate) background steering vectors denoted as $\mathbf{g}_0(\mathbf{X}_s,\omega) = [G_0(\mathbf{X}_s, \mathbf{R}_1,\omega),.., G_0(\mathbf{X}_s, \mathbf{R}_N,\omega)]^T$ where G_0 is the Green's function of the homogeneous medium. In inhomogeneous media, $\mathbf{g}(\mathbf{X}_s)$ is replaced with $\mathbf{g}_0(\mathbf{X}_s)$. Since $\mathbf{g}_0^*(\mathbf{X}_s)$ cannot fully cancel the phase variations of $\mathbf{g}(\mathbf{X}_s)$ (as $\mathbf{g}^*(\mathbf{X}_s)$ would) during TR, the resulting images might differ for each different realization leading to (statistically) unstable images. In order to obtain more stable images for a given inhomogeneous medium, one can take advantage of frequency decorrelation and employ the whole bandwidth available. This is achieved by UWB-MUSIC that combines the null space at each frequency along the bandwidth as follows:

$$\mathbf{M}_{UWB}(\mathbf{X}_s) = \left(\int \sum_{i=M_s(\omega)+1}^{N} \left| \left\langle \mathbf{g}(\mathbf{X}_s,\omega), \mathbf{v}_p^*(\omega) \right\rangle \right| d\omega \right)^{-1} \quad (4)$$

The relative performances of these four TR-based imaging methods, i.e. CF-DORT, TD-DORT, CF-MUSIC and UWB-MUSIC will be considered next.

3. SIMULATION SCENARIO AND RANDOM MEDIA MODELS

Two-dimensional finite-difference time-domain (FDTD) simulations with grid size $N_x \times N_y$ = 200×220 are used to obtain the TRO random media (with uniform spatial cell size of Δs=1.37 cm). TM_z polarization is considered. Continuous random medium has spatially fluctuating permittivity $\varepsilon(\vec{r}) = \varepsilon_m + \varepsilon_f(\vec{r})$ where $\vec{r}$ denotes the spatial position, and ε_m is the average relative permittivity equal to 2.908 (dry sand). The fluctuating permittivity $\varepsilon_f(\vec{r})$ is a zero mean Gaussian random variable which has a Gaussian correlation function $C(\vec{r}_1-\vec{r}_2) = \delta\exp(-|\vec{r}_1-\vec{r}_2|^2/l_s^2)$ with variance δ and correlation length l_s.

The linear TRA consists of N=7 dipole transceivers. Each dipole is initially fed by an UWB time-domain excitation[7] centered at f_c=400 MHz. The TRA lies parallel to the *x*-direction and the dipoles are distributed evenly along an aperture length of $a=3\lambda_c$, with the central dipole located at the origin. To isolate volumetric scattering effects, surface roughness and mutual interactions among array elements are not included.

4. CLUTTER AND NOISE PERTURBATIONS

In this section, we apply TR-imaging for a metallic cylinder with radius r=4.1 cm buried 1.71 m below (i.e. along *y* direction) the central antenna either under clutter (volumetric secondary scattering) or additive noise conditions. Clutter is produced by the background inhomogeneities (generally) arising from the random dielectric contrast. Compared to additive system noise, clutter is multiplicative in nature and exhibit spectral signatures more similar to those of targets. The following signal-to-clutter ratio (SCR) is used to define the amount of clutter received by all MDM elements:

$$\mathrm{SCR} = \left(\frac{\sum_{l,m=1}^{N} \int \left| \vec{E}_{lm}^{\mathrm{HOM}}(t) \right|^2 dt}{\sum_{l,m=1}^{N} \int \left| \vec{E}_{lm}^{\mathrm{RM}}(t) - \vec{E}_{lm}^{\mathrm{HOM}}(t) \right|^2 dt} \right)^{1/2} \tag{5}$$

where $\vec{E}_{lm}^{\mathrm{RM}}(t)$ and $\vec{E}_{lm}^{\mathrm{HM}}(t)$ are the scattered fields received by the lth antenna due to initial excitation applied at the mth antenna in random (RM) and homogeneous media (HM), respectively. Note that SCR depends on the scatterer dimension and distance from the TRA elements as well as fluctuations of the background media. For the range of δ and l_s considered here, increasing δ at fixed l_s or decreasing l_s at fixed δ tends to increase SCR[7]. Similarly, signal-to-noise ratio (SNR) is defined as the ratio between the received signal power and additive white Gaussian noise (AWGN) power over the whole MDM elements. In this case, new MDM is defined as $\mathbf{K}_{noisy}=\mathbf{K}+\mathbf{N}_{SNR}$ where $\mathbf{N}_{SNR}$ is the AWGN matrix.

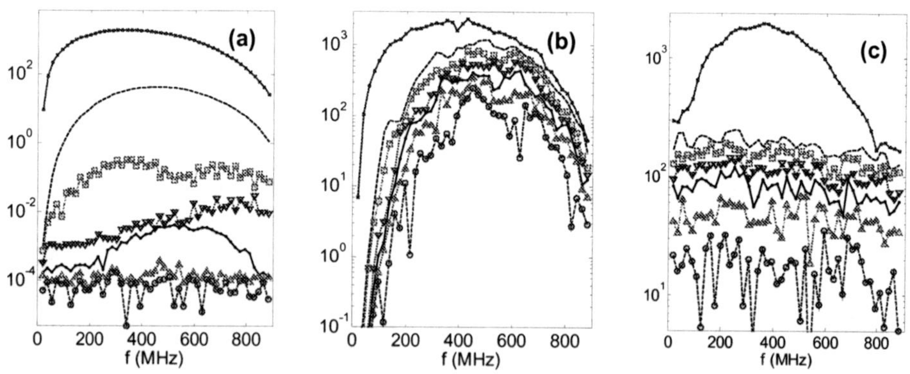

Figure 1. All seven singular values for different clutter and noise conditions (a) noiseless homogeneous medium, (b) noiseless random medium with $\delta=0.01$, $l_s=\lambda/8$, (c) noisy homogeneous medium with SNR=10.0

Effects of both clutter and noise on the singular values are shown in Fig. 1, where singular values obtained in noiseless and noisy homogeneous media and noiseless random medium are compared. For the noise-free homogeneous medium case, one singular value clearly dominates over the others. With increasing noise or clutter, more singular values become significant, resulting in a potentially wrong estimate of the number of discrete scatterers. In the noisy case, singular values corresponding to receiver noise contributions assume a (relatively) white distribution along the bandwidth, whereas clutter singular values have signatures more similar to the target one. In the results that follow, a threshold of 10% of the maximum singular value at each frequency is used to distinguish NS and SS. Resulting images obtained from each method applied in noise-free HM and RM are shown in Fig. 2 and Fig. 3. Corresponding cross-range plots are provided in Fig. 4. Both MUSIC methods provide the best range and cross-range resolution in the HM case. Although TD-DORT has better cross-range as compared to CF-DORT, it underperforms those of MUSIC. In the RM case, clutter affects the resulting images as follows: For CF-MUSIC, there are multiple peaks at different parts of the images which could be misinterpreted as scatterer locations, and with different distributions depending on the RM realization. Therefore, CF-MUSIC does not provide statistically stable images. On the other hand, UWB-MUSIC combines images obtained at different frequencies samples to yield a statistically stable image. This comes at the expense of poorer range and cross-range resolutions (blurring). For both MUSIC methods, increased clutter (i.e. decreased SCR) yields wider cross-range resolutions and higher noise floors (Fig. 4). Also, TR-MUSIC fails to work in cases where the NS is null. This happens, for example, when all clutter eigenvalues are greater than the chosen threshold.

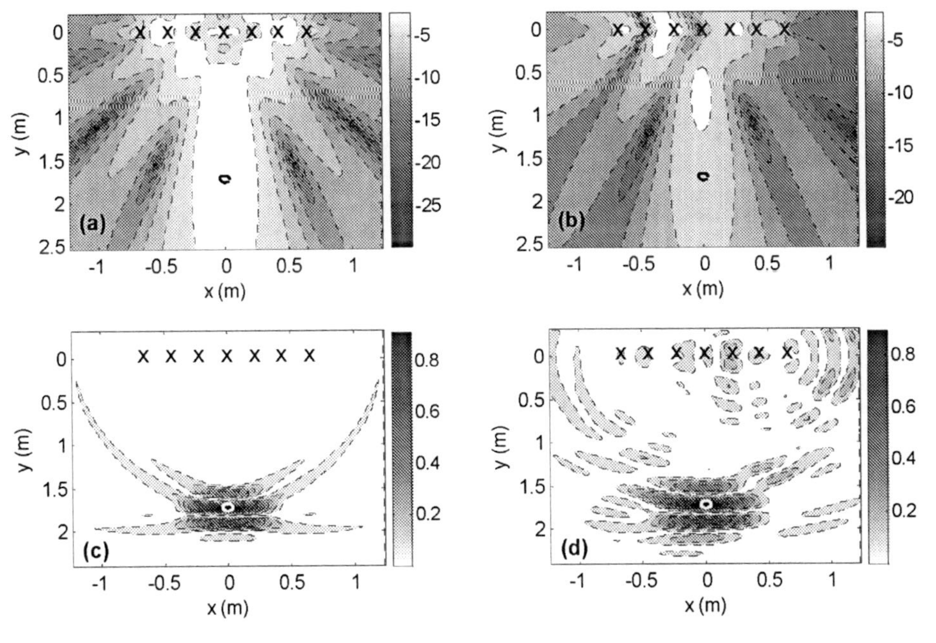

Figure 2. Images obtained using CF-DORT in (a) HM, (b) RM (both normalized and dB format) and TD-DORT in (c) HM, (d) RM (Note that RM defined by δ=0.01 and $l_s \approx \lambda_c/10$).

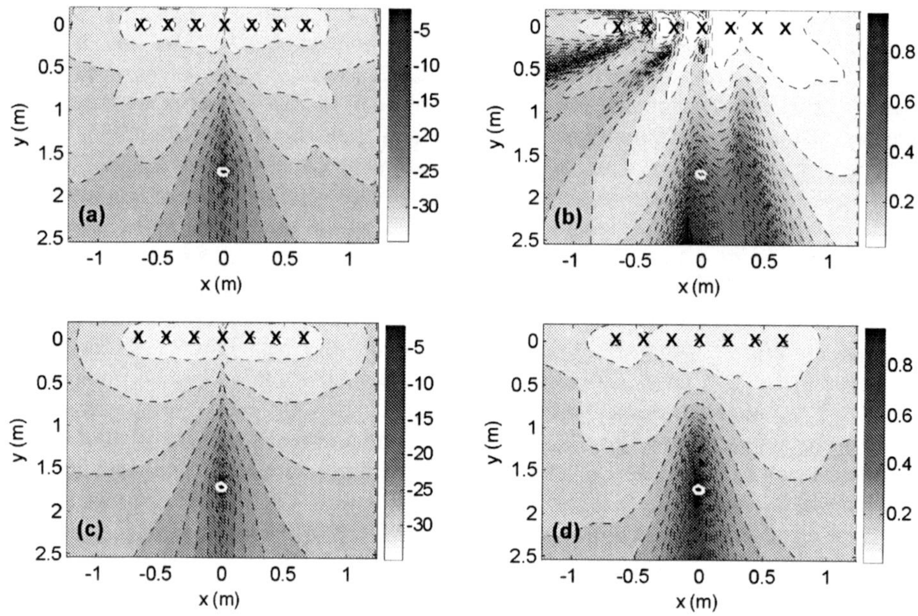

Figure 3. Images obtained using CF-MUSIC in (a) HM, (b) RM and UWB-MUSIC in (c) HM (d) RM (Note that RM defined by δ=0.01 and $l_s \approx \lambda_c/10$). All plots are normalized, (a) and (c) in dB format

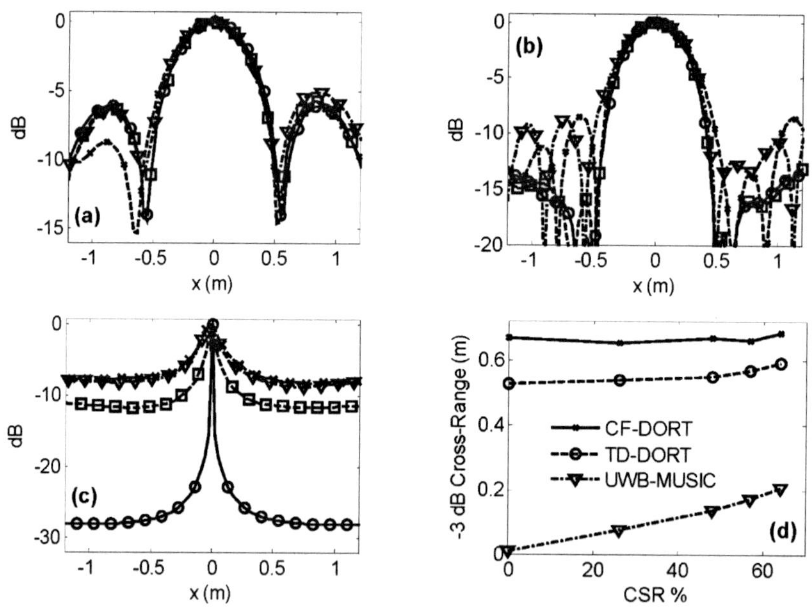

Figure 4. Cross-range resolutions under increasing SCR for (a) CF-DORT, (b) TD-DORT and (c) UWB-MUSIC. (Solid line with circles: SCR=∞, dashed line with squares: SCR=3.64, dot-dashed line with triangles: SCR=1.44, dashed line with crosses: SCR=1.21) (d) -3 dB resolutions vs CSR (clutter to signal ratio= SCR^{-1})

Both DORT images have similar range and cross-range information even for increasing clutter (Fig. 4). In other words, DORT is robust against clutter and provides stable images. Sidelobe levels for TD-DORT are less than CF-DORT in general, but for smaller SCR, sidelobe levels increase for both methods. We get similar results for smaller white noise, which are not shown here for the sake of brevity.

5. PERTURBATION ON TR INVARIANCE (LOSSES)

Many realistic scenarios involve dispersive and/or lossy media where the TR invariance is broken. This constitutes yet another perturbation of the ideal TR scenario. The performance of any imaging method depends on the frequency of operation, scatterer depth and the TRA characteristics. Therefore, to isolate the effects of losses, all other parameters are kept constant in the following simulations.

We apply CF-DORT and CF-MUSIC in the case of a cylindrical object buried in homogeneous soil with different (static) conductivities. Background steering vectors are also evaluated for the given conductivity values. Cross-range resolutions of the obtained images are depicted in Fig. 5. With increasing conductivity, received signals at the TRA and background steering vectors undergo more attenuation for the same physical distances. As a result, the CF-DORT cross-range shows decreased amplitudes and a corresponding degradation on the focusing resolutions. However, CF-MUSIC performance seems to be more stable against increasing conductivities as long as the received signals from the scatterers are not attenuated below the noise level. This can be observed when cross-ranges obtained for different medium conductivities are compared. For conductivities up to $\sigma = 0.04$ S/m, CF-MUSIC cross-range patterns and noise floors are very similar to each other, whereas CF-DORT is gradually degraded. But when CF-DORT cross-range becomes very weak due to strong attenuation ($\sigma > 0.04$ S/m), CF-MUSIC cross-range patterns start having higher noise floors and slightly worse focusing resolutions. Note that these discussions can be extended to dispersive media where the loss effects are distributed along the whole bandwidth. Compensation methods can also be employed for better performance in lossy and dispersive cases[8].

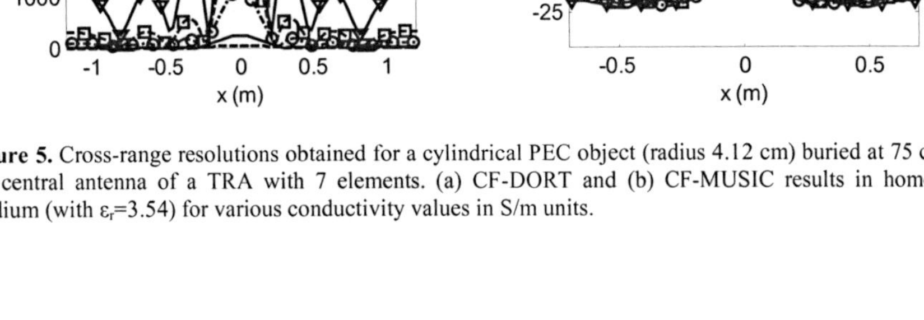

Figure 5. Cross-range resolutions obtained for a cylindrical PEC object (radius 4.12 cm) buried at 75 cm below the central antenna of a TRA with 7 elements. (a) CF-DORT and (b) CF-MUSIC results in homogeneous medium (with ε_r=3.54) for various conductivity values in S/m units.

6. SUMMARY AND CONCLUSIONS

We have considered the effect of perturbations on narrowband and UWB TR imaging methods. It was shown that in noiseless homogeneous media, TR-MUSIC outperform DORT both in co-range and cross-range resolutions. However, as the amount of clutter increases, eigenvalues that correspond to random background scattering centers become more significant. The new signal space is also corrupted by such background inhomogeneities, and CF-MUSIC images may have spurious peaks which are realization dependent. By utilizing the null-space over a wider bandwidth, UWB-MUSIC yields statistically stable images. In addition, it was observed that DORT cross range resolution is more stable against clutter as compared to UWB-MUSIC. MUSIC methods require the existence of a non-null NS, but for highly cluttered environments and/or small number of TRA antennas, this requirement may not be satisfied. As for the effect of losses, it was observed that DORT cross-ranges degrade with increasing conductivities. However, as long as the received signals from the scatterers are distinguishable, MUSIC results are relatively more stable.

7. ACKNOWLEDGMENTS

This work has been supported in part by National Science Foundation under grant ECS-0347502, and Ohio Supercomputer Center under grants PAS-0061 and PAS-0110.

8. REFERENCES

1. M. Fink, D. Cassereau, A. Derode, C. Prada, P. Roux, M. Tanter, J. Thomas, and F. Wu, Time-reversed acoustics, *Rep. Prog. Phys.* **63**, 1933–1995 (2000)
2. C. Prada and M. Fink, Eigenmodes of the time reversal operator: a solution to selective focusing in multiple-target media, *Wave Motion* **20**, 151–163 (1994)
3. C. Prada, S. Mannevile, D. Spoliansky, and M. Fink, Decomposition of the time reversal operator: detection and selective focusing on two scatterers, *J. Acoust. Soc. Am.* **99**(4), 2067–2076 (1996)
4. H. Lev-Ari and A. J. Devaney, The time-reversal techniques re-interpreted: subspace-based signal processing for multistatic target location, *Proc. IEEE Sensor Array and Multichannel Signal Processing Workshop*, 509-513, (March 2000)
5. M. E. Yavuz and F. L. Teixeira, Full time-domain DORT for ultrawideband electromagnetic fields in dispersive, random inhomogeneous media, *IEEE Trans. Antennas and Propag.* (accepted)
6. P. Blomgren, G. Papanicolaou, C. Tsogka, and J. Berryman, Super resolution in time-reversal acoustics, *J. Acoust. Soc. Am.*, **111**(1), 230-248 (2002)
7. M. E. Yavuz and F. L. Teixeira, A numerical study of time reversed UWB electromagnetic waves in continuous random media, *IEEE Antennas and Wireless Prop. Lett.*, **4**, 43–46 (2005)
8. M. E. Yavuz and F .L. Teixeira, Frequency dispersion compensation in time reversal techniques for UWB electromagnetic waves, *IEEE Geosci. Remote Sensing Lett.* **2**(2), 233–237 (2005)

MODELLING OF REFLECTION OF UWB PULSES FROM TRAPPED HUMAN BEINGS

Amer Nezirovic, Mingying Liu and Alexander G. Yarovoy*

1. INTRODUCTION

In emergency situations, particularly within smoke-filled, partially or completely collapsed large buildings, knowledge of the location and the state of the victims is largely limited, due to various obstacles that are result of an earthquake or fire. Therefore a reliable tool is needed which would have good victim detection and localization, but also material penetration capabilities.

Currently, there are several approaches for solving the problem:

1. Sound detection equipment, which has a limited use considering the potential weakened state of the victims as well as high sound attenuation of the rubble,
2. IR camera, which exhibits limitations in NLOS situations or in presence of strong surrounding heat, and
3. Use of radars: Doppler[1], interferometric[2] or UWB radar[3].

The use of impulse UWB radar, which makes use of high temporal and thereby also spatial resolution by using ultra-short pulses, is considered to be a potential solution, mainly because of its detection, localization and penetration capabilities. Such pulses have a wide frequency spectrum that can easily pass through obstacles, such as rubble or thick smoke. Furthermore, short pulses lead to low energy consumption, which makes UWB radars power efficient. Being power efficient, UWB systems intended for victim search can therefore be implemented in hand-held devices, which would be carried by the operators.

In order to detect a victim in a background of stationary interference, the detection techniques that we have considered so far will rely on the chest movements due to breathing. Breathing motion causes changes in frequency, phase, arrival time and, to a certain extent, amplitude of the electromagnetic wave reflected from a human being.

---

* International Research Centre for Telecommunications-Transmission and Radar (IRCTR), P.O. Box 5031, 2600 GA Delft, The Netherlands

One of the main obstacles faced in human breathing detection using UWB radar is the low SNR values associated with it, which makes target detectability challenging. A way to improve the SNR is to perform matched filtering in the receiver. It is therefore desirable to have a template which matches the scattered pulse as much as possible.

In this paper we analyse how changes in thickness and conditions of body layers influence the shape of the scattered pulse and effectively the need of accurate matched filter template. On the other hand, obviously, more bandwidth results in higher resolution and thus facilitates breathing detection, but at the same time increases interference to other systems, thus, it is further important to investigate what the minimal bandwidth and optimal centre frequency should be for successful breathing detection. The detectability will be assessed by a developed human breathing detection algorithm.

2. MODELLING REFLECTION OF THE HUMAN BODY

2.1 Human body model

In order to analyse the reflection properties of the human body, we have developed a one-dimensional, multi-layered human body model in MATLAB, representing the cross-section of the chest area. This model consists of 9 tissue layers with their respective thicknesses modelled realistically, see figure 1.

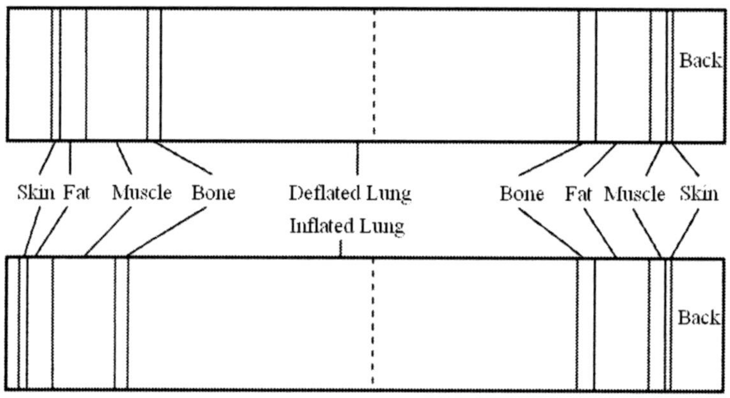

Figure 1. Modeled cross-section of the human chest for deflated and inflated case

In implementing this model we assume that:

1. A TEM plane wave propagates perpendicularly to the orientation of a horizontally positioned human being.
2. All layers are linear, homogeneous, isotropic media with frequency-dependent permittivity (since a large bandwidth is occupied).
3. The pulse travels from the chest to the back of the human body and is reflected back.
4. Multiple reflections occur between the tissue layers.

2.2 Global transfer function

In order to obtain the global transfer function, we need to apply a recursive formula which incorporates the transfer function of each of the involved tissues in the human body model[4]:

$$R_{i-1}(\omega)=\frac{R_{i-1,i}+R_i(\omega)\cdot e^{-2jk_i(\omega)L_i}}{1+R_{i-1,i}\cdot R_i(\omega)\cdot e^{-2jk_i(\omega)L_i}} \tag{1}$$

where $c=3\cdot 10^8$ m/s is the speed of light in free space, ε_i is the permittivity of the *i*-th layer, $k_i=\omega\cdot\sqrt{\varepsilon_i}/c$ and L_i are the wave number and the thickness of the i-th layer, respectively and $R_{i-1,i}$ is the reflection coefficient at the boundary between i−1-th and i-th layer: $R_{i-1,i}=\left(\sqrt{\varepsilon_{i-1}}-\sqrt{\varepsilon_i}\right)/\left(\sqrt{\varepsilon_{i-1}}+\sqrt{\varepsilon_i}\right)$. Equation (1) can be simplified as follows:

$$R_{i-1}(\omega)=R_{i-1,i}+\frac{R_i(\omega)\cdot(1-R^2_{i-1,i})e^{-2jk_i(\omega)L_i}}{1+R_{0,1i-1,i}\cdot R_i(\omega)\cdot e^{-2jk_i(\omega)L_i}} \tag{3}$$

The global transfer function is obtained from (3) by putting i = 1:

$$H(\omega)=R_0(\omega)=R_{0,1}+\frac{R_1(\omega)\cdot(1-R^2_{0,1})e^{-2jk_1(\omega)L_i}}{1+R_{0,1}\cdot R_1(\omega)\cdot e^{-2jk_1(\omega)L_i}} \tag{4}$$

The impulse response can be obtained from (4) by taking the inverse Fourier transform:

$$h(t)=2\cdot\mathrm{Re}\left\{\mathbb{F}^{-1}\left[H(\omega)\right]\right\} \tag{5}$$

2.3 Permittivity modelling

Since the propagating TEM wave into the human body model occupies large bandwidth, the respective permittivity of the involved tissues cannot therefore be assumed constant. In order to model the permittivity as a function of frequency, 4-Cole-cole method is used[5]:

$$\varepsilon(\omega)=\varepsilon_\infty+\sum_{m=1}^{4}\frac{\Delta\varepsilon_m}{1+(j\omega\tau_m)^{(1-\alpha_m)}}+\sigma_i/j\omega\varepsilon_0 \tag{6}$$

where $\varepsilon(\omega)$ is the complex permittivity, ε_∞ is the permittivity in the terahertz frequency range, $\Delta\varepsilon_m$ is the drop in permittivity in the frequency range, ω is the angular frequency, τ is the relaxation time for each dispersion region, α is the exponent, σ_i is the ionic conductivity and $\varepsilon_0 = 8.8542 \cdot 10^{-12}$ F/m is permittivity of free space. The modeled permittivity values as function of frequency can be seen in figure 2 for comparison.

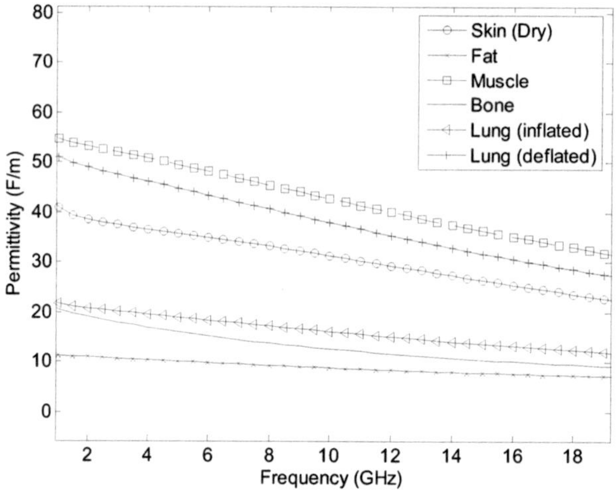

Figure 2. Modeled permittivity of the human body tissues

2.4 Reflection properties of the human body

Since the anatomy of all human beings differs slightly from individual to individual, we cannot assume the same scattered response from all human beings being equal, given otherwise identical conditions (orientation and position of the body, aspect angle, etc). Our interest is in finding out how much and in what way does the thickness of the different layers in the human body model influence the backscattered response, both in frequency and time domain.

The considered body tissues and their respective thicknesses are set as following:

1. Skin: 1.5 mm, 2.0 mm and 2.5 mm.
2. Fat: 8.5 mm, 9.5 mm and 15 mm.
3. Deflated and inflated lung layer: 128 mm and 148 mm, respectively.

3. SIMULATION PROCEDURE

3.1 Human breathing detection algorithm

The most important purpose of using the human breathing detection algorithm is to reduce the noise as much as possible but at the same time preserve the energy of the backscattered response. A beneficial feature of the breathing motion is that it is periodic. Thus, instead of processing each of the acquired range profiles in range dimension and looking for periodic time shifts of the backscattered response, the processing is done in the time dimension for each of the range profile samples. The reason is that the breathing motion in time dimension shows up as $b(t) = -|\sin(2\pi f_b t)|$. So therefore we only need to apply a filter in time dimension with the passband centred at the expected breathing frequency $\overline{f}_b = E\{f_b\}$, which we assume is known and constant during the observation time. Thus we perform filtering of the signal vector $\mathbf{s}_k = [s_k(1)\, s_k(2) \dots s_k(N)]^T$ for k = 1:M, i.e. in total M filter iterations. The spectrum of the breathing motion contains harmonics, so a first order harmonic is included in the filtering process as well. The benefits of the filtering operation in time dimension are twofold: reduction of noise and suppression of stationary clutter. The clutter is assumed to be stationary during the observation time, thereby constituting the DC component of the signal vector $\mathbf{s}_k$. After the filtering process, the energy vector $\boldsymbol{\varepsilon}_k$ of the filtered signal vectors $\hat{\mathbf{s}}_k$ is computed and summed in time dimension, thus obtaining an energy profile as a function of range:

$$\boldsymbol{\varepsilon}_n = \sum_{k=1}^{N} \varepsilon_{k,n} = \sum_{k=1}^{N} \hat{s}_{k,n}^2 \tag{7}$$

In order to determine the range of the breathing victim a mean of the energy profile $\overline{\boldsymbol{\varepsilon}}_n$ is computed to obtain the threshold. The criterion for detection is: *any peak that is L dB above the threshold is considered a breathing victim.* The value of L should be statistically evaluated based on the desired false alarm rate. In this work, though, we will present the variation of L of the peak in the energy profile, with bandwidth and centre frequency of the backscattered response.

3.2 Transmitted bandwidth vs. centre frequency evaluation

In this part we are simulating the process of data acquisition for human being detection, for various values of bandwidth B and centre frequency f_c of the backscattered response from the body. The acquired data consists of an M number of range profiles containing the appropriately filtered backscattered responses stacked in matrix form overlaid with AWGN, having the same B and f_c values. The clutter is disregarded in this simulation because it is assumed stationary and thus can easily be suppressed by the human breathing detection algorithm.

During one of our measurement campaigns we have found that the human breathing motion is proportional to $-|\sin(2\pi f_b t)|$, where f_b is the breathing frequency. Therefore the thickness of the lung layer in the model is modulated in time accordingly. The chest movement due to breathing occurs in forward direction only, i.e. the back part of the body is motionless, see figure 1. Since the front half of the chest is the one that is moving, the reflected pulse will exhibit periodic differences in time-of-arrival at the receive side according to $-|\sin(2\pi f_b t)|$ pattern. The receive matrix can now be expressed:

$$\mathbf{R} = \mathbf{s}_n + \mathbf{n}_n = [\mathbf{s}_1\, \mathbf{s}_2 \dots \mathbf{s}_N]^T + [\mathbf{n}_1\, \mathbf{n}_2 \dots \mathbf{n}_N]^T \tag{8}$$

where $\mathbf{s}_n = [s_n(1) s_n(2) \dots s_n(M)] = \left[0 \dots 0 \; r\left(k - |\sin(2\pi f_b n)|\right) \; 0 \dots 0\right]$ is a 1×M vector containing the backscattered response $\mathbf{r}$ having dimension 1×K, where K<M, and

$\mathbf{n}_n = [n_n(1)\, n_n(2) \ldots n_n(M)]$ is the vector containing additive white Gaussian noise. Signal-to-noise ratio is defined as: $\mathrm{SNR} = \varepsilon_s / E\{\varepsilon_n\} = \varepsilon_s / E\{\varepsilon_n\} \cdot (M/K)$. The scaling coefficient $(M/K) > 1$ is included since the length of vector the backscattered response and the noise vector are not the same. The bandwidth B and centre frequency f_c of both **r** and **n** are set the same and varied in the interval of 0.5-3.2 GHz for B and 1.8-8.0 GHz for f_c. The receive matrix **R** is processed using the human breathing detection algorithm and the value of L is plotted for each value of B and f_c.

4. RESULTS

4.1 Reflection properties of the human body

The influence on the varying thickness of the skin layer in time domain can be seen in figure 3. As the thickness increases the response of the body in time domain becomes increasingly delayed and attenuated. In frequency domain we see changes in attenuation for frequencies above 1GHz, which suggests that frequencies lower than that pass the skin layer virtually unaffected, see figure 4. The responses in frequency domain do not differ more than 2dB at most.

The influence of the the changes in thickness of the fat layer are more prominent than for the skin layer, especially for frequencies lower than 1GHz, see figure 5-left. In time domain, figure 5-right, the changes are visible following a short delay, which is reasonable since the fat layer comes after the skin layer whose thickness does not change.

Inhaling and exhaling produces minor changes in the backscattered response from the modelled human chest. In time domain the difference is minor and is observed in the late-time response, see figure 6-left. The changes in frequency domain are visible mostly in the region below 2GHz, see figure 6-right. That allows for preserved backscattered waveform during breathing if the frequencies used are higher than 2GHz since these frequencies are unaffected. That is explained by the fact that such high frequencies do not penetrate the chest plate and reach the lung layer, whereas frequencies below 2GHz do.

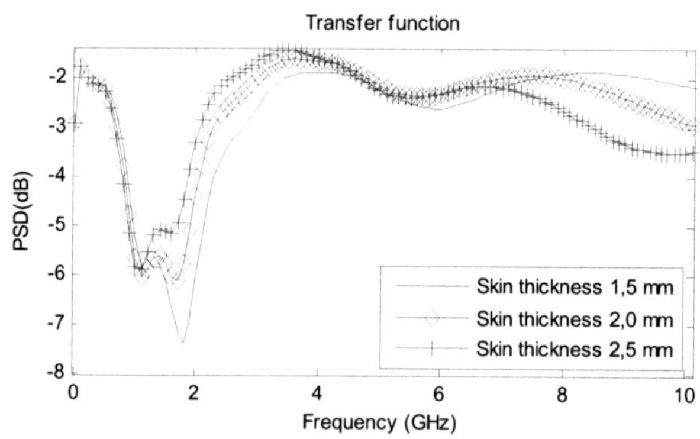

Figure 3. Frequency response of the model for various skin layer thicknesses.

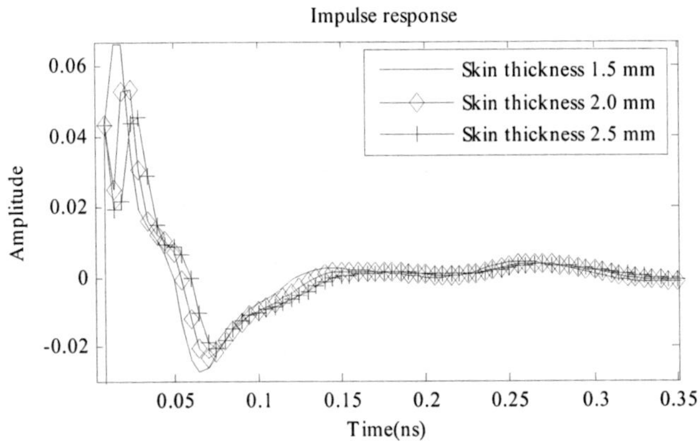

Figure 4. Time response of the model for various skin layer thicknesses.

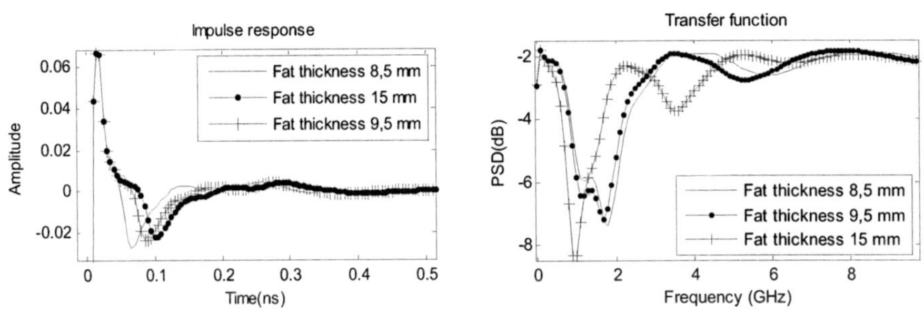

Figure 5. Frequency response of the model for various fat layer thicknesses in time domain

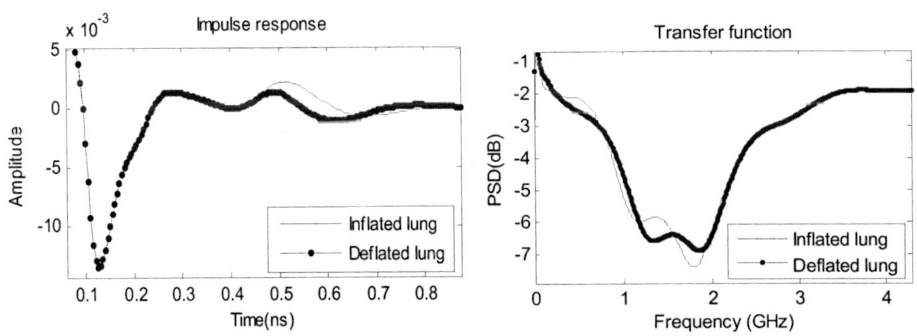

Figure 6. Response of the model for inflated and deflated lung case in time domain (left) and frequency domain (right).

Probability of detection of breathing victims using UWB radar increases with increasing carrier frequency and bandwidth, while keeping the SNR value constant, see figure 7. From that point of view it is desirable to use as high bandwidth and centre frequency as possible in order to increase probability of detection.

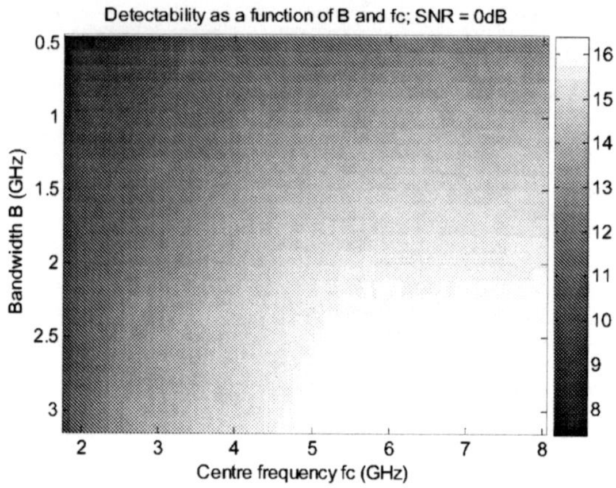

Figure 7. Breathing motion detectability as a function of bandwidth and centre frequency measured in the number of dB over the threshold

5. CONCLUSION

In this paper reflective properties of the human body are investigated using a developed one-dimensional human body model. It is shown that the change in thickness of skin layer affects frequencies higher than 1GHz. For fat layer even lower frequencies are affected and for the case of inflated and deflated lung, changes can only be seen for frequencies lower than 2GHz, since only those frequencies penetrate the chest plate and reach the lung layer. Strictly looking from a point of view of increasing the probability of detection of a breathing victim, the transmitted waveform should have as high centre frequency and bandwidth as possible.

6. REFERENCES

1. I. Arai, Survivor Search Radar System for Persons Trapped under Earthquake Rubble, *Proceeding of the IEEE Microwave Conference*, vol. 2, p. 663-668 (2001).
2. Chen, K.-M. and Chuang, H.-R. Measurement of Heart and Breathing Signals of Human Subjects Through Barriers with Microwave Life-Detection Systems, Proceeding of the IEEE International Conference on Engineering in Medicine and Biology Society, vol. 3, p.1279-1280 (1988).
3. G. Ossberger; T. Buchegger; E. Schimback; A. Stelzer and R. Weigel, Non-Invasive Respiratory Movement Detection and Monitoring of Hidden Humans Using Ultra Wideband Pulse Radar, *Processing of the International Workshop on Ultrawideband Systems and Technologies*, p. 395-399 (2004).
4. P. Lombardo, A. M. Magenta and E. Pettinelli, Multichannel fusion of subsurface radar images at different resolutions, *IEE Proc.-Radar, Sonar Navig.*, **147**(3), p. 121-133 (2000).
5. C. Gabriel and S. Gabriel, Compilation of the Dielectric Properties of Body Tissues at RF and Microwave Frequencies, (Texas, June 1996); http://www.brooks.af.mil/AFRL/HED/hedr/reports/dielectric/home.html.

SOURCES

GLOBAL LIMITATION ON FAST SWITCHING BY SEMICONDUCTOR DEVICES

Alexei F. Kardo-Sysoev [1]

1. INTRODUCTION

One of the important problems of UWB short pulse Electromagnetic is the pulse generation. The core of a pulse generator is a fast switch, a device, which could be switched between two states "off" and "on" with low and high conductance. Difference between the conductances must be as large as possible. Many types of semiconductor devices could be used for the switching. For many of them figures of merit, which shows their limitations, are well known (for an example – time of flight limitations).

In this work global limitations, which could be applied to any kind of a switch are considered. Due to generalization of the approach, we make very rough estimations of the limits, just to show orders of magnitudes. In reality, any limitation is connected with some physical process or combination of several processes and with the geometry under consideration. Switching processes consist in the filling of semiconductor volume by electron-hole plasma (turn on) and its removal (turn off). Plasma gets in and gets out by only two ways:

- Injection/extraction through the borders of the volume;
- generation/recombination inside the volume.

For mentioned above ways the next physical processes could be considered, which leads to a variety of limitations. Corresponding figures of merit could be estimated as well.

1. Space charge builds up.
2. Sharply changed current and corresponding change of electromagnetic field limits the working width of the area where the current flows. For linear case this leads to well known skinning in conductors.
3. Fast heating due to transient dissipation of energy and corresponding shock wave generation in semiconductor limits current density.
4. Heating plus skinning limit current density per length of the switch perimeter.
5. During switch off and on processes EM field changes its configuration: gets inside and out the semiconductor volume, which leads to "time of EM wave flight" limitation.

An analysis of the limitations makes it possible to understand and evaluate the degree of approximation of the real device to the ideal one: that is to evaluate the main

[1] Ioffe PTI Russian Academy of Science, St. Petersburg, Russia 194021

possibility of improving the parameters, as well, as to show what technical steps (if any) should be done for the improving of the device.

2. A GENERAL APPROACH

A generator of short powerful pulses could be split into two main parts: a storage and a switch (Fig.1a). Either Electric or Magnetic fields could be used to store energy. The work of any switch is based on the change of material parameters such as dielectric permittivity (ε), magnetic permittivity (μ), conductivity (σ) or any their combination. From practical point of view, the range of change of the parameter should be as large as possible, preferably many orders of magnitude, and the change should be easily triggered, predictable and stable. At present time, the most of these requirements could be satisfied by use of semiconductors with changing conductivity. A semiconductor switch may be considered as a bulk of semiconductor placed between two electrodes (Fig. 1.b). The bulk is filled by electron-hole plasma (turn-on) and set free from it (turn-off) in series.

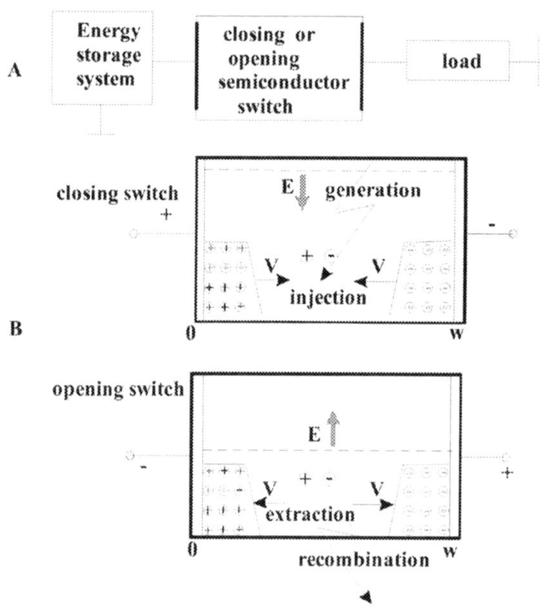

Figure 1. Switch as a modulated conductivity of a seniconductor bulk.

To fill the bulk there are just two ways:

a. To inject carriers from borders.

b. To generate electron-hole pairs in the volume.

To set free there are the reverse processes:

a. To extract carriers into the borders.

b. To recombine electron – hole pairs.

Injection, extraction could be externally triggered (as well as generation in case of external irradiation by light, particle beam) independently of the process in bulk (for example, in three electrode transistors).There is no practical way to trigger recombination – it is self organizing process. For injection/ extraction process there is well-known "time of flight" limitation on the length of turn-on and/or turn-off (τ) times

$$\tau > W / V_s \qquad 1)$$

Where W – bulk width; Vs – maximal "saturated velocity of electrons/holes under high electric field. Maximal field strength is determined by impact/tunnel ionization E_α, so Eq. (1) can be rewritten as

$$\tau > U_m / E_\alpha V_s \quad , \quad dU/dt \approx U_m/\tau \approx E_\alpha V_s, \qquad (2)$$

i.e. switching time is proportional to the blocking voltage (U_m), maximum voltage change rate is a constant for given semiconductor with given E_α, V_s.

3. SPACE CHARGE LIMITATION

Continuity and Poisson equations for the case of saturated velocity lead to the equation for current density rise

$$j'_m = dj/dt = -V_s^2\,(d\rho/dx) + 2qV_sG, \qquad (3)$$

where j = qVs(p+n) conducting current density; p,n – hole, electron concentrations, $\rho = q(p-n)$ – space charge, G – carriers generation rate, q – electron charge. Here we assume saturated velocities of electrons and holes being equal.

For the case of only injection/extraction, G =0, assuming zero fields on the borders, we find from Eq. (3) in respect to U

$$j'_m = (12\varepsilon V_s^2 U) / W^3 \qquad (4)$$

Equation (4) has clear physical meaning (Fig. 2). Accumulation of carriers is connected with gradient of the carriers flow. The gradient leads to space charge build. When voltage drop on space charge region (SCR) reaches applied voltage source farther current rise rate decreases.

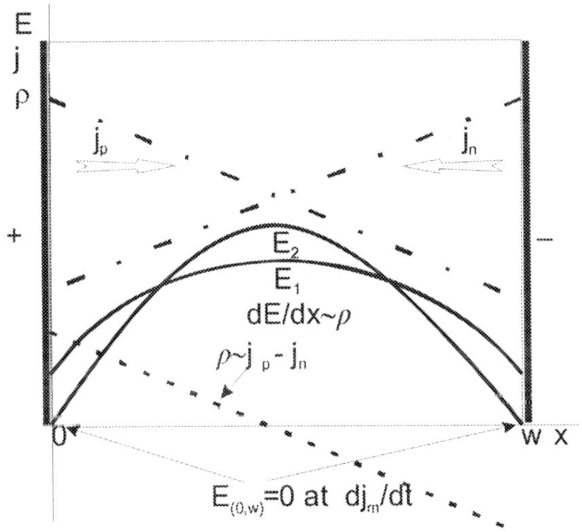

Figure 2. Field E and space charge ρ distributions due to injection

In the case of generation of carriers, they appear in pairs. Space charge appears due to separation of these pairs in the field and described above the sequence start to work again. However, the distance of pair separation (δ), which determines the SCR size is $\delta \approx V_s t_\delta$, where t_δ is the time of separation.

Equation (3) shows, that the first component in Eq. (3) compensates the second (generation) and j'_m goes to zero. So the maximum of j'_m is at the beginning of the process of concentration build up.

$$j'_m < 2qV_sG \qquad (5)$$

From Eq. (3) and Poisson equation it may be shown, that $j' = 0$, when current density reaches the maximum value

$$j_m \le \varepsilon V_s U/\delta^2 \ . \qquad (6)$$

In the case of external irradiation, $G = \Phi\gamma / \lambda$, where Φ – irradiation flux, γ – quantum efficiency, λ – extinction depth.

In the case of impact ionization

$$G < \alpha(E)\cdot j/q = \alpha(E)(p+n)V_s\ , \qquad (7)$$

Where α(E) – ionization coefficient strongly depending on the field.

From Eq. (6), (7) taking into an account $\alpha = 2\delta$, we get $G_m \le 8\varepsilon V_s \alpha_m^2 U$. Limitations to j' with numerical estimations are shown in Table 1.

4. ELECTROMAGNETIC SKINNING

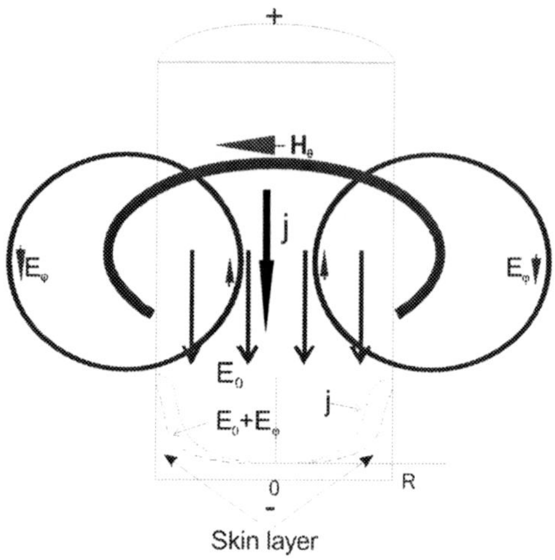

Figure 3. Skin effect in a switch

The fast change of current in the bulk induces changing magnetic field (Fig. 3). Changing magnetic field induces curling electric field. The last in its turn changes initial electric field, which produces initial current. As it could be seen from Fig. 3, the induced electric field decreases initial field in the volume of semiconductor. For linear case, where j~E, Maxwell equations lead to diffusion equation for E with diffusivity coefficient $D \approx 1/\sigma\mu$. So the depth of the field penetration is $\delta_s = (\tau/\mu\sigma)^{1/2}$, where τ is the switching time. The relation for δ_s is the time domain presentation for the very well known skin layer thickness frequency domain, where τ is substituted by $(\omega)^{-1}$and ω is frequency. As it was mentioned above, during transient process in a semiconductor switch, carriers drift with saturated velocity, which is independent of field intensity, for the case we have from Maxwell equations for the Fig. 3 geometry:

$$j'_k = 4E_m/\mu R^2, \quad I'_k = j'_k \pi R_0^2 = 4\pi E_m/\mu. \tag{8}$$

When the current rise rate reaches I'_k value, field at the centre goes to zero. The rise at the centre stops. If $j' > j'_k$, the part of volume where field and current goes to zero widens and current concentrates near the surface. This can be interpreted as a skin effect for nonlinear conductivity. From Eq. (8) we get estimation for skin layer depth (δ)

$$\delta \approx (E_m / \mu j')^{1/2}. \tag{9}$$

5. THE HEATING OF THE MATERIAL

The density of heating energy due to conductivity current j under applied electric field E is

$$Q_T = \int_0^T jE dt = \int_0^j (jE/j')dj \tag{10}$$

The maximum of energy density, which can be dissipated in a semiconductor, is limited by melting or the mechanical stress, which arises from the thermal expansion. In the first case, the maximal energy is the sum of the energy needed to heat up to melting point and specific melting heat.

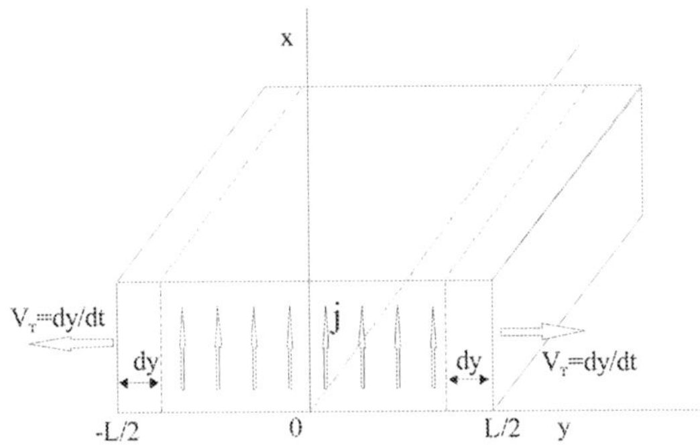

Figure 4. Thermal Mechanical stress.

The mechanical stress is minimal for uniform heating of the bulk with mechanically free boundaries. In stationary case, there are no stress at all, but this is not so for transient process.

Let us consider the simplest case of parallelepiped (Fig. 4). Increase of temperature by dT causes expansion along y axis $dy = \alpha_T \cdot y\, dT$, where α_T – is the thermal expansion coefficient. At point y pressure $P_T \approx m(d^2y/dt^2)$ arises, where $m = g(L - y)$ is the inert mass for a unit area, g – is the material density. Finally we have

$$P_T = \alpha_T \cdot gy(L - y)d^2T/dt^2. \qquad (11)$$

The maximum pressure is in the center. It could be found from Eq.(10), (11)

$$P_m \approx (\alpha_T \cdot gL^2)/4C_T\,(d^2Q/dt^2) \approx (\alpha_T \cdot gL^2\,\hat{E}\,/4C_T)j', \qquad (12)$$

Where C_T - heat capacity, $\hat{E}$ – some average field intensity.

The speed of boundaries widening (V_T) is $V_T \approx dy\,/\,dt$, but the maximum for V_T is the speed of sound C_S. In the case of short switching transient time τ_t, when τ_t is shorter, than time of flight of sound wave ($\tau_t << L/C_S$), the boundaries could be considered to be fixed and we get “dynamic” stress

$$P_d \approx (\alpha_T \Delta T)/\chi = (\alpha_T Q_T)/\chi\, C_T, \qquad (13)$$

where χ – is the compressibility of the semiconductor.

Estimation Eq. (12) is valid for cylindrical geometry as well with the substitution of 2R for L.

Substituting into Eq. (13) values for Q_T, derived from Eq. (10) for different processes of switching we get limitations on current density j, shown in Table 1.

For cylindrical geometry under skinning condition, the total current is $I_m \approx 2\pi R\delta \cdot j_m$. From Eq. (9) and Eq. (10) we get

$$I_m/2\pi\, R \lesssim (Q_T/\mu) \qquad (14)$$

Expression Eq. (14) shows the limitation of linear current density, taking into an account nonlinear skinning and thermal heating. It could be shown, that for silicon, the energies Q_T needed to melt and to reach dynamic mechanical breakdown threshold $P_m = 10^5$ N/cm^2 are very close $Q_{Tm} \approx 10^4$ J/cm^3.

Estimations for silicon for all considered above limits are shown in Table 1.

Table 1 shows, intuitively expected, result – in theory carrier generation, especially impact ionization could provide many orders magnitude higher current density and current density rise rate. The main practical problem is how to trigger such high ionizing rate? The best know solution is the use of the delayed ionization. However, the process is inherently unstable, what manifests itself in filamentation and worsens reliability.

The high dj/dt decreases skin layer and as a result, the maximum current linear density (current per length of circumference) does not depend on the plasma filling process. To overcome the time of flight limitation Eq. (2) for injection/extraction processes, which are stable contrary to impact ionization, it is possible to use many stacked in series

Table 1. Limits for different Plasma filling processes

Limiting process	**Plasma filling process.** Estimations are for Silicon devices		
	Double injection from borders	Carries generation by external radiation	Carriers generation by impact ionization
Space charge build up limits current density rise rate $dj/dt=dj/dt<$	$(12\varepsilon V_S^2 \cdot U)/W^3$ $1.2 \cdot 10^3$ A/cm^{2}nsec $U \approx 1$ kV; W=100 mkm	$(2qV_S \gamma \Phi)/\lambda$ 10^3 A/cm^{2}nsec $\Phi \sim 1$ kW/cm^2, $\lambda=0.3$cm	$8\varepsilon V_S^2 \alpha^3 U$ 10^9 A/cm^{2}nsec $\alpha \approx 10^4$ cm^{-1}, $U \approx 1$kV
Skinning nonlinear case skin layer $\delta \approx$ linear case skin layer $\delta \approx$	$\sqrt{\frac{E}{j'\mu}}$ 1 cm for $E \approx 10^5$ V/cm $\sqrt{\tau/\mu\sigma}$	$\sqrt{\frac{E}{j'\mu}}$ 1 cm for $E \approx 10^5$ V/cm $\sqrt{\tau/\mu\sigma}$	$\sqrt{\frac{E}{j'\mu}}$ 10^{-3} cm for $E \approx 5*10^5$ V/cm $\sqrt{\tau/\mu\sigma}$
Heating limits current density $j<$ $Q_T \sim 10^4$ J/cm^3	$\frac{4V_S(\varepsilon Q_T)^{1/2}}{W}$ $j_m \approx 10^5$ A/cm^2	$\frac{(2\varepsilon Q_T)^{1/4}(qV_S\gamma\Phi)^{1/2}}{\lambda^{1/2}}$ $j_m \approx 10^5$ A/cm^2	$4V_S\, \alpha\, (\varepsilon Q_T)^{1/2}$ $j_m \approx 10^7$ A/cm^2
Heating + skinning limits linear current density $I_m/L<$	$4\,(Q_T/\mu)^{1/2}$ $I_m/L \approx 10^6$ A/cm	$4\,(Q_T/\mu)^{1/2}$ $I_m/L \approx 10^6$ A/cm	$4\,(Q_T/\mu)^{1/2}$ $I_m/L \approx 10^6$ A/cm

p-n junctions. For n junctions dU/dt may be increased n-times. Practically for fast switching only two electrodes, devices could be stacked.

The best results, very close to Eq. (2) limit are achieved in the case of Drift Step Recovery Diodes (DSRD). Still there is the Electromagnetic limit for total power both voltage and current increase by stacking in series and increasing circumference of a switch.

6. ELECTROMAGNETIC TIME OF FLIGHT LIMIT.

Let us consider the EM field picture of a pulser in case of cylindrical (coaxial) geometry. Initially (off state) energy is stored in static electric field on the left part of a wave guide (Fig. 5). The electric field lines penetrate semiconductor volume in direction from anode to cathode, which is perpendicular to the field in storage part. In final "on state" the field is forced out of the volume to build TEM configuration. The semiconductor becomes like a conductor. It is evident, that the change of the EM field configuration between "off" and "on" states, takes time not less, than the "time of flight" of EM wave (T_{EM}) along the largest size (l_m) of the semiconductor bulk (W,R) or the waveguide δR, where field changes the geometry.

$$\tau > T_{EM} = l_m / c_{EM} , \quad (15)$$

where c_{EM} – electromagnetic wave velocity in semiconductor or waveguide dielectric accordingly.

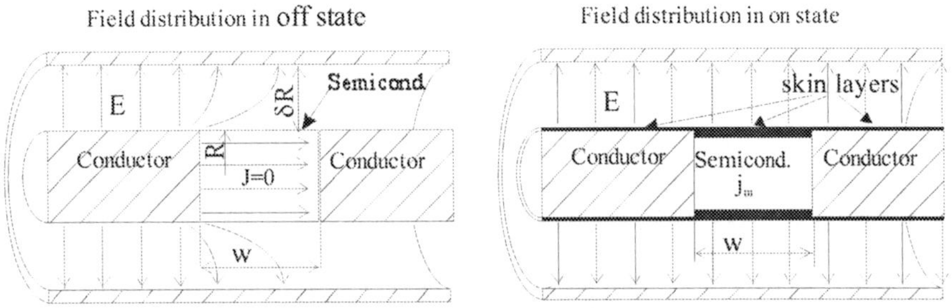

Figure 5. EM field reconstruction during switching process

From Eq. (2) and Eq. (15) we get maximum switching voltage (U_{ms}) for a switch with switching time τ, independently of a number of stacked p-n junctions:

$$U_{ms} < c_{EM} \cdot E_\alpha \cdot \tau \,. \qquad (16)$$

For a silicon switch with switching time ~0,1 ns, Eq. (16) gives $U_{ms} \approx 100$ kV.

7. INJECTION/EXTRACTION DEVICES

As it was mentioned, the most promising from practical point of view are devices based on fast voltage restoration – Drift Step Recovery Diodes, which exploit double injection – extraction processes. Beyond considered above limits some additional relations could be evaluated.

From practical point of view, diameter of a device should be less than skin layer δ, which provides uniform current distribution and the whole area use. So the maximum total current rise rate could be derived from Eq. 9

$$dI_m/dt \le \delta^2 \, dj_m/dt \approx E_\alpha/\mu \qquad (17)$$

The maximum switched (during time τ), current is

$$I_m \approx E_\alpha \cdot \tau/\mu \qquad . \qquad (18)$$

Maximum for one p-n junction voltage is $U_m \approx E_\alpha W$. We can determine from Eq. 18. the ratio $\rho_m = U_m/I_m$, which is the output impedance corresponding the maximum output power

$$\rho_m = U_m/I_m \approx V_s \cdot \mu \qquad . \qquad (19)$$

It should be mentioned, that the impedance is nearly the same for any semiconductor $\rho_m \approx 0.10$ Ohm and is very low value. Because the mot of UWB applications are connected with pulse radiation, the output impedance should be compared with a typical UWB antenna impedances dozens - hundreds Ohms. Such large difference could be treated from both ends:

a. To increase a number of stacked p-n junction up to the limit Eq. (16), which leads to $\rho_{ms} \approx (\mu/\varepsilon)^{1/2}$ – wave impedance of the material. However, the disadvantages are: complicated mechanical design and proportional thermal conductance decrease, which worsens reliability, average power and duty factor.

b. To decrease antenna input impedance, which requires new antenna and waveguide approaches.

8. CONCLUSION

It was shown, that in general a number of fundamental physical effects determine the limitations for performance of switching devices. The limitations set boundaries and show guidelines for improvements in the art of design of switches and generators of power short pulse generation. All considerations have been made for a single switch, although it could consist of many p-n junctions.

One of the most important factors, limiting the performance of a switch is the delay of electromagnetic wave propagation in the bulk of the switch as well as in the surrounding volume of the wave guide structure.

The propagation process could change from wave–like to diffusion-like (in the last case it close to well known skinning), depending on the conductivity of semiconductor (change from "off" to "on" state). In any case, the parts of the bulk separated by distance, which is larger, than the distance passed by the wave during switching could be considered causally separated and so could be separated physically. The switched power of the parts still could by combined by use of additional electromagnetic structures. The combining is possible in space (in case of radiation) or in the load. For combining, it is necessary to synchronize all generated wave fronts. In the case due to the absence of internal interaction between parts of the device, which could provide self synchronization (as in DSRD), special means are necessary for the synchronization, like PLL widely used in microwaves. Due to very small inherent jitter in semiconductors switches, PLL is needed to compensate only slow thermal and aging temporal drift.

A LOW-COST, COMPACT PLANAR INTEGRATED-CIRCUIT TUNABLE MULTI-PULSE TRANSMITTER FOR UWB RADAR AND COMMUNICATION SYSTEMS

Jeongwoo Han, Rui Xu, and Cam Nguyen*

1. INTRODUCTION

Ultra-wideband (UWB) radar and communication systems are attractive for various applications, such as pavement assessment, bridge-deck inspection, geophysical explorations, collision avoidance, fluid level sensing, detection and classification of unexploded ordnance (UXO) and land-mines, and short-range in-building communications [1]-[9].

In UWB systems, transmitters capable of generating multiple pulses with various durations are very desirable. Pulses of wide duration contain large low-frequency components, enabling the pulse signals to propagate deeply into a medium because of the relatively low propagation loss of their low-frequency components. Pulses of shorter duration, on the other hand, have wider frequency bandwidth, making feasible higher range resolution. A tunable multipulse transmitter would therefore have both the advantages of increased penetration (or range) and fine range resolution for UWB applications. Electronically tunable pulse generators are also desired for measurement equipment. The polarimetric video impulse radar described in [7]-[8] is a good example showing the usefulness of the pulse's tuning capability. Pulse generators with tunable step-functions and impulses have been reported [10]-[12]. Recently, tunable monocycle pulse generators have also been developed [13]-[14]. Step-function, impulse and monocycle pulse are typical waveforms used in UWB systems. Step-function and impulse contain DC and large portion of low-frequency spectral components, which can not be transmitted through a practical antenna. Monocycle pulse, on the other hand, has no DC component and narrower bandwidth, facilitating its transmission using a practical

* Jeongwoo Han, Danam Systems, Inc., Anyang, South Korea. Rui Xu, Texas A&M University, College Station, Texas 77843. Cam Nguyen, Texas A&M University, College Station, Texas 77843

antenna. Furthermore, using monocycle pulse facilitates the design of other components including antenna in the system. Monocycle pulses are thus typically preferred for UWB systems.

In this paper, we present the development of a new transmitter for UWB radar and communication systems. The developed transmitter produces multiple monocycle pulses with 10-MHz PRF, 0.4-1.2 ns pulse duration, and 0.2-3.9 GHz operating frequency range. The measured output pulse amplitude is from 6.5-9V peak-to-peak voltage.

2. DESIGN OF TUNABLE MULTIPULSE TRANSMITTER

2.1. Impulse Generator

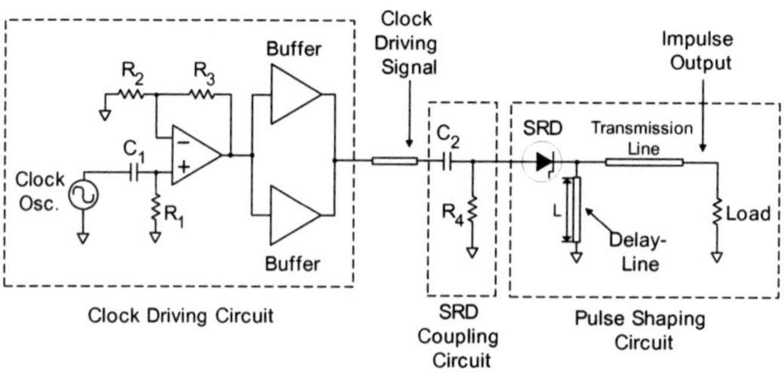

Figure 1. Circuit diagram for the designed SRD delay-line impulse generator.

The basic element in the developed tunable multipulse transmitter is the delay-line impulse generator. Formation of an impulse using delay lines is a classical technique used in the digital and pulse circuit areas. Some step recovery diode (SRD) impulse or monocycle-pulse generators have been designed using the same delay-line principle and implemented on microstrip or CPW structures [15]-[19]. Fig. 1 shows the overall circuit diagram of the designed impulse generator.

The delay-line SRD impulse generator, was designed, based on the circuit diagram of Fig. 1, and fabricated using microstrip lines on RT/duroid 6010 substrate having a relative dielectric constant of 10.2 and a thickness of 0.127 cm. Fig. 2 shows the measured output pulse of the fabricated circuit. The measured impulse has 8-V peak amplitude, 160-ps FWHM (Full Width Half Maximum), 300-ps pulse width (defined at 10 % of the peak amplitude).

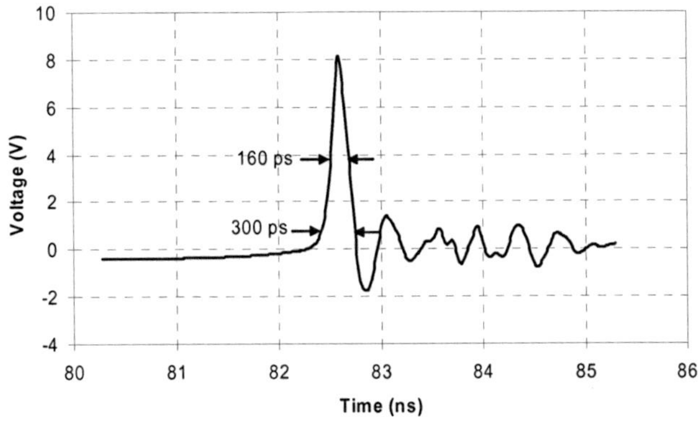

Figure 2. Measured impulse output of the fabricated impulse generator.

2.2. Tunable Monocycle Pulse Generator

The basic idea for forming the multiple monocycle pulses relies on using PIN-diode switches to realize synthetic short-circuited delay transmission lines. The underlying principle of the tuning method using PIN diode switching is relatively simple. Fig. 3 shows an instance of the distributed delay-line for tuning of the impulse duration. The original delay-line, seen in Figs. 1 is subdivided into several transmission-line sections, separated by DC blocking capacitors, to form a distributed delay-line with each section containing an anti-parallel PIN-diode pair and a biasing circuitry. A synthetic short-circuit can be created at each diode-pair connection point by turning on the PIN diodes through a DIP switch. By changing the DIP switch connections alternately, various delay-lines of different lengths can be effectively made, hence generating different impulse durations corresponding to different round-trip times of the step signal propagating on the delay-line.

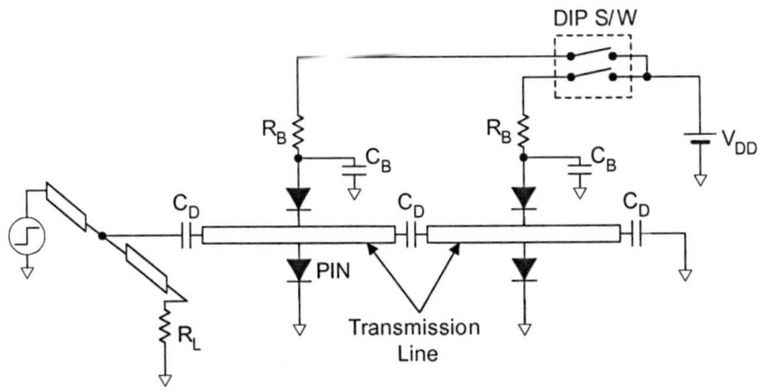

Figure 3. Circuit diagram for a distributed delay-line using anti-parallel PIN diode pairs.

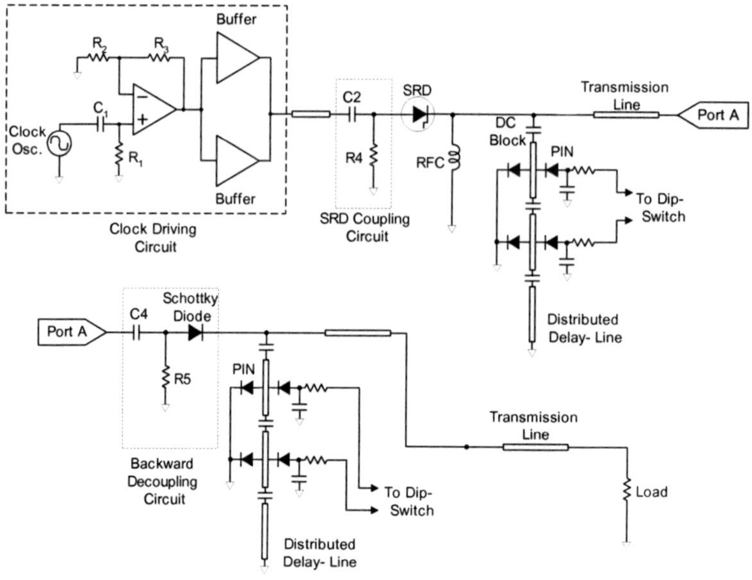

Figure 4. Circuit diagram of the designed tunable multipulse transmitter.

Fig. 4 shows a simplified circuit diagram of the designed multipulse transmitter. The clock signal, produced by the Clock Driving Circuit, drives the SRD, via the SRD Coupling Circuit, to generate a step function. This step function is converted into impulses of various durations by turning on and off the PIN diodes located along the first Distributed Delay Line using a DIP switch. A second identical Distributed Delay Line is then used to convert these impulses into monocycle pulses of different durations by means of PIN diodes. A direct coupling of the two distributed delay-lines may cause problem of backward transmission of the reflected impulse from the second to the first delay-line, making large multiple reflections and eventually causing large ringing on the output signal. The Backward Decoupling Circuit is used to reduce the backward coupling effect.

3. PERFORMANCE OF TUNABLE MULTIPULSE TRANSMITTER

The tunable multi-pulse transmitter was fabricated using microstrip structure on the same substrate as that used for the impulse generator. Its photograph is shown in Fig. 5. The transmitter employs two identical distributed delay-lines, each divided into four sections to generate four different monocycle pulse durations. By turning on a branch of the DIP switches, the corresponding two PIN-diode pairs located on the two distributed delay-lines are turned on simultaneously, generating an output monocycle pulse with a particular duration. The pulse duration, corresponding to the delay-line length, can be varied according to the selected turn-on position of the PIN-diode pairs.

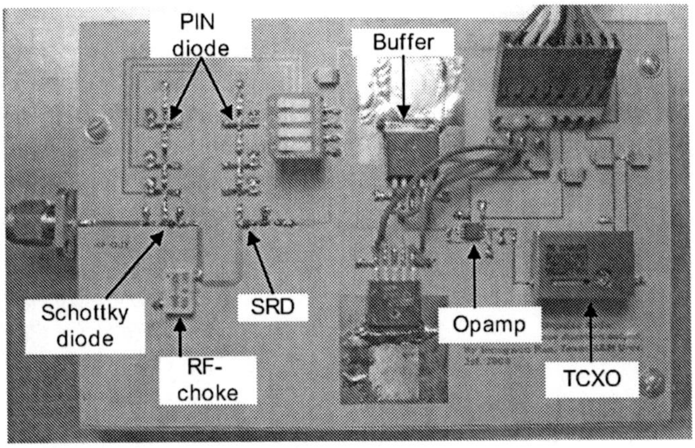

Figure 5. Photogrtaph of the tunable multi-pulse transmitter integrated with the clock driving circuit.

Fig. 6 shows the measured output monocycle pulses having four different pulse durations. The characteristics of each generated pulse are summarized in Table I along with the designed values for comparison. The measured pulse width at 10% in Table I is the pulse width measured at 10 % level of the peak amplitude. As can be seen, the measured pulses show practically good waveforms with acceptable ringing levels.

Spectrum analysis has also been done for the generated monocycle pulse signals to find their frequency bandwidths. Fig. 7(a) shows the frequency-spectrum data for the ideal 420-ps-duration monocycle pulse and the measured 400-ps-duration monocycle. The ideal monocycle is represented as a perfect single cycle of sinusoid. In the figure, it is found that both of the main- and side-lobes of the ideal monocycle pulse are higher than those of the measured pulse. The reason is that the ideal pulse has abrupt transitions at the beginning and ending parts of the single-cycle duration. Nevertheless, the spectrum of the measured pulse follows that of the ideal monocycle pulse reasonably well for most engineering purposes. The available bandwidth (BW) is defined in Fig. 7(a) as the instantaneous bandwidth at 10-dB level. For instance, the 400-ps-duration monocycle pulse has about 3.5-GHz available BW. Note that an UWB impulse GPR having more than 3-GHz available BW for the transmitting pulse signal is usually known as a high-resolution GPR, which can be used for high-resolution applications such as land-mine or UXO detection. Available BW data are also needed for the design of the system's components. Both the transmit and receive antennas, and the receiver need to be designed to cover the entire available BW. The effective pulse width in Table I is directly related to the available bandwidth and is defined as the pulse width of the ideal monocycle that has the same available bandwidth as the measured pulse. In view of the close resemblance of the pulse widths between the ideal and actual pulses for a particular bandwidth, it is considered that the effective pulse width defined in here is a more accurate characterization parameter than the commonly used pulse width at 10% to represent the performance of the range resolution in impulse radar applications.

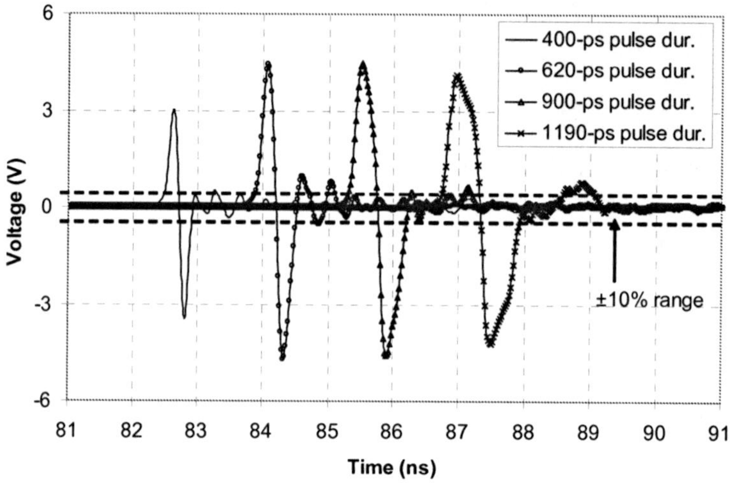

Figure 6. Measured output monocycle pulses having four different pulse durations controlled by PIN diode switching.

Table 1. Designed and measured performance of the UWB tunable multipulse transmitter

Designed Pulse Width (ps)	Measured Pulse Width@10% (ps)	Effective Pulse Width (ps)	Operating Frequency Band (GHz)	Pulse Amplitude (Vpp)
350	400	420	0.50-3.90	6.5
550	620	630	0.35-2.60	9
850	900	850	0.20-1.95	9
1150	1190	1170	0.20-1.40	8.3

Spectrum analysis results for all the four generated monocycle pulses are shown in Fig. 7(b) and the measured operating frequency bands at 10-dB level are specified in Table I. The operating frequency bands of these generated pulses are from 0.2-3.9 GHz. Measured pulse amplitudes are in the range of 6.5-9 V peak-to-peak voltage as seen in Fig. 6.

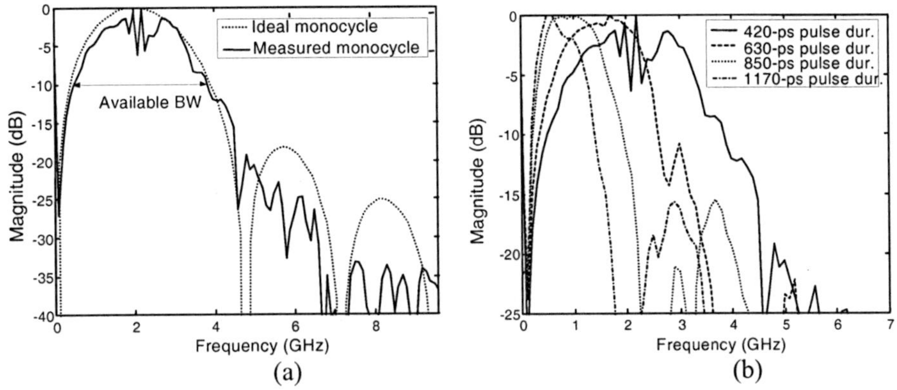

Figure 7. (a) Spectrums of the ideal monocycle pulse with 420-ps duration and the measured monocycle pulse with 400-ps duration. (b) Spectrums of all the monocycle pulses generated from the multipulse transmitter.

4. CONCLUSION

A new UWB transmitter capable of generating multiple monocycle pulses has been developed for UWB applications. The multipulse is achieved by alternately switching on and off the PIN-diode pairs spatially located along the delay lines, which is easy to be implemented and results in a compact circuit structure. The transmitter produces monocycle pulses of varying duration from 0.4-1.2 ns, corresponding approximately to the operating frequency range of 0.2-3.9 GHz, and 6.5-9 V of peak-to-peak voltage. These results demonstrate that the designed transmitter with multipulse capability can be used for most short-range UWB applications, including high-resolution radar applications such as UXO and land-mine detection. Particularly, the *effective pulse width* derived via the pulse's spectrum analysis is deemed as a useful and accurate parameter for calculating the range-resolution of UWB impulse radar.

5. ACKNOWLEDGEMENT

This work was supported in part by the National Science Foundation, the Texas Transportation Institute, the Southwest Region University Transportation Center, the Texas Advanced Research Program and Dell Computer Corporation.

6. REFERENCES

1. R. J. Fontana. (2000, May). "Recent applications of ultra wideband radar and communications systems," in *Ultra-Wideband, Short-Pulse Electromagnetics 5*, P. D. Smith and S. R. Cloude, Ed. New York: Kluwer Academic/Plenum Publishers, 2002, pp. 225-234.
2. S. Azevedo and T. E. McEwan, "Micropower impulse radar," *Science & Tech. Review*, pp. 17-29, Jan./Feb. 1996.
3. J. S. Lee, C. Nguyen and T. Scullion, "Impulse ground penetrating radar for nondestructive evaluation of pavements," in *2002 IEEE MTT-S Int. Microwave Symp. Dig.*, 2002, pp. 1361-1363.

4. J. P. Warhus et al. (1994, Sep.). Advanced ground-penetrating, imaging radar for bridge inspection. Lawrence Livermore National Laboratory, CA. [Online]. Available: http: // www-eng. llnl. gov / dsed / documents / em / jwpctta93.html
5. A. Duzdar and G. Kompa, "Applications using a low-cost baseband pulsed microwave radar sensor," in *Proc. 18th IEEE Instrumentation and Measurement Technology Conf.*, 2001, pp. 239-243.
6. R. Tielert, "Subnanosecond-pulse generator employing 2-stage pulse step sharpener," *Electron. Lett.*, vol. 12, pp. 84-85, Dec. 1976.
7. A. Yarovoy and L. Ligthart, "Full-polarimetric video impulse radar for landmine detection: Experimental verification of main design ideas," in *Proc. 2nd Int. Workshop on Advanced Ground Penetrating Radar*, 2003, pp. 148-155.
8. L. Ligthart and A. Yarovoy, "STW Project: Advanced relocatable multi-sensor system for buried landmine detection," in *Proc. 2nd Int. Workshop on Advanced Ground Penetrating Radar*, 2003, pp. 5-8.
9. Time Domain Corp., AL. PulsON Technology: Time modulated ultra-wideband for wireless applications. [Online]. Available: http://www.timedomain.com/Files/downloads/techpapers/PulsONoverview.pdf.
10. K. Daneshvar and L. Howard, "High current nanosecond pulse generator," in *Proc. IEEE Southeastcon '89*, 1989, pp. 572-576.
11. U. A. Uhmeyer and J. C. Libby, "A fast variable transition time pulse generating circuit," in *Proc. 9th IEEE Instrumentation and Measurement Technology Conf.*, 1992, pp. 152-157.
12. J. Han and C. Nguyen, "Ultra-wideband electronically tunable pulse generators," *IEEE Microwave Wireless Compon. Lett.*, vol. 14, no. 3, pp. 112-114, Mar. 2004.
13. J.W. Han and C. Nguyen, "On the Development of a Compact Sub-Nanosecond Tunable Monocycle Pulse Transmitter for UWB Applications," *IEEE Trans. on Microwave Theory and Techniques*, Vol. MTT-54, No. 1, Jan. 2006, pp. 285-293.
14. J.W. Han, R. Xu and C. Nguyen, "A Multipulse Transmitter for UWB Radar and Communication Systems," *IEEE AP-S International Symposium & USNC/URSI National Radio Science Meeting*, Albuquerque, New Mexico, July 9-14, 2006
15. G. D. Cormack and A. P. Sabharwal, "Picosecond pulse generator using delay lines," *IEEE Trans. Instrum. Meas.*, vol. 42, no. 5, pp. 947-948, Oct. 1993.
16. J. S. Lee and C. Nguyen, "Uniplanar picosecond pulse generator using step-recovery diode," *Electron. Lett.*, vol. 37, no. 8, pp. 504-506, Apr. 2001.
17. J. S. Lee, C. Nguyen and T. Scullion, "New uniplanar subnanosecond monocycle pulse generator and transformer for time-domain microwave applications," *IEEE Trans. Microwave Theory and Techn.*, vol. 49, no. 6, pp. 1126-1129, Jun. 2001.
18. J. S. Lee and C. Nguyen, "Novel low-cost ultra-wideband, ultra-short-pulse transmitter with MESFET impulse-shaping circuitry for reduced distortion and improved pulse repetition rate," *IEEE Microwave Wireless Compon. Lett.*, vol. 11, no. 5, pp. 208-210, May 2001.
19. J. Han and C. Nguyen, "A new ultra-wideband, ultra-short monocycle pulse generator with reduced ringing," *IEEE Microwave Wireless Compon. Lett.*, vol. 12, no. 6, pp. 206-208, Jun. 2002.

UWB Antenna Surrogate Design

Paul Robert Hayes, Luis Hernandez, and Andrew Austin**

1 Introduction

Ultra Wide Band (UWB) antenna systems have a wide variety of uses including impulse radiation for communication, material property determination, structural feature viewing in building trades, buried object detection and other uses. In each case, the UWB nature (significant frequency content from kHz through multi-GHz) of the overall system combined with the near- and mid-field aspects of the operation present unique challenges. Previous EMP/HEMP work where far-field approximations apply and objects are significantly shielded [1] may not be applicable [2]. Often very little or no shielding is present with complex structures (Figure 7) invalidating cavity/perturbational approximations [3] and the antenna systems are often near, mid and far-field within the same time-domain impulse drive due to the UWB nature thereby invalidating far-field approximations [4]. Major sub-systems for a successful UWB system design include the radiating antenna, the power supply, the power source and the near-field structural environment.

In the process of developing a production system, a "closed-loop" process of physical testing and analysis, combined with modeling and simulations, provides a more complete perspective on the operational aspects permitting significant optimizations in a cost-effective manner. Lengthy physical testing, as well as manufacturing multiple full-scale test antenna systems, can be time-consuming and cost prohibitive. In other situations, creating a complete full-scale mockup of the antenna system may not be possible due to cost constraints. In a study of capabilities, an antenna system may not initially be available for physical testing. In each case, creating a suitable surrogate for preliminary physical testing and simulations is advantageous. [2]

** Paul Robert Hayes, The CEMTACH Group, King George, VA 22485. Luis Hernandez, BAE Systems, Minneapolis MN 55421. Andrew Austin, L3 Communications, Salt Lake City, Utah 84116-0850.

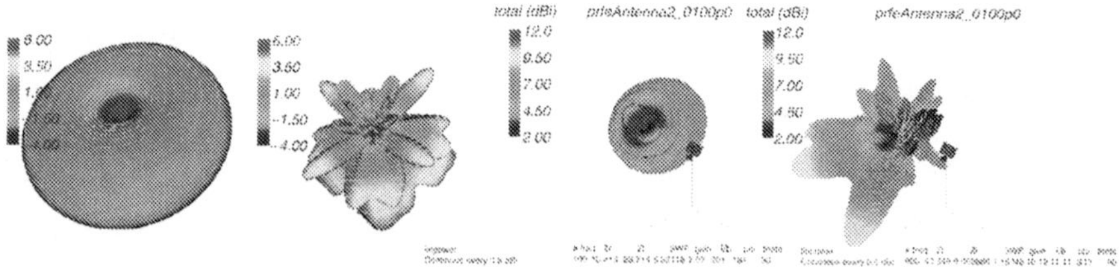

Fig. 1. Two free-field radiation patterns (left) from the many for an example boxCone UWB operational range presenting the challenge in radiation pattern surrogate symmetry. At low frequency, agreement with cylindrical systems is maintained (far-left), yet in the higher operational bands (mid-left) the radiation pattern no longer resembles cylindrical discone or cylCone designs. On the right are 100 MHz (mid-right) and 900 MHz (far-right) radiation patterns above a ground for a tilted boxCone.

When designing a surrogate antenna system for physical testing, maintaining symmetry eventually present in a final design is important.[5] Surrogate antenna systems are valuable when experiments with the real antenna system are not possible due to cost, design complexity, design maturity, non-disclosure concerns or other issues. Exigencies of a physical test may lead to surrogate design features that adversely affect results as shown in Figure 1 where a box was chosen for drive components over a cylinder. The radiation pattern changes significantly through the operational band causing wide variations in the direction of energy flow. Simulations capabilities and experience can aid in surrogate antenna system design yielding rapid prototyping while maintaining accurate symmetries with respect to final design guidelines. Several examples are discussed including discone (cylindrical) versus box discone (cylindrical/cartesian) antenna patterns, frequency characteristics and pulsed results.

1.1 UWB Radiating Antenna Characteristics

Impulse radiating antennas (IRA), bowties, folded dipoles, discones and cylindrical dipoles are examples of broadband radiating antennas each with unique characteristics and applications. For this work, we consider the basic discone as shown in Figure 2 with four variants with the design and analysis concepts for surrogates relevant to other UWB systems.

From the input impedance of the antenna, which can be measured in physical tests or simulations, we can derive SWR, reflected power and other drive characteristics which will affect the ability of the power supplies to effectively radiate UWB. Figure 3 shows that below 4 MHz none of the 4 discones radiates effectively and thus should the power supply contain most of it's energy content below this, the overall system will be significantly degraded. One can also note that the discone and cylCone designs are relatively close

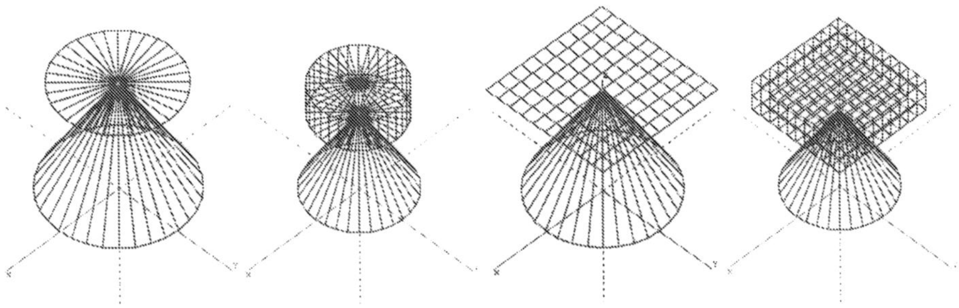

Fig. 2. Four prospective designs [discone, cylinder cone (cylCone), square plate cone (squareCone) and box cone (boxCone)] for use as surrogates shown on ortho view.

drive surrogates while the squareCone and boxCone are significantly different leading to a different radiated spectrum. Thus, the squareCone and boxCone would not serve as adequate surrogates for a discone design.

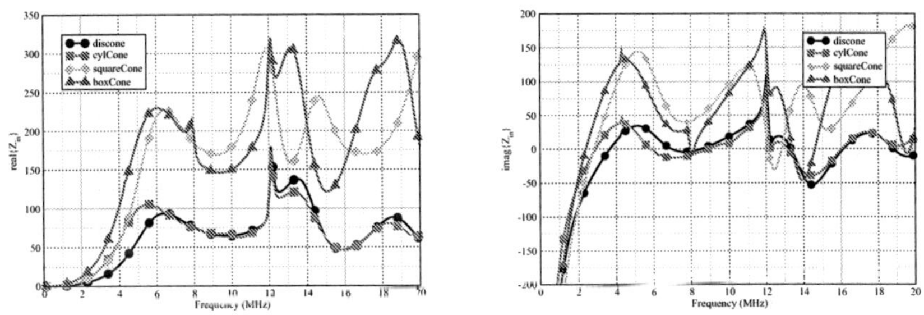

Fig. 3. Portion of drive characteristics for the four sample discone antenna systems illustrating differences and similarities in design for surrogates where the systems should resemble each other as closely as possible.

Figure 4 shows a flexible test probing methodology for UWB systems. Caution must be exercised as noticed in Figure 1 especially when surrogate systems are under study as radiation patterns can significantly skew field probe data. Due to the UWB nature of the field probe information, coupled with the wide variations in radiation patterns, gathering a few (5-10) point probes (time-domain d-dot E field, for example) does not yield enough information to reverse engineer how well the antenna is operating. For example, while the field probe and antenna are set at a fixed physical height, the UWB nature changes the electrical height from a fraction of a wavelength through many wavelengths resulting in peaks and nulls in the field probe spectra, Figure 4. Thus, even though the antenna radiated significant energy within a band, there is no guarantee that a distant field probe will observe that. For

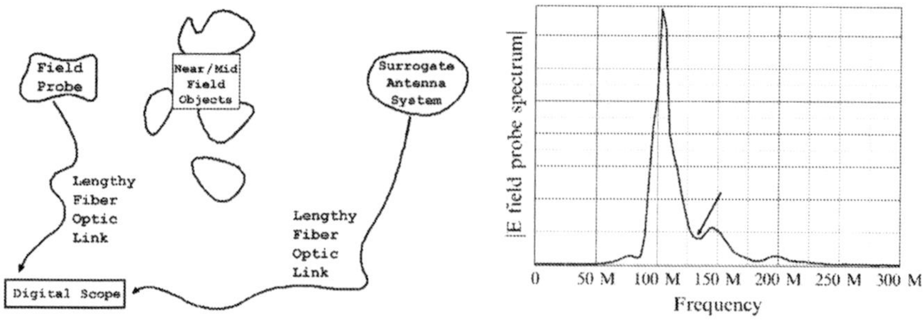

Fig. 4. Test probing example for UWB systems illustrating fiber optic links for isolation with near/mid field objects, antenna feedback probing and test field probes all connected back to a sampling digital oscilloscope (left). At right is a sample of the spectra from a field probe which is affected by the antenna design where the arrow may indicate a radiated field null.

an accurate surrogate, the field probes at all positions should ideally match in time-, frequency-domain. The most accurate strategy comes directly from the drive waveform and the antenna itself, with probes placed directly on the antenna systems and targeted distant field probing.

1.2 UWB Power Supplies

UWB pulses are generally characterized by subnanosecond risetimes and nanosecond pulse widths. The final pulse generator employed for an antenna system is a matter of the specifics to be studied. Pulse generating systems capable of generating hundreds of kV in this time frame include Marx generators, spark gap switched transmission lines and blumleins. Of course it would be ideal if one could study the system without operating at full voltage permitting the researcher to use simple, reliable, low-cost systems to generate surrogate pulse waveforms. Simple relays can be used up to a few kV as can simple commercial off-the-shelf (COTS) based step recovery diode pulsers. This additionally makes time domain pulse shaping a trivial matter. Stubs can be used to produce bipolar pulses as can various transmission line techniques [6]. This may or may not be a critical step depending on the actual antenna to be used however can be advantageous to shift the energy spectra where it will more efficiently radiate from the antenna, Figure 5.

Figure 6 shows four waveforms from relatively "slow" through "fast" risetimes with the corresponding Fourier spectra shown below. The Pulse 1 "slow" risetime drive contains significant energy at lower frequency bands. As the risetime diminishes to the "fast" Pulse 4, energy content is distributed in an increasingly wider band. Pulse width is also a critical parameter to consider given that an antenna will radiate the derivative of a pulse so long as the

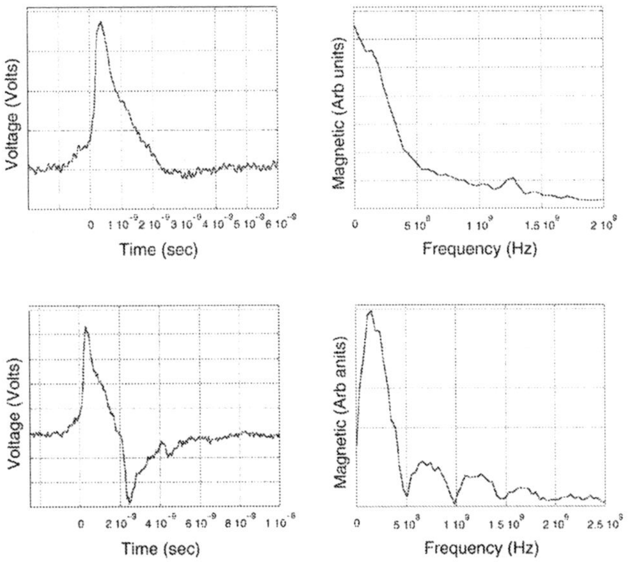

Fig. 5. A typical voltage drive pulse (top left) and spectra (top right) with significant energy content in low frequencies can be shaped into a bipolar drive pulse (lower-left) and spectra (lower right) with energy shifted up in frequency.

pulse width is shorter than the antenna transit time. [7] With the non-trivial nature of impulse radiating antennas, it is clear that care must be taken if a test with any fidelity is to be conducted.

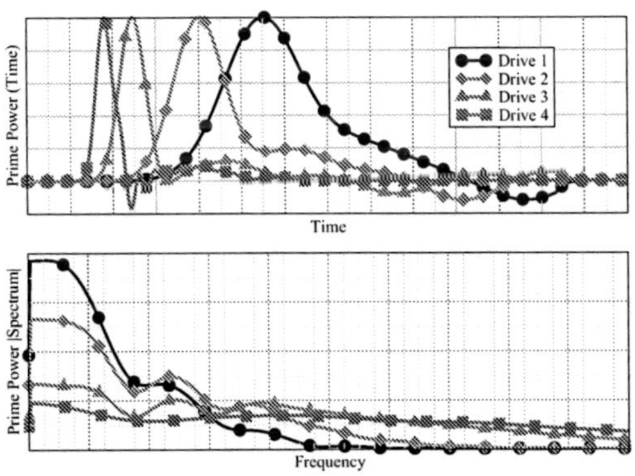

Fig. 6. Four representative impulsive drive waveforms from a proposed UWB power supply for a discone-type radiating antenna.

Using a surrogate in the field may include the use of a battery pack to provide prime power. While this seems a trivial addition, great care must be made to ensure that there is little change to the final desired radiation pattern. Significant battery subsystems that are in the near field can cause major changes to the radiation pattern, Figure 1, and antenna characteristics, Figure 3. The cause of such a non-symmetric pattern is a cubic battery/drive housing directly beneath the antenna, or more accurately put, the cubic battery housing elements of the antenna.

Rapid physical prototyping may also utilize a wire feed for power which, if placed in a null of the antenna radiation patterns, will have adverse effects (surrogate is no longer valid). Careful simulations and physical testing reveals dead-zones, or nulls in the radiation of the antenna, where relatively minimal effects of connecting a wire-feed are possible. Due to the UWB nature, the "dead-zone" must occur across the entire operational range of the proposed system and surrogates. While nulls are challenging to design around with physical testing, simulations are a cost-effective method for analysis and feedback during the "closed-loop" design process.

1.3 Near-field structural environment

UWB systems may not be powerful enough to operate in a far-field manner and some are designed to specificly operate in a near-field low-power mode (structural or buried anomaly detectors). In these situations, a wide range complex material objects from electrically small through electrically large are present and must be considered for operational design.

One underlying challenge is the relative uncertainty of objects (wires, field probes, et cetera) placement in complex systems for simulations and physical experiments such as shown in Figure 7. [8] [9] In order to generalize beyond a specific test system, simulations and tests can be performed with a variety of object placements emphasing the need for a statistical approach to analyzing the results. A statistical, generalized approach for simulation and experimental analysis also permits a more thorough sampling of the wealth of information [8] [3] while avoiding the pitfalls of small sample size in which results could be inadvertently shifted by chance to higher or lower statistics resulting in over or under compensation of design features for the antenna surrogate along with resulting costs or susceptibilities.

Higher level validations and analysis are possible for physical testing and simulations as well. In some instances, probing the antenna system directly may not be viable. Figure 8 illustrates a statistical process for validation. In a real world physical test, only a few probes are available due to costs and the field perturbations that would occur. In simulations, thousands of non-perturbing field probes (Figure 8 left) are available distributed randomly throughout a test structure (Figure 7). The maximum electric field magnitude achieved at each simulation probe in space for all time ($|E|_{max} \forall t$) (Figure 8 middle) were collected to create a cumulative distribution function (CDF)

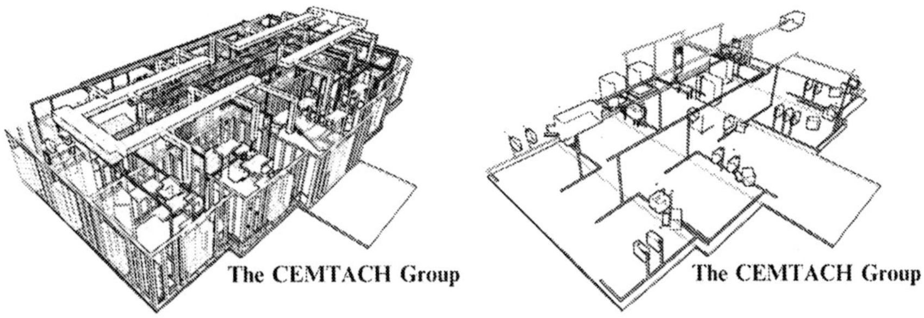

Fig. 7. Interior geometry of a representative brick structure illustrating interior 2×4 wood framed walls, doors, windows, HVAC system, electrical power system, overhead lights, Ethernet network, office furniture and other common components appropriately placed.

(Figure 8 right) allowing prediction of the peak achieved at other "typical" probe locations.

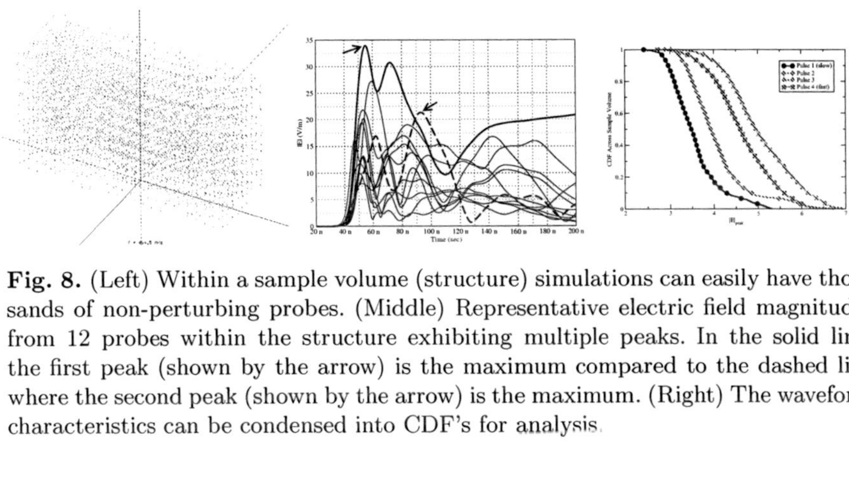

Fig. 8. (Left) Within a sample volume (structure) simulations can easily have thousands of non-perturbing probes. (Middle) Representative electric field magnitudes from 12 probes within the structure exhibiting multiple peaks. In the solid line, the first peak (shown by the arrow) is the maximum compared to the dashed line where the second peak (shown by the arrow) is the maximum. (Right) The waveform characteristics can be condensed into CDF's for analysis.

Figure 8 also illustrates why physical test probing is challenging. With the logistics of few physical probes available, those probes may be inadvertently placed at positions resulting in the lower right portion of the CDF graph yielding an atypical "high". This could result in excessive shielding for a structure or a false sense of the effectiveness of the antenna. Shielding strategies based on the 25 highest probes ($100\% \rightarrow 300\% E_0$) may be cost prohibitive compared to the average throughout the structure for all incident angles. Placing the probes inadvertently at physical positions resulting in the lower CDF values (upper left) results in an atypical "low". This could result in too little shielding for a structure or a false sense of the ineffectiveness of an antenna system.

The most promising strategy is to combine the wealth of information from simulations with targeted physical testing, "closed-loop" analysis, at low- and high-level validation points.

1.4 Conclusions

In conclusion, the design and analysis of UWB systems is challenging with surrogates and models no less difficult. Surrogates systems should match real and proposed systems at low levels (drive characteristics, radiation patterns, point field probe, et al) and high levels (statistical means, multiple locations of antenna, probe and nearby objects, et al). In order to facilitate physical testing, simulations can aid in test design avoiding "Quick" & "Cheap" fixes which may make a physical test happen but adversely affecting results yielding false positives or negatives. Simulations are a valuable asset in analysis, aiding cost-efficient and effective physical testing. Designing UWB surrogates should be approached with a "closed-loop" design-verify process considering many factors.

References

1. P. R. Hayes, "Transmission-line matrix modeling," Master's thesis, Auburn University, December 1990.
2. Y. T. Lo and S. W. Lee, eds., *Antenna Handbook*, vol. 4. Van Nostrand Reinhold, 1993.
3. P. Hayes and B. Hankla, "Numerical estimates of electromagnetic environments inside complex commercial information system installations," in *Proceedings of The International Conference on Electromagnetics in Advanced Applications (ICEAA 01)*, (Turin, Italy), ICEAA, 2001.
4. P. R. Hayes, B. J. Hankla, W. C. Baedke, and D. C. Stoudt, "Basic issues of field sampling for em environment descriptions," in *2002 AMEREM*, (Baltimore, MD), AMEREM, 2002.
5. P. R. Hayes, "Uwb antenna surrogate design," in *2006 AP/URSI/AMEREM*, (Albuquerque NM), AP/URSI/AMEREM, July 2006.
6. Z. Kaplan, "Simple, fully isolated, pulse transformer," *Review of Scientific Instruments*, vol. 55, pp. 1355–1356, Aug 1984.
7. A. Austin, D. Marvin, M. Moroz, A. Yuenger, and L. Hernandez, "Development of an ultra-wideband (uwb) radiating system for a solid-state pulse generator," Tech. Rep. 8620, NAWCWD, China Lake, CA 93555-6100, June 2006.
8. M. P. Bernardin, "Scaling laws for coupling quantities on emp," *IEEE Transactions on Electromagnetic Compatibility*, vol. 30, pp. 48–53, February 1988.
9. C. A. Ropiak, R. L. Gardner, and D. C. Stoudt, "Electromagnetic susceptibility analysis using multivariate logistic regression," in *Proceedings of the General Assembly meeting of the International Union of Radio Science*, (Maastricht, Netherlands), URSI, August 2002.

INDEX

Antennas
- array, 101-103, 143-151, 153-160
- conical, 25-32, 63
- descriptors, 75-82
- dispersion, 63-68
- folded monopole, 49-56
- impulse radiating, 9-16, 117-120
- loop, 33-40
- omnidirectional, 41-48
- pattern variation, 57-62
- self complementary, 17-24
- slot, 1-8
- TEM 6-8
- tuning, 83-87
- UWB (general), 251-258
- Vivaldi (phase center), 69-73
- VSWR, 41-48

Clutter, 222-225
Communication, 243-250

Data processing, 113-120, 121-125
Deconvolution 129
Dielectric, 187-193, 196, 229-230
Directional coupler, 114-117, 179-188
DORT, 203-210, 220-221

Electromagnetic implosion, 97-104
E-pulse, 199
E-wave (TM), 92-93

FFT (fast Fourier transform), 197-199

High-voltage, 179-186
H-wave (TE), 5-6
Human beings, 227-234

Imaging, 153-160
Impulse response, 127
Integrated circuit, 161-170
IRA, (impulse radiating antenna)
– see antennas

Lens, 89-96

M-Sequence, 125-142
Multipath, 1i72

Radar, 153-186, 243-250,
ground penetrating, 153-160
Ray tracing, 176
Reflector (prolate spheroidal), 98-101

Sampling, 161-170
Scattering, 197-104, 187-234
Scattering center,s 127-134
Semiconductor, 234-242
SEM (singularity expansion method),
187-193, 195-202
Spherical modes, 85
Subregions, 211-218
SVM regressor, 143-150
Switching, 235-242

Target detection, 148
Time intregrals, 105-112, 121-125,
219-226
Time reversal, 105-112, 171-178,
203-210, 219-226

Wireless, 105-112

Printed in the United States
90434LV00001BB/5-6/A

9 780387 730455